Canine and Feline
Behavior and Training

A COMPLETE GUIDE
TO UNDERSTANDING
OUR TWO BEST FRIENDS

Join us on the web at

agriculture.delmar.cengage.com

Canine and Feline
Behavior and Training

A COMPLETE GUIDE
TO UNDERSTANDING
OUR TWO BEST FRIENDS

By Linda P. Case

With Illustrations by Bruce MacAllister

DELMAR
CENGAGE Learning™

Australia • Brazil • Japan • Korea • Mexico • Singapore • Spain • United Kingdom • United States

**Canine and Feline Behavior and Training:
A Complete Guide to Understanding Our
Two Best Friends**
Linda P. Case

Vice President, Career and Professional
Editorial: Dave Garza

Director of Learning Solutions: Matthew Kane

Acquisitions Editor: Benjamin Penner

Managing Editor: Marah Bellegarde

Product Manager: Christina Gifford

Editorial Assistant: Scott Royael

Vice President, Career and Professional
Marketing: Jennifer McAvey

Marketing Director: Debbie Yarnell

Marketing Manager: Erin Brennan

Marketing Coordinator: Jonathan Sheehan

Production Director: Carolyn Miller

Production Manager: Andrew Crouth

Senior Content Project Manager: James Zayicek

Senior Art Director: David Arsenault

Technology Project Manager: Tom Smith

Production Technology Analyst: Thomas Stover

© 2010 Delmar, Cengage Learning

ALL RIGHTS RESERVED. No part of this work covered by the copyright herein may be reproduced, transmitted, stored, or used in any form or by any means graphic, electronic, or mechanical, including but not limited to photocopying, recording, scanning, digitizing, taping, Web distribution, information networks, or information storage and retrieval systems, except as permitted under Section 107 or 108 of the 1976 United States Copyright Act, without the prior written permission of the publisher.

> For product information and technology assistance, contact us at
> **Cengage Learning Customer & Sales Support, 1-800-354-9706**
> For permission to use material from this text or product,
> submit all requests online at **www.cengage.com/permissions**.
> Further permissions questions can be e-mailed to
> **permissionrequest@cengage.com**

Library of Congress Control Number: 2009922730

ISBN-13: 978-1-4283-1053-7

ISBN-10: 1-4283-1053-3

Delmar
5 Maxwell Drive
Clifton Park, NY 12065-2919
USA

Cengage Learning is a leading provider of customized learning solutions with office locations around the globe, including Singapore, the United Kingdom, Australia, Mexico, Brazil, and Japan. Locate your local office at: **international.cengage.com/region**

Cengage Learning products are represented in Canada by Nelson Education, Ltd.

To learn more about Delmar, visit **www.cengage.com/delmar**

Purchase any of our products at your local college store or at our preferred online store **www.ichapters.com**

Notice to the Reader

Publisher does not warrant or guarantee any of the products described herein or perform any independent analysis in connection with any of the product information contained herein. Publisher does not assume, and expressly disclaims, any obligation to obtain and include information other than that provided to it by the manufacturer. The reader is expressly warned to consider and adopt all safety precautions that might be indicated by the activities described herein and to avoid all potential hazards. By following the instructions contained herein, the reader willingly assumes all risks in connection with such instructions. The publisher makes no representations or warranties of any kind, including but not limited to, the warranties of fitness for particular purpose or merchantability, nor are any such representations implied with respect to the material set forth herein, and the publisher takes no responsibility with respect to such material. The publisher shall not be liable for any special, consequential, or exemplary damages resulting, in whole or part, from the readers' use of, or reliance upon, this material.

Printed in Canada
1 2 3 4 5 6 7 12 11 10 09

dedication

in loving memory of

My Father, Robert Palas
Whose wisdom, compassion, and positive outlook continue to guide me
AND
The animal companions who are no longer with us,
but live forever in our hearts:
Dogs – Fauna, Stepper, Roxie, Gusto, Sparks, and Nike
Cats – Mac and Nipper

table of contents

PREFACE / xv
ACKNOWLEDGMENTS / xvii
ABOUT THE AUTHOR / xix

SECTION 1 – DOMESTICATION, SOCIAL BEHAVIOR, & COMMUNICATION

CHAPTER 1
The Beginning of the Friendship – Domestication 2

The Evolutionary History and Taxonomy of Dogs and Cats / 2

    Domestication: Scavenging Wolves and Opportunistic Wildcats / 5

Social Organization: Why Dogs Are Not Cats and Cats Are Not Dogs / 8

    The Social Groups of Dogs / 9

    Social Behavior of Cats / 12

History of Dog Breeds and Breed-Specific Behaviors / 15

History of Cat Breeds and Breed-Specific Behaviors / 19

References and Further Reading / 22

CHAPTER 2
Behavior through the Life Cycle: Newborns to Seniors 26

The Early Weeks: Neonatal and Transition Periods / 27

    Neonatal Period / 28

    Transition Period / 28

    Practical Applications / 29

Sensitive Period of Socialization / 30

    Learning to Be a Dog / 30

    Learning to Be a Cat / 32

Puppy Play vs. Kitten Play / 33

Socialization with Humans / 33

Weaning / 36

Juvenile (Adolescent) Period and Adulthood / 38

Senior Pets / 39

References and Further Reading / 41

CHAPTER 3
Social Behavior and Communication in Dogs and Cats 44

Olfactory Communication / 45

Urine Marking / 46

Feces Marking / 49

Allorubbing in Cats / 49

Scratching in Cats / 50

Auditory Signals / 50

Dog Vocalizations / 51

Cat Vocalizations / 53

Visual Signals / 55

Distance-Reducing Signals – Greeting / 56

Distance-Reducing Signals – Play Solicitation / 60

Distance-Increasing Signals – Offensive Threat / 63

Distance-Increasing Signals – Fear / 65

Distance-Increasing Signals – Defensive Threat / 67

Predatory and Feeding Behaviors of Dogs and Cats / 68

Predation Sequence and Feeding Behavior of Dogs / 69

Predation Sequence and Feeding Behavior of Cats / 70

Social Relationships of Dogs / 72

Social Relationships of Cats / 75

References and Further Reading / 77

SECTION 2 – HOW DOGS AND CATS LEARN – TRAINING TECHNIQUES AND PRACTICAL APPLICATIONS

CHAPTER 4
How Dogs and Cats Learn: Principles of Learning Theory 82

Classical (Pavlovian) Conditioning / **82**

Operant Conditioning (Instrumental Learning) / **87**

    Reinforcement vs. Punishment / **89**

    Types of Reinforcing Stimuli for Dogs and Cats / **92**

    Primary and Secondary (Conditioned) Reinforcers / **94**

    Value of Reinforcing Stimuli / **95**

Habituation and Sensitization / **96**

Social Learning / **97**

References and Further Reading / **101**

CHAPTER 5
Practical Applications: Training and Behavior Modification Techniques 104

The Roles of Classical and Operant Conditioning / **104**

Positive vs. Negative (Aversive) Control of Behavior / **106**

    An Historical Perspective / **106**

    More about Aversive Stimuli / **106**

    Maximizing Pleasant and Minimizing Aversive Stimuli / **109**

Training Preparations / **110**

    Identify Objectives and Goals / **110**

    Select a Training Program or Behavior Modification Technique / **111**

    Develop a Reasonable and Achievable Training Schedule / **111**

    Evaluate Progress / **111**

Selecting Primary Reinforcers / **112**

Using a Conditioned Reinforcer / **112**

    Clicker as a Conditioned Reinforcer / **113**

Getting Started / 114

Timing and Schedules of Reinforcement / 114

Successive Approximation (Shaping) / 119

 Prompting vs. Free-Shaping / 120

Behavior Chains and the Premack Principle / 122

 Fading Cues and Attaining Stimulus Control / 124

Behavior Modification Techniques / 127

 Extinction / 127

 Systematic Desensitization / 129

 Counter-Conditioning (and Counter-Commanding) / 129

 Flooding / 130

References and Further Reading / 131

CHAPTER 6
Training and Problem Prevention for Puppies and Kittens 133

The Newly Adopted Puppy and Kitten / 133

 The First Day / 134

  Establishing a Regular Daily Schedule / 136

Socialization / 136

House-training Puppies / 139

Litter Box Training Kittens / 142

Teaching Puppies to Accept Isolation / 145

Preventing Nipping in Puppies / 147

Teaching Kittens to Play Gently / 149

Teaching Puppies to Chew Appropriate Items / 151

Preventing Objectionable Climbing and Clawing in Kittens / 152

Introducing a New Puppy or Kitten to Other Pets / 154

 Introducing a Puppy to a Resident Adult Dog / 154

 Introducing a Puppy to a Resident Adult Cat / 155

 Introducing a Kitten to a Resident Adult Cat / 155

 Introducing a Kitten to a Resident Adult Dog / 156

References and Further Reading / 157

CHAPTER 7
Teaching Dogs and Cats Desirable Behaviors and Good Manners............... 159

Preventing Problems and Building Bonds / **159**

Basic Manners Training for Dogs / **160**

    Teaching Sit / **161**

    Teaching Down / **163**

    Teaching Stay / **164**

    Teaching Wait / **171**

    Teaching Walk on a Loose Lead / **172**

    Teaching Come When Called / **176**

Training a Well-Mannered Cat / **179**

    Training Cats to Use Specific Sleeping Areas / **180**

    Training a Cat to Come When Called / **181**

    Training a Cat to Walk on a Harness / **181**

References and Further Reading / **183**

SECTION 3 – UNDERSTANDING AND SOLVING COMMON BEHAVIOR PROBLEMS

CHAPTER 8
Elimination Problems in Dogs and Cats.. 186

Elimination Problems in Dogs / **187**

    Incomplete House-Training / **188**

    Marking Behavior / **192**

    Submissive/Excitable Urination / **193**

    Other Causes of House Soiling in Dogs / **195**

Elimination Problems in Cats / **195**

    Inappropriate Elimination / **195**

    Treatment for Inappropriate Elimination / **199**

    Marking Behavior in Cats / **202**

    Treatment for Marking / **203**

    Pharmacotherapy for House Soiling in Cats / **206**

References and Further Reading / **207**

CHAPTER 9
Unruly and Disruptive Behaviors in Dogs and Cats............................ 210

Dogs: Jumping Up / 211
- Factors to Consider / 211
- Training an Alternate and Incompatible Behavior / 213
- Approaches to Avoid / 214
- Shaping and Practicing / 215

Dogs: Destructive Behaviors / 215
- Exploratory Chewing / 215
- Chewing as a Result of Boredom / 219
- Chewing (or Stealing) as Attention-Seeking Behaviors / 221

Dogs: Problem Barking / 222
- Repetitive Barking / 223
- Attention-Seeking Barking / 225
- Punishment and Problem Barking / 227

Dogs: Problem Digging / 229

Dogs: Overly Active (Hyperactive) / 230

Cats: Furniture Clawing / 232

Cats: Jumping Up on Counters/Furniture / 233

Cats: Nocturnal Activity / 235

Cats: Excessive Vocalization / 236

Cats: Plant Eating and Pica / 239
- Eating Houseplants / 239
- Pica (Eating Non-Nutritional Items) / 239

References and Further Reading / 240

CHAPTER 10
Separation, Fear, and Anxiety Problems in Dogs and Cats...................... 244

Normal vs. Problematic Anxiety and Fear Reactions / 245

Separation Anxiety / 246
- Risk Factors and Predisposing Temperament Traits / 247

Behavioral Signs and Diagnosis of Separation Anxiety / 249

Treatment of Separation Anxiety (Dogs) / **250**

    Reducing Dependency on Owner / **250**

    Counter-Conditioning Pre-Departure Cues / **253**

    Desensitization to Isolation / **254**

    Management Approaches / **256**

    Adjunctive Drug Therapy / **256**

    Ineffective Treatments for Separation Anxiety / **259**

Prevention of Separation Stress / **260**

A Cautionary Note: Separation Anxiety or Something Else? / **261**

Treatment Protocol for Cats with Separation Anxiety / **264**

Fear-Related Behavior Problems / **265**

    Common Fears in Dogs and Cats / **266**

    The Role of Avoidance Learning in Maintaining Fear-Related Behaviors / **268**

    Treatment of Fear-Related Behavior Problems / **269**

References and Further Reading / **273**

CHAPTER 11
Problem Aggression in Dogs and Cats 277

Overview of Aggression / **279**

Problem Aggression in Dogs / **281**

    Dominance-Related Aggression / **282**

    Possessive Aggression / **289**

    Territorial Aggression / **293**

    Fear-Related Aggression / **295**

    Ineffective Treatments for Aggression Problems in Dogs / **300**

Problem Aggression in Cats / **301**

    Inter-Cat Aggression in Multiple-Cat Homes / **302**

    Aggression toward a Newly Introduced Cat / **303**

    Aggression toward Humans / **304**

References and Further Reading / **308**

APPENDIX 1 – Recommended Books / **311**

APPENDIX 2 – Resources and Professional Associations / **313**

APPENDIX 3 – Sample Dog Behavior History Profile / **315**

APPENDIX 4 – Sample Cat Behavior History Profile / **320**

GLOSSARY / **324**

INDEX / **327**

preface

We are a nation of dog and cat lovers. Never before in our history have we spent more time, emotional energy, and money on a group of animals who are kept solely for companionship. In the United States alone, we share our homes with over 65 million dogs and more than 75 million cats. Americans spend more than 12 billion dollars per year on pet food and 9 billion dollars on veterinary care. This devotion is further evidenced by the increasing popularity of pet superstores, dog parks, training centers, luxury boarding kennels, dog sports, and doggie daycare centers. Moreover, the relationships that we have with our dogs and cats are not inconsequential. During the 1980s, recognition of the human-animal bond led to serious research into the roles that pets play in our lives. These studies have shown that pets provide significant benefits to our emotional, physical, and social well being. It is clear that dogs and cats are here to stay, and learning more about their behavior, communication, and response to training is of interest to owners, students, and pet professionals in a variety of fields.

Working with dogs and cats in any capacity requires an understanding of species-specific behavior patterns, learning theory, and training techniques. Although dogs and cats are equally matched in popularity as pets, their social behavior, relationships with people, and the types of challenges they present can differ dramatically. *Canine and Feline Behavior and Training: A Complete Guide to Understanding Our Two Best Friends* is written to provide well-researched and accessible information about how dogs and cats behave, learn, and respond to training. The format that is used allows comparison between dogs and cats and promotes understanding and respect for each species in their own right. My primary intent with this book is to provide in-depth information about the behavior and communication patterns of domestic dogs and cats, and to promote positive approaches to training and behavior modification. Content focuses on the development of well-adjusted and well-behaved companion pets and on the use of appropriate and humane training techniques to prevent and solve problem behaviors.

The organizational structure of the book examines dogs and cats sequentially within topical sections. This allows readers who have an interest in both dogs and cats to compare and contrast their social behaviors, communication patterns, and problem behaviors. For those who are interested primarily in either dogs or cats, clear delineations between the two species are provided within chapter subheadings to allow for quick reference. The book is divided into three sections. The first explores domestication, developmental behavior, social behavior, and species-specific behavior patterns of dogs and cats. Emphasis is placed on learning to recognize and correctly interpret communication signals of dogs and cats. Because so many homes today include multiple pets, social relationships within species and between pets and their human caretakers are explored in detail. The second section of the book examines learning theory and its application to companion animal training and problem prevention. Topics in these four chapters include the use of learning theory in training programs, principles of behavior modification, and manners training for young and adult pets. Training instructions incorporate the principles of classical and operant conditioning and are designed for teaching desirable behaviors to dogs and cats and preventing problem behaviors. The underlying theme throughout the book is the promotion of training to build and maintain strong and positive bonds between owners and their pets. This paradigm recognizes and addresses the behavioral needs of dogs and cats, as well as owner responsibilities to reliably provide for those needs. Problem behaviors and their solutions are addressed in depth in the final section. Content in these three chapters emphasizes understanding the underlying motivations and causes of undesirable behaviors, the importance of exercise, mental stimulation, and companionship in pets' lives, and the use of methods that can successfully modify unwanted behaviors while still preserving the pet-owner relationship.

This book is written for both professional and hobbyist dog trainers, cat fanciers, and pet owners who are interested in understanding their pets better and learning how to best care for their companion animals. It can also serve as a textbook for a variety of professional programs in companion animal care and training. Chapters include more than 50 illustrations and photographs depicting communication signals of dogs and cats and various training techniques. Numerous sidebars are included to spotlight new information and to outline training steps, while boxes present personal case studies. This format is accessible and user-friendly and will encourage the practical use of the principles and training techniques that are outlined in the book. Finally, it is my sincere hope that the reader comes away from this book not only with an increased understanding of behavior and training, but also with the book's intended message of appreciation, tolerance, and empathy for the many dogs and cats who contribute so very much to the quality of our lives.

Linda P. Case
Mahomet, Illinois
June 2008

acknowledgments

This book would not have been possible without the help of numerous friends and colleagues who read and critiqued drafts of the manuscript. Their suggestions and help have been invaluable and have contributed significantly to the content and clarity of the book. I am deeply grateful to Rebecca Buraglio, Jill Cline, April Hammer, Susan Helmink, Erica Jewell, Jessica Lockhart, Sandy Myer, and Pam Wasson. Special thanks to my mom, Jean Palas, who read every word of the manuscript, and provided me with many insights and countless hours of enjoyable and challenging discussions about behavior, training, and our relationships with dogs and cats. All of the illustrations in this book are the work of Bruce MacAllister, who is not only an outstanding artist, but a longtime friend. Individual photographs were generously contributed by acquaintances, family, and friends; my appreciation for these goes to Meg and Paul Bender, Idil Bozkurt, Diana Chaytor, Justin Frisch, Jon Mnemonic, and Glen Peterson. Many of the photographs are the work of Heather Mohan-Gibbons, whose understanding of dog behavior and ability to think "one second before the dog" allows her to capture often-missed intricacies of behavior. Special mention also to Jim Robertson for his stunningly beautiful wildlife photography and for being a kindred spirit in his respect for all of the animals with whom we share this world. Finally, credit must go to David Rosenbaum, my acquisitions editor, whose enthusiasm and dogged (pun intended) persistence got me up off of the couch with the dogs to get started on this project. My thanks also to Chris Gifford, my editor, for her positive attitude, infinite attention to detail, and willingness to graciously consider my endless modifications and cogitations as we progressed through the editing process together.

And, as always, boundless gratitude goes to my husband and best friend, Mike, who continues to provide enthusiasm and support for all of my writing projects, invaluable computer help, and, most importantly, that needed time together, running and hiking the trails with our family of dogs. I hope that everyone enjoys reading the results of our hard work as much as I enjoyed the process.

In addition, Delmar Cengage Learning and the author would like to thank the following individuals for their review of the manuscript throughout the development process:

Anita Oberbauer
UC Davis

Gary L. Wilson
Moorpark College

Janice Siegford
Michigan State University

Dr. Glenn Wehner
Truman State University

about the author

Linda P. Case owns AutumnGold Consulting and Dog Training Center in Mahomet IL and teaches canine and feline behavior and training at the College of Veterinary Medicine at the University of Illinois. She is the author of numerous articles on companion animal behavior, nutrition, and healthcare and has written three other books: *The Dog: Its Behavior, Nutrition, and Health*, *The Cat: Its Behavior, Nutrition and Health*, and *Canine and Feline Nutrition: A Resource for Companion Animal Professionals*. Linda and her husband share their lives with three dogs, Cadie, Vinny, and Chip, and two cats, Tara and Pumpkin Joe, and enjoy running, hiking, and traveling with their dogs.

SECTION 1

Domestication, Social Behavior, & Communication

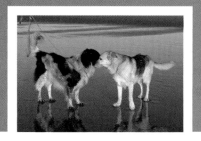

CHAPTER 1

The Beginning of the Friendship – Domestication

Without a doubt, the dog and cat are the most popular companion animals in the United States. According to a biannual survey conducted by the American Pet Product Manufacturers Association, there are more than 65 million dogs and 77 million cats living in homes today. Moreover, approximately one-third of dog lovers share their lives with two or more dogs, and almost half of pet-owning homes include multiple cats! So what exactly was it that brought humans and dogs, and humans and cats, together so many years ago? And, more importantly, what characteristics of these two species have enabled them to forge the strong and ongoing bonds with their human caretakers that are so important to us today?

THE EVOLUTIONARY HISTORY AND TAXONOMY OF DOGS AND CATS

Anyone who lives with both dogs and cats will emphatically affirm that the dog and the cat have some significant and important behavioral differences. In fact, it is often said that some people are "dog people" while other folks are "cat people" – a distinction that has much to do with respective differences between the natural behavior patterns of the two species. Despite these differences, however, when viewed in the context of evolutionary history, dogs and cats are actually quite closely related. Both dogs and cats are classified within the taxonomical order "Carnivora," a group of mammals that evolved approximately 40 to 60 million years ago and which today includes 17 families and about 250 species. The very first carnivores were collectively referred to as the Miacidae family. This was a very diverse group of small, slender, tree-dwelling predators, somewhat weasel-like in appearance. About 30 million years ago, the viveravines branched off from the miacines. Viveravines are now considered to be the oldest ancestor of the domestic cat, and miacines are the ancestors of our present-day dogs. Included with the dog in this group are the ancestors of the other extant (presently existing) canid species (wolves, jackals, coyotes, and foxes), as well as the bear, raccoon, and weasel. The viveravines further branched into two primary lines. One produced several of the large and now-extinct prehistoric cats, including *Smilodon*, the sabertooth tiger. The second line included *Dinictis*; a small cat that later evolved into several distinct cat species. *Dinictis* is considered to be the primary cat ancestor of all cat species alive today, including our domestic cat.

Today, along with our domestic dog, the other canid species that are found in North America include wolves (two species), coyotes (one species), and foxes (five species) (Figures 1.1 through 1.3). No present-day wild cousins of the domestic cat are found in North America, but subspecies of the wildcat live freely in Europe, Northern Africa, and parts of Asia. Canid and felid species are classified within separate families, the Canidae and Felidae, respectively, within the order Carnivora (Sidebar 1). Carnivores are so named because of a set of enlarged teeth (the carnassials) which comprise the enlarged upper fourth

figure 1.1 Timber Wolf (*Canis lupus*); The domestic dog's wild progenitor species

figure 1.2 Coyote (*Canis latrans*)

figure 1.3 Red Fox (*Vulpes vulpes*)

Sidebar 1 | CLASSIFICATION OF THE DOG AND CAT

Taxonomy refers to the present-day classification of a species. The domestic dog's genus is *Canis* and its species is *familiaris*, correctly expressed as *Canis familiaris* or *C. familiaris*. The Canidae family also includes the wolf, coyote, dingo, fox, jackal, and Cape hunting dog.

The domestic cat is classified as a member of the Felidae family. Like the dog, the cat (*Felis catus*) is considered to be a domesticated species that is taxonomically distinct from its progenitor species, the African wildcat (*Felis silvestris lybica*). This family includes the four genera *Felis, Lynx, Panthera,* and *Acinonyx*. The species included in these genera are considered to be the true cats, all existing as carnivorous predators. The *Felis* genus is comprised of 26 cat species, including the domestic cat, *Felis catus*.

TAXONOMY	CAT	DOG
Phylum	Animalia	Animalia
Class	Mammalia	Mammalia
Order	Carnivora	Carnivora
Family	Felidae	Canidae
Genus	*Felis*	*Canis*
Species	*catus*	*familiaris*

premolar and the lower first molar on each side of the mouth. Both dogs and cats have these dental adaptations, which are efficient for shearing and tearing prey. Carnivores also have small, sharp incisors at the front of the mouth for holding and dissecting prey, and four elongated canine teeth that evolved for predation and defense.

Interestingly, despite these dental modifications, not all of the present-day species that are included in the order Carnivora are strict carnivores. Some, such as bears and raccoons, are omnivorous and at least one species, the Giant Panda, is primarily vegetarian. This diversity is of practical significance when comparing the feeding behavior and dietary preferences of our present-day dogs and cats. Although both evolved as predatory species, the dog is decidedly more omnivorous than the cat, which is classified as an obligate carnivore. As we will see, these differences significantly affect the predatory and feeding behaviors, and the type of food-related behavior problems that each species tends to develop.

Domestication: Scavenging Wolves and Opportunistic Wildcats

Both dogs and cats are considered to be "domesticated" species, meaning that they are genetically and behaviorally distinct from their wild progenitor (ancestor) species. The phenomenon of domestication can be viewed as an evolutionary process in which the affected animals have been selectively bred over many generations to adapt to a new **ecological niche** – that of living in close association with humans. Domesticated animals rely almost exclusively upon human caretakers for survival and for the opportunity to reproduce. Specific behavioral adaptations that are common among all species of domestic animals include the absence of a fear of humans, enhanced adaptability and acceptance of handling and control, and increased sociability with humans and with other of their own species (conspecifics). For the dog and the cat, this enhanced sociability is often referred to as the human–animal bond. Indeed, it is the ability of dogs and cats to develop strong attachments to their human caretakers that is in many ways responsible for their popularity as animal companions today.

Changes of Domestication: Many of the physical and behavioral changes of domestication are explained by alterations in the rate of development of the young animal, a process known as *paedomorphism*. Body features and behaviors that are normally expressed only during the animal's juvenile period are retained into adulthood. Paedomorphosis can be achieved through changes in the onset, rate, or completion of various types of physical development. One subtype of paedomorphosis, called **neoteny**, is defined as a reduced *rate* of development, resulting in the persistence of juvenile characteristics. For the domestic dog, the selection for traits that occur during different points of the juvenile period may be one source of the wide variation in size, body type, and behavior that are seen in different breeds. This diversity is not as obvious in the domestic cat, which as a species shows much less variation in appearance than does the dog.

Dogs as Village Scavengers: Domestication of the dog is believed to have begun during the latter part of the Mesolithic period, 12,000 to 15,000 years ago, on the continent of Eastern Asia (Sidebar 2). During this time, humans were changing from being completely nomadic hunter-gatherers to living in semi-permanent settlements. For a number of years, a theory that was used to explain the domestication of the dog relied upon the assumption that hunters of the Mesolithic period coexisted with wild wolves and often competed for the same prey species. This explanation further assumed (since archeological evidence does not preserve attitudes or belief systems), that Mesolithic hunters recognized the superior hunting abilities of wolves and exploited those abilities by capturing, raising, and taming individual wolf pups, who were then used as hunting aids. Although this theory gained enormous popular acceptance, in recent years it

Sidebar 2: TRACING THE DOG'S ORIGINS IN THE NEW WORLD

For many years, it was believed that the dog, similar to pigs and horses, had been domesticated on several occasions, in different geographic regions of the world. The theory of multiple domestication events helped to explain the diversity that we see today in different breeds of dogs as well as their presence throughout the globe. However, a group of evolutionary biologists at the Sweden's Royal institute of Technology recently provided evidence for a single domestication event in the dog occurring approximately 15,000 years ago in Eastern Asia, while another group has been able to explain how the dog first arrived in the Americas.

Dr. Peter Savolainen and his colleagues studied mitochondrial DNA mutations in present-day dogs and wolves to estimate evolutionary changes over time in each species. The evidence showed that the domestic dog split off from wolves about 15,000 years ago and, showed no evidence of any other domestication event in either the Old or New World after that point in time. Savolainen's group even suggests that the most influential genetic contributions to our present-day dogs came from just three female Asian wolves! (More recent studies have increased this estimate to five). The implication of this research is that although dogs were found all around the world 9,000 years ago, this must have been because they were already traveling with their human companions. Indeed, Vila's group's examination of the genetic makeup of the Mexican hairless dog (or Xoloitzcuintel), which is considered to be one of the oldest identifiable breeds of America, showed that this ancient dog is closely related to several modern European breeds of dogs and were not at all related to the American grey wolf.

Although the previous results supported a single, Eurasian domestication event, they could not be considered conclusive because genetic analysis of present-day American dogs could reflect inter-breeding between native American dogs and dogs who had been brought to the New World with European explorers. To solve the mystery for once and for all, Jennifer Leonard examined genetic material recovered from the ancient, pre-Columbian remains of 17 dogs found in archeological digs of Latin America and Alaska. Her results showed unequivocally that while the dogs were all closely related to each other, they were not related to American wolves.

Collectively, these studies show that a single domestication event occurred for the dog, somewhere in Europe or Asia, from Old World grey wolves, and a separate domestication did not occur in the New World involving the North American wolf. This means that humans brought the domesticated dog with them when they colonized America from Asia approximately 15,000 years ago. Although inter-breeding between dogs and wolves may occur occasionally (and between dogs and coyotes, as well), there is no evidence of enduring genetic contributions of the American wolf in any breeds that have been studied to date. All of the breeds that we know today are descended from a single sub-species of wolf living many years ago in Eastern Asia.

Leonard JA, Fisher SC. The origin of the American dogs. *30th WSAVA Conference*, Mexico City, Mexico, May 11–14, 2005.

Savolainen P, Zhang Y, Luo J, Lundeberg J, Leitner T. Genetic evidence for an East Asian origin of domestic dogs, *Science*, 298:1610–1613, 2002.

Vila C, Maldonado J, Wayne RK. Phylogenetic relationships, evolution, and genetic diversity of the domestic dog. *J Heredity*, 90:71–77, 1999.

---

has been successfully challenged by evolutionary biologists. An alternate, more defensible theory proposes that the early domestication of the dog was largely unintentional and occurred as a result of adaptive radiation and natural selection as wolves adapted to a new ecological niche – the village dump.

At the end of the last Ice Age, humans gradually became less nomadic and began to live in semi-permanent villages. This new way of life not only benefited people, but also created a new ecosystem into which wolves could adapt. Specifically, the outskirts of permanent villages provided a steady source of food from human waste and garbage. These dump areas also provided relative safety from other predator species and the potential for new and protected nesting sites. Although popular mythology surrounding wolves often depicts them as efficient predators, they are also highly opportunistic scavengers, capable of consuming and thriving on a highly varied and omnivorous diet. Therefore, as a species, the wolf was already well-suited to feed at these newly formed "dump sites," which contained a wide variety of waste and food scraps.

The selective pressures of this new environment favored less timid wolves with a higher tolerance (less fear) of human proximity. Quite simply, the less fearful individuals had increased opportunities to feed because they stayed longer and fled less readily than the more timid animals. The social behavior of the wolf was also affected by this new way of life. Selective pressure for social hierarchies and strict pack order relaxed as pack-hunting behaviors were replaced by semi-solitary or group-scavenging behaviors. As this *proto-dog* became more adapted to eating and reproducing in the presence of humans, isolated sub-populations became "naturally" domesticated. In this branching of the dog and wolf's evolutionary tree, the wild wolf, *Canis lupus*, remained a pack-living predator, while the dog evolved specialized adaptations for living in close proximity to humans. It is hypothesized that it is from these semi-domestic village scavenger populations that individual dogs were eventually chosen for further taming. Eventually (many generations later) it was the selective breeding of these dogs that led to the working partners and companions who we know today.[1]

Cats as Opportunistic Mousers: The African wildcat has a similar early domestication story to that of the dog, although it occurred more recently. Just as village dump sites attracted scavenging wolves, the grain storage barns of the ancient Egyptians living about 4,500 years ago (and possibly several hundred years earlier in Cyprus) attracted the African wildcat. Granaries naturally are infested with mice and other rodents, which happen to be one of the preferred prey species of the African wildcat. The barns of agricultural communities also provided protected nest sites for female cats to raise their kittens. The African wildcat is a solitary species and by nature is extremely shy of humans. Adults live completely separate lives and use established territories to advertise their presence and prevent contact with others. With the exception of coming together during mating season, males and females do not form lasting pair bonds. Males have no involvement at all in raising kittens and, once kittens have dispersed from their mother and littermates, adults avoid contact with others of their species except when breeding.

So, just as the exploitation of village dump sites by wolves led to changes in their social behavior, so too did the ecosystem provided by barns and granaries exert new selective pressures on the African wildcat. As wildcats began to exploit the surplus mice population of Egyptian barns, selective pressure favored those individuals who were more tolerant of the presence of other cats. In the presence of plentiful and easily available food, the ability to peacefully share food and nesting sites enhanced an individual's chance of surviving and producing kittens. And, because there would also have been human presence around these sites, those cats who were less fearful of humans would also be more "fit" in this new ecological niche. Over many generations, this evolving subpopulation became reproductively isolated from the African wildcat and group-living cats of barns began to exhibit modified social behaviors. Females shared protected nesting sites within barns and communally nursed their kittens. Loosely organized group-living communities evolved, a way of life that was decidedly different from the African wildcat's solitary lifestyle. Natural selection for a more communal-living mouser changed several important aspects of the cat's behavior patterns, leading to a domesticated animal who is actually much more social than is often assumed. The importance of the domestic cat's sociability becomes clear when we examine the cat's interaction with human caretakers and with other cats in multiple-cat homes.

SOCIAL ORGANIZATION: WHY DOGS ARE NOT CATS AND CATS ARE NOT DOGS

Although dogs evolved from an obligatory pack-living species (the wolf) and cats from a highly solitary species (the wildcat), the process of domestication along with the inherent flexibility of social behaviors within species has changed the types of social groups that are common to companion animals today. In addition to being part of human society (and human social groups), dogs and cats live in a variety of social environments – as single pets, with others of their own species, and, very often today, with one another as household companions. Recent studies of dogs and cats in homes and as strays or feral animals have led to an increased understanding of the types of social relationships that our two best friends can develop.

The Social Groups of Dogs

It is well-established that the dog's closest wild relative, *Canis lupus*, lives as a highly social, predatory species. In the wild, wolf packs consist of small groups of related individuals who remain together throughout the year to hunt, rear young, and protect a communal territory. Social ranking within packs is important, as it facilitates the cooperation that is needed for hunting, raising young, and protecting territories together. The pack hierarchy is maintained using highly ritualized behaviors that signal an individual wolf's intentions and rank. Having ritualized signals that communicate dominance, submission, and appeasement enhance the survival and reproductive chances of each individual and the pack as a whole, and also serve to minimize aggressive interactions between members of the group.

Flexibility of Social Groups: Although the pack is the first type of social organization that comes to mind when most people think of wolves, as a species the wolf is capable of forming a variety of different types of social groups. Biologists studying wild canids have found that the type of social organization that wild canids form within a region is influenced by the type of prey species available, and its abundance and distribution within their territory. Wolves have been observed living in unusually large packs of 25 or more, as small family groups, as pairs, and as solitary animals. Studies of free-ranging domestic dogs in rural and urban environments report similar variations in the social behavior of domestic dogs, ranging from solitary living or loose pair-bonds to stable but loosely organized groups and to groups demonstrating some hierarchical ranking (Sidebar 3).

The domestic dog has inherited the social nature of his original wolf ancestor. Dogs have also retained parts or all of the wolf's ritualized behavior patterns that function to signal dominance and submission. However, just as domestication has caused the dog to diverge in physical appearance from the wolf, so too has social behavior been modified. Specifically, the provision of food, shelter, and protection from other predators over hundreds of generations removed the selective pressure for the ranked social groups that were needed by wolves to hunt large prey and raise the pups of the breeding pair. The domestic dog is highly social and demonstrates the same social flexibility of his wild cousins. Dogs form close relationships with human family members, and many dogs living in multiple pet households also develop primary social attachments with the other dogs or pets who are present. Conversely, the presence of rigid dominant/submissive rankings or the identification of a particular "alpha" dog within a group is less

Sidebar 3: SOCIAL BEHAVIOR IN FREE-RANGING DOGS

The earliest studies of free-ranging dogs were conducted with dogs who may have been owned, but were allowed to run free in urban and suburban settings. These studies, conducted in the 1970s and 1980s, reported that free-ranging dogs were solitary or, less frequently, paired with one or two other dogs. The formation of groups of dogs was rare and only tended to occur when dogs congregated for short periods of time around a food source or protected resting place. In addition, the dogs did not show any signs of territoriality and **agonistic** (aggression) encounters were rare or not reported. These studies concluded that free-ranging dogs in cities and suburbs do not form stable social groups, probably because many had owners who regularly fed them and provided some level of care.

In contrast, other researchers found that stray (or feral) dogs in urban areas will form stable social groups and demonstrate long-term affiliative relationships when there is no human care or intervention. In these studies, individuals tended to forage alone for food, but small stable groups defended common territory, raised puppies cooperatively, and demonstrated behaviors consistent with ranked social hierarchies. Together, these studies suggest that the domestic dog, just like his wolf cousin, modifies social behavior in response to the distribution and abundance of food and other resources.

In short, although the identification of the dog as our "pack-living" companion animal has been highly popularized, in reality the dog is capable of forming a variety of different types of social relationships.

Berman M, Dunbar I. The social behavior of free-ranging suburban dogs. *Applied Animal Ethology*, 10:5–17, 1983.

Font E. Spacing and social organization: Urban stray dogs revisited. *Applied Animal Behaviour Science*, 17:319–328, 1987.

Fox MW, Beck AM, Blackman E. Behavior and ecology of a small group of urban dogs (*Canis familiaris*). *Applied Animal Ethology*, 1:119–137, 1975.

Pal SK, Ghosh B, Roy S. Agonistic behaviour of free-ranging dogs (*Canis familiaris*) in relation to season, sex and age. *Applied Animal Behaviour Science*, 59:331–348, 1998.

---

obvious. Although dogs do express signals that communicate dominance and submission, these communication signals should not be confused with the existence of hierarchical roles of individuals within a strictly ranked social group.

Dogs Are Not Wolves: Generations of selective breeding to develop dogs for different functions diversified the dog with regard to the ways in which individuals form and maintain pair relationships. While some breeds and individual dogs do not generally form rigid dominant/subordinate relationships, others more readily seek out and adopt ranked relationships. In some cases, these relationships are obvious, with one dog displaying offensive threats towards other dogs, guarding access to food and toys, and consistently gaining access to choice resting places. However, in most homes, dominant/subordinate interactions between dogs are less obvious and, when they do occur, are situational and context-specific. For example, one dog may guard his food bowl from other dogs in the home, but not show any response if another dog takes his toy or sleeps in his favorite spot. In addition, established rank orders, when they are evident, are not characterized by aggressive fights or even excessive posturing. More commonly, they can only be discerned through observing which dog consistently gains access to desired resources and which dog or dogs display appeasement or submissive postures most frequently during interactions.

In many multiple-dog homes, rank order is not clearly evident, the dogs show little or no competition for resources, and displays of conflict are rare or absent. For this reason, describing all dog social groups in terms of who is "dominant" and who is "subordinate" is misleading and probably inaccurate for most inter-dog relationships. In most multiple-dog homes, the majority of dog-to-dog interactions are deferential in nature, and communication tactics that signal appeasement, conciliation, and invitations to play are much more common than are agonistic (conflict-related) interactions. Many dogs living together behave in a peaceful and affectionate manner toward each other and the occurrence of aggressive displays and fights are rare. This fact should not be interpreted as denying the existence of seriousness of inter-dog aggression problems within homes. When these conflicts occur, they can be very difficult problems to treat. (See Chapter 11, pp. 287–288 for a complete discussion.) However, the highly popularized but inaccurate portrayal of normal dog-to-dog relationships as consisting of endless rounds of scheming and battling to achieve "alpha" status does not generally apply and ignores the context-specific nature of most inter-dog relationships. Because relationships between dogs in multiple-dog homes are more fluid and less hierarchical, and do not directly impact a dog's ability to obtain food or survive, the dog's social organization is more accurately defined as a social group rather than the highly popularized and value-laden label of "pack."

For the same reasons, the "pack" model of social organization is an inaccurate model for describing social relationships that dogs have with their human caretakers. Concepts of pack behavior and ranked social groups became conventional as a way of explaining relationships between dogs and their owners during the latter half of the 20th century. This resulted from an extrapolation of that period's understanding of wolf pack behavior onto the social behavior that is observed between dogs and their human owners. Because social ranking was emphasized, dogs were considered to all be "naturally" dominant and were expected to constantly challenge their human caretakers in an effort to achieve "alpha" status. As a result of this highly popularized (but incorrect) concept, almost any behavior that a dog offered that was not in compliance with their owner's wishes often came to earn the label of "dominant."

However, unlike wolves, the dog is the result of generations of purposeful selective breeding for behaviors that enhance trainability and sociability with humans. Domestic dogs are well-adapted to forming strong and naturally deferential relationships with their human caretakers. Although dogs are still capable of displaying dominant and submissive signals, the expression of these communication patterns should not be confused with unrelenting attempts to gain social status over their owners. Although dominance challenges between dogs and their owners can occur in some dogs, the use of a dominance model for describing the normal social relationships between all owners and their dogs has been largely discarded. Many behaviors that in the past were interpreted as a dog "being dominant" are more often simply unruly or attention-seeking behaviors in dogs who have not been trained to behave differently.[2] These behaviors and their interpretations are discussed in detail in future chapters.

Social Behavior of Cats

It is true that the cat is not as highly social as the dog. And, of course, we simply need to look to the natural behavior of the African wildcat to understand why. However, during domestication, the African wildcat slowly adapted to live in higher densities with other cats and to tolerate human presence. These changes provided selective pressure for the cat to develop communicative signals that facilitate living in groups. Similar to dogs, studies of free-living and feral cats have found that the type of feline social system that prevails in a particular location is related to the availability of food or prey, the number of other cats who are sharing the area, and the frequency and types of interactions that occur with humans. For example, feral cats tend to live solitary lives around farmed areas containing abundant and well-dispersed populations of mice and voles. Conversely, when a centralized and reliable food source is present, free-living cats readily form social groups. These usually occur around waste areas or garbage dumps or when someone regularly provides food to stray cats (Sidebar 4).

In homes, the domestic cat is observed living as an only cat (solitary), as part of a pair or group of cats, and, not uncommonly, with one or more companion dogs (Sidebar 5). The cat's flexible nature and ability to adapt to various types of social groups is demonstrated by the fact that many cats living together in homes exhibit relatively peaceful coexistence, with affiliative bonds forming between pairs or groups. However, within groups of cats, the existence of established hierarchies has not been demonstrated and social rankings do not appear to be an important component of cat relationships. Although one male or female in a group may be more aggressive than others, interactions between cats within groups do not conform to the accepted definition of ranked orders in which a dominant animal controls access to resources such as food, resting places and interactions with others in the group (e.g., opportunities to mate). This type of

SOCIAL BEHAVIOR IN FREE-LIVING AND FERAL CATS — Sidebar 4

Studies of free-living cat colonies have shown that communal access to a concentrated and stable food source supports the development of stable cat social groups. Also important is the availability of protected shelter and nest sites. Conversely, free-living cats will adopt solitary living when there is a surplus of well-dispersed prey or food. In these cases, which are almost exclusively rural, adult cats live and hunt alone, coming together only to mate.

The most frequently observed examples of group living in cats are barnyard cats who are fed regularly by the human residents, or stray/feral cats who congregate around a garbage site or who are fed by a human caretaker. Studies of barnyard cats show that these social groups consist of related adult queens and their offspring. Adult male cats are found living on the periphery of groups and often travel between several groups of females to mate. The spontaneous movement of queens between groups is rare and familiar queens within a group show hostility toward strange queens who try to join an established group. Individual bonds form between cats and it is not unusual for two queens to raise their kittens together in a single nest site. **Allogrooming, allorubbing** and other affiliative behaviors are commonly observed between cats within groups. While a particular group of females usually has one adult tomcat who mates most often with the females and will fight with other adult males who visit, there is no distinct social hierarchy between breeding males, and intact males rarely behave as full members of the group.

Calhoon RE, Haspel C. Urban cat populations composed by season, sub habitat and supplemental feeding. *Journal of Animal Ecology*, 58:321–328, 1989.

Genovesi LA, Besa M, Toso S. Ecology of a feral cat *Felis catus* population in an agricultural area of northern Italy. *Wildlife Biology*, 1:333–337, 1995.

Liberg O, Sandell M. Spatial organization and reproductive tactics in the domestic cat and other felids. In: *The Domestic Cat: The Biology of its Behaviour*, 1st edition, DC Turner and P Bateson, editors, pp. 83–98, Cambridge University Press, Cambridge, UK, 1988.

Warner RE. Demography and movements of free-ranging domestic cats in rural Illinois. *Journal of Wildlife Management*, 49:340–346, 1985.

---

resource guarding and control occasionally is seen between individual cats, but is generally not a regular component of cat social groups.

Domestication created a cat with an enhanced tolerance of and affiliation for others compared with its wild ancestors, perhaps as a result of paedomorphism. Affiliative behaviors such as care-soliciting, mutual grooming, and playing are all seen in young African wildcats but are infrequent or totally absent in adults.

Sidebar 5: CATS AND DOGS LIVING TOGETHER – CAN THEY COMMUNICATE?

It is often thought that communication between dogs and cats is difficult or impossible because of their differing evolutionary histories and social behaviors. However, the increased frequency with which people keep dogs and cats together in homes without problems led investigators to study the relationships between dogs and cats and their abilities to correctly interpret each other's communication signals (Feuerstein and Terkel, 2008).

The investigators interviewed 170 owners who shared their homes with both a dog and a cat and also observed behavior of the pets in a subset of 45 homes. Greater than 60 percent of owners in the study reported that their dog or cat was amicable toward their pet of the opposite species, and only 9 percent stated that their pet showed aggression toward the alternate-species housemate. The remaining pets were largely indifferent to each other. Interestingly, the type of relationship that existed was not affected by species. Dogs were just as likely to be friendly or aggressive toward their cat housemate as were cats likely to be friendly or aggressive toward the dog in their home. This result is in contrast to widely held beliefs that cats instantly dislike dogs and that dogs are more likely to want to either befriend or to chase or kill a cat.

When the pets' behaviors were observed, mutual play made up a substantial proportion of interactions, and maintaining close proximity by staying in the same room or resting together was also common. Cats offered significantly more play behaviors towards dogs than vice versa, but also were more likely to be fearful or aggressive than dogs. The order of adoption and age at which the pet was adopted also were important factors affecting pets' relationships. In homes in which the cat had been adopted first, dogs tended to be more likely to have friendly relationships with the cat than when the dog had been adopted prior to the cat. However, for cats, order of adoption did not affect the cat's relationship with the dog. As expected, both dogs and cats who were adopted as young animals were more likely to develop affectionate relationships with the other pet than were animals adopted when older.

The investigators were also interested in the use of communicative body signals between dog and cat housemates. Of the 45 pairs observed, interactions in which the body posture of one species had an unrelated or opposite meaning to the other species were still correctly interpreted by the receiving animal in the majority of interactions. Another interesting finding was that dogs and cats living together in homes often greeted using a feline-specific pattern, the nose-touch, rather than canine-specific greeting signals (Figure 1.4 and Figure 1.5).

figure 1.4 Two familiar cats greeting with "nose touch"

figure 1.5 Dog and cat housemates greeting with "nose touch"

This study suggests that dogs and cats who share homes often develop close affiliative relationships and show this through proximity, mutual play, and greeting behaviors. In addition, both dogs and cats are able to learn to understand the other species' communication signals. Owners may facilitate the development of these positive relationships by adopting the cat first, and introducing pets when they are young (preferably six months or younger for cats and one year or younger for dogs).

Feuersten N, Terkel J. Interrelationships of dogs (*Canis familiaris*) and cats (*Felis catus*) living under the same roof. *Appl Anim Behav Sci*, 113:150–165, 2008.

In contrast, adult domestic cats continue to seek out petting and affection, are often very playful and loving, and commonly develop strong and enduring bonds of attachment and affection with their human owners and with other cats in their social group. While some cats are social only with their human family (and sometimes even only with one person in the home), others are virtual social butterflies, ready to greet and interact with any visitor. Similarly, while some cats adapt very well to living with other cats, others do not and are best kept in single-cat homes.

HISTORY OF DOG BREEDS AND BREED-SPECIFIC BEHAVIORS

Of all of the domesticated species of animal that exist today, the dog has probably been subjected to artificial selection for the longest period of time. Since domestication, at least 4,000 generations of dogs have been selectively bred for various functions, producing a diverse number of breeds. Most of the functional groups that we recognize today (hunting, guarding, and herding dogs) were already in existence during the Roman period almost 2,500 years ago. Since that time, selective breeding of dogs to meet the functional needs of humans took place during two major periods. The first of these was during the Middle Ages, when hunting became associated with aristocracy and was restricted to the land-owning nobility. Different types of dogs were developed and bred for hunting different game species. Examples of breeds of that period include deerhounds, beagles or harriers, foxhounds, various types of terriers, and, later in the period, various types of gundogs. Although the landed gentry paid attention to pedigrees and prided themselves on the working prowess of their kennel of dogs, selective breeding and breed development focused upon working ability and behavior and was not restricted, as are purebreds today, within a set of animals who were identified as registered members of their breed.

The second wave of breed development occurred during the mid-19th century and represented a new approach to breed development. Breeds gradually became defined less by function and more by their uniformity in appearance and genetic relatedness. The current creation of a purebred breed of dog requires four essential elements: a set of founder animals, reproductive isolation of those animals from the general population of dogs, generations of inbreeding within that group to stabilize physical and behavioral attributes that define the breed, and, finally, selection of breeding animals who most closely conform to the prescribed "breed type." At some point during this process, the new breed is recognized and accepted by an external purebred registry organization. The majority of our modern-day breeds have been developed within the last 150 years by restricting breeding to animals with verifiable lineages (i.e. "pedigreed" dogs). Prior to that time, breeding was not restricted within purebred lineages and the name of a breed simply reflected the function of a loosely related group of dogs who were all used in a similar manner for either hunting, herding, or guarding.

The first breed clubs of the 1800s created most of the purebreds that are in existence today. The growing dog fancy of that period placed unprecedented emphasis on the purity of dogs' lineages and began to intensively select for dogs who conformed to a standard appearance described as "ideal" for the breed. To codify this practice, kennel clubs governing the first dog shows during the late 1800s established "breed barrier" rules. These regulations maintained that only dogs who were the offspring of a registered dam and sire were recognized as registered members of the breed and likewise eligible for exhibiting and breeding. Therefore, from each breed's point of origin and creation of a stud book, all future breeding was limited to descendents of the breed's founding dogs. These rules ensured that the genetic pool for each breed was reproductively isolated from that of the general dog population (much as the ruling aristocracy of that period isolated themselves from the working classes). This projection of the values of the upper classes is reflected in the cultural importance assigned to dogs' pedigrees and lineages. Breeding related individuals to one another became a common practice that was used to rapidly create a uniform appearance and to enhance the expression of desirable traits within a line of dogs.

Within each of these newly emerging and reproductively isolated breeds, the number of the founding dogs, their genetic diversity (heterozygosity), and the vigor with which the prohibition against breeding outside of registered lines was enforced would eventually impact the degree of inbreeding and genetic homogeneity for all future generations. Unfortunately, while this approach to selective breeding initially enhanced the ability of dogs to perform (in the short term), and resulted in the highly uniform physical appearance of breeds that we recognize today, the imposed genetic isolation and the inclusion of relatively small numbers of founding individuals in the original breed gene pools has also contributed to the many genetically influenced diseases occurring in purebred dogs today.[3]

Breeding for a specific type of working function impacts both physical traits of the dog and the behaviors that are necessary to carry out that function. For example, the long, slender legs and deep chest of the Greyhound contribute to its ability to hunt using its eyesight (Figure 1.6). This breed also possesses a very strong chase instinct, which is considered to be a modification of predatory behavior. By comparison, the short, thick legs of the Basset Hound contribute to this breed's talent as a scent trailer, along with behaviors that many owners can find frustrating – such as a propensity to keep its nose to the ground when out walking (Figure 1.7). Therefore, depending upon the original function for which a breed or breed-type was developed, different dogs display certain behavior patterns in variable manners or to varying degrees of intensity. Today, although

figure 1.6 Greyhound (Sighthound Breed)

figure 1.7 Bassett Hound (Scent Hound Breed)

the majority of dogs are kept primarily as companions and are not used for the function of their breed, they still inherit the behavior patterns and predispositions that were strongly selected for in the development of that breed (Sidebar 6 and Sidebar 7). These breed-specific functions and their associated behavior patterns must always be considered when teaching new behaviors and when attempting to understand and address behavior problems in individual dogs.

Sidebar 6: COMMON BREEDS OF DOGS

Sporting Breeds	*Pointers:* German Shorthair Pointer, Pointer, Wirehaired Pointing Griffon *Setters:* English Setter, Gordon Setter, Irish Setter *Retrievers:* Golden Retriever, Labrador Retriever, Flat-coated Retriever *Spaniels:* Brittany, Cocker Spaniel, English Springer Spaniel
Hounds	*Scent Hounds:* Basset Hound, Beagle, Bloodhound
	Sight Hounds: Borzoi, Greyhound, Saluki, Whippet
Working Breeds	Akita, Alaskan Malamute, Boxer, Great Dane, Mastiff, Newfoundland, Rottweiler, Siberian Husky
Terriers	Border Terrier, Bull Terrier, Miniature Schnauzer, Parson Russell Terrier, Smooth and Wire Fox Terrier
Toy Breeds	Cavalier King Charles Spaniel, Chihuahua, Maltese, Pekinese, Pomeranian, Poodle, Pug, Shih Tzu, Yorkshire Terrier
Herding Breeds	Australian Shepherd, Belgian Tervuren, Border Collie, Collie, German Shepherd Dog, Pembroke Welsh Corgi, Shetland Sheepdog

Sidebar 7: GENERAL BREED-SPECIFIC BEHAVIORS IN DOGS (AMERICAN KENNEL CLUB CLASSIFICATIONS)

- *Sporting Breeds (Gundogs):* These breeds were developed to aid hunters by locating, flushing, and retrieving game on land and in water. Sporting dogs are energetic and active, and require regular vigorous exercise. They are generally highly trainable and social, and low in aggressive reactivity.

- *Hounds:* The two primary types of hound were both developed for hunting. The scent hounds follow a scent trail to find game, while sight hounds use eyesight and speed to chase and capture quarry. Hounds work well ahead of the hunter and, as a result, are relatively independent or even aloof in nature. Some sight hounds, such as the Greyhound and Whippet, are known for their extremely gentle and quiet dispositions.

- *Working Breeds:* Dogs classified as working breeds were bred to guard property or livestock, pull sleds, or perform water rescues. Because they were often required to actively protect by warning or even attacking intruders, the working breeds are high in reactivity and moderate to high in aggression. These dogs tend to bond strongly to one person or family and, when raised in a structured environment, are highly trainable.

- *Terriers:* Terriers were developed to find and kill small rodents and other animals that were considered to be pests. These breeds worked with little or no direction from their handler and were required to immediately kill their prey upon catching it. These two requirements resulted in breeds who have low-to-medium trainability and very high reactivity. In general, terriers show increased inter-dog aggression as well as a strong predatory response.

- *Toy Breeds:* Many of these dogs represent miniaturizations of other breeds. In some cases, they retain behaviors similar to that of their larger forefathers. In others, a more subordinate nature was selected along with the neotenized features. The toys were probably the first true companion dogs, and many of these breeds reflect this in their strong predisposition to bonding to humans, puppy-like behaviors, and high trainability.

- *Herding Breeds:* Herding breeds were developed to move livestock. They are considered to be highly trainable and will bond very strongly to their human caretakers. Because of their need to respond quickly to the movements and changes in the behavior of the herd, herding dogs are also usually highly reactive and have a strong chase instinct.

HISTORY OF CAT BREEDS AND BREED-SPECIFIC BEHAVIORS

Although the cat is considered to be a domesticated species, the breeding of cats has historically been under very little human control. Cats are still very capable of living on the peripheries of human communities, mating and raising kittens without human care or interference. As a result, barn cats, stray cats, and even feral cats have been a primary source of companion cats for many generations. It is only within the last 150 years that purebred cats have been developed through artificial selection and strict controls over breeding. Even today, the majority of cat owners still share their lives with "mix-breed" cats (typically referred to as domestic shorthairs or domestic longhairs), and only a relatively small proportion select a purebred cat as a companion.

Even without human interference, a number of mutations occurred early in the cat's domestic history that led to a variety of new coat colors. The cat's wild type of coloring, seen in *Felis silvestris*, is the striped tabby (also called mackerel). Another type of tabby, called the blotched or classic tabby, is considered to be an early mutation, and today is commonly seen in domestic cats of many breeds and breed mixes (Figure 1.8). Other coat colors that emerged early and which

figure 1.8 Cat showing the "wild type" of striped or mackerel tabby color pattern (right) and the blotched (also called classic) tabby color pattern (left)

now are common in cats include black, orange, white-spotted, and all-white. Once human intervention in breeding became established, additional colors, coat patterns, and coat types began to emerge as cat fanciers selected for unusual and often bizarre characteristics in their quests to create new breeds of cat.

The practice of keeping cats as companions first became popular during the early 1800s. Selective breeding for "pedigreed" cats began during the latter part of that century. The first recorded cat shows were held during the 1870s in England and in the United States, and official studbooks for the registry of pedigree (purebred) cats were established about 20 years later. Today, more than 400 cat shows are held annually in the United States. These are organized and regulated by several purebred cat registry organizations, such as the Cat Fanciers Association, the International Cat Association, and the Canadian Cat Association. These organizations differ somewhat in the way that they classify cats, but generally accepted breed categories include shorthaired breeds, longhaired breeds, Rex cats, spotted cats, and tailless cats (Sidebar 8). In addition, cats are often described as being of either "cobby" or "foreign" body type. The cobby breeds include cats who have a compact, heavy-boned and sturdy body type, with a deep chest, broad shoulders and hindquarters, short legs, and a short, round head. Although there are many variations in this type, it is exemplified by the British Shorthair and Persian breeds (Figure 1.9). Conversely, cats described as having a foreign or oriental body type are light-boned, with a narrow, wedge-shaped head, long legs, and a long, slender body. The Siamese is considered to be an extreme version of this body type (Figure 1.10).

Although cat breeds differ very dramatically in appearance, differences in behavior between breeds are less evident. This may be in part because, unlike the dog, different cat breeds were not developed for different working functions. Rather,

BREEDS OF CATS — Sidebar 8

Abyssinian, American Shorthair, British Shorthair, Burmese, Colorpoint Shorthair, Russian Blue, Siamese, Tonkinese	**Shorthair Breeds**
Angora, Balinese, Birman, Himalayan, Javanese, Maine Coon, Persian, Turkish Van, Norwegian Forest Cat, Somali, Turkish Van	**Longhair Breeds**
Cornish Rex, Devon Rex, Selkirk Rex	**Rex Cats**
Bengal, Egyptian Mau, Ocicat	**Spotted Breeds**
American Bobtail, Cymric (longhaired Manx), Japanese Bobtail, Manx	**Tailless and Bobtail Breeds**
American Curl, Japanese Bobtail, Munchkin, Ragdoll Cat, Scottish Fold, Sphynx	**Rare and Unusual Cat Breeds**

figure 1.9 Manx cat with cobby body type

figure 1.10 Siamese cat with oriental body type

cats exhibiting an unusual coat color, coat type, or even an anatomical mutation such as shortened legs or a folded ear were selected to create new breeds of cat. The absence of strong selective pressure for cats to perform different functions has resulted in much less diversity in behavior patterns within the population of domestic cats, when compared with the diversity of behaviors that we see among dog breeds. Still, many breeders and purebred proponents maintain that some behavioral differences are found among cat breeds. For example, the Siamese cat is commonly thought to be highly affectionate and vocal when compared with Persians, who are typically thought to be inactive and quiet.

Several survey studies of cat show judges and veterinarians have supported the general temperament characteristics that are reported in most cat books about popular breeds of cats. A recent study surveyed cat owners' perceptions of their cats' behaviors and found that two popular breeds of cat (Persian and Siamese) were perceived by their owners as being more affectionate, vocal and friendly than non-pedigreed cats.[4] However, because this was a survey study of owner attitudes that did not directly assess cat behavior, these differences may have been due as much to the expectations and beliefs of the owners as to true differences between purebred and mixed-breed cats. The study's authors suggested that selective breeding of cats in general seems to have selected for a generalized increase in sociable and friendly cats, regardless of breed. However, because so few studies of this type have been conducted, it is still difficult to make qualified conclusions about the effect of breed on the behavior of individual domestic cats.

REVIEW QUESTIONS

1. What factors in human evolution facilitated the domestications of the dog and cat, respectively?
2. During the domestication process, what changes occurred in canid social behavior? What specific selective pressures caused these changes?
3. How did the social behavior of the wildcat change during domestication? What specific selective pressures caused these changes?
4. In what ways does a dog's breed (or breed type) influence his or her temperament and behaviors?
5. Describe the various types of social groups to which cats living in homes may belong.

REFERENCES AND FURTHER READING

Adams JR, Leonard JA, Waits LP. **Widespread occurrence of a domestic dog mitochondrial DNA haplotype in southeastern US coyotes.** *Molecular Ecology,* 12:541–546, 2003.

Beadle M. **The cat: History, biology, and behavior,** Simon and Schuster, New York, NY, 1977.

Byrne RW. **Animal communication: What makes a dog able to understand its master?** *Current Biol*, 13:R3467–R348, 2003.

Clutton-Brock J. **A review of the family Canidae with a classification by numerical methods.** *Bull Brit Museum Nat Hist Zoology*, 29:117–199, 1976.

Clutton-Brock J. **Man-made dogs.** *Science*, 197:1340–1342, 1977.

Clutton-Brock, J. **A natural history of domesticated mammals.** Cambridge University Press, Cambridge, UK, pp. 133–140, 1999.

Coppinger R, Coppinger L. **Dogs: A startling new understanding of canine origin, behavior, and evolution.** Scribner, New York, NY, 352 pp., 2001.

Davis SJ, Valls FR. **Evidence for domestication of the dog 12,000 years ago in the Natufian of Israel.** *Nature*, 276:608–610, 1978.

Driccoll CA, Menotti-Raymond M, Roca AL, et al. **The Near Eastern origin of cat domestication.** *Science*, 317:519–523, 2007.

Feuersten N, Terkel J. **Interrelationships of dogs (*Canis familiaris*) and cats (*Felis catus*) living under the same roof.** *Appl Anim Behav Sci*, 113:150–165, 2008.

Fiennes R, Fiennes, A. **The natural history of dogs.** The Natural History Press, Garden City, NY, 1970.

Fox MW. **Socio-ecological implications of individual differences in wolf litters: a developmental and evolutionary perspective.** *Behav,* 41:298–313, 1972.

Fox, MW. **Behaviour of wolves, dogs and related canids.** Harper and Row, New York, NY, 1971.

Fox, MW. **The dog: Its domestication and behavior.** Garland STPM Press, New York, NY, 1978.

Frank H, Frank MG. **On the effects of domestication on canine social development and behavior.** *Appl Anim Ethology*, 8:507–525, 1982.

Goodwin D, Bradshaw JW, Wickens SM. **Paedomorphosis affects agonistic visual signals of domestic dogs.** *Anim Behav*, 53:297–304, 1997.

Hare B, Tomasello M. **Human-like social skills in dogs?** *Trends in Cog Sci*, 9:439–444, 2005.

Hoage, RJ (Editor). **Perceptions of animals in American culture.** Smithsonian Institution Press, Washington, D.C., 151 pp., 1989.

Irion DN, Schaffer AL, Famula TR. **Analysis of genetic variation in 28 dog breed populations with 100 micro satellite markers.** *J Heredity*, 94:81–87, 2003.

Kretchmer KR, Fox MW. **Effects of domestication on animal behaviour.** *Vet Rec*, 96:102–108, 1975.

Leonard JA, Wayne RK, Wheeler J. **Ancient DNA evidence for Old World origin of New World dogs.** *Science*, 298:1613–1616, 2002.

Lorenz K. **Man meets dog.** Kodansha International, New York, NY, 211 pp., 1953 (reprint 1994).

Miklosi A, Kubinyi E, Topal J. **A simple reason for a big difference: Wolves do not look back at humans, but dogs do.** *Curr Biology*, 13:763–766, 2003.

Morey DF. **Size, shape and development in the evolution of the domestic dog.** *J Arch Sci*, 19:181–204, 1992.

Morey DF. **The early evolution of the domestic dog.** *Amer Scientist*, 82:336–347, 1994.

Morris D. **Cat breeds of the world.** Viking Press, New York, NY, 256 pp., 1999.

Odendaal JSJ, Meintjes RA. **Neurophysiological correlates of affiliative behaviour between humans and dogs.** *Vet J*, 165:296–301, 2003.

Olsen SJ. **Origins of the domestic dog.** University of Arizona Press, Tucson, AZ, 1985.

Olson PN, Hall MF, Peterson JK, Johnson GS. **Using genetic technologies for promoting canine health and temperament.** *Anim Reprod Sci*, 82:225–230, 2004.

Overall K. **Update on canine behavioral genetics: What vets should know to help breeders and clients who love purebred pets.** *Proc NAVC 2006*, 168–170, 2006.

Parker HG, Kim LV, Sutter NB, et al. **Genetic structure of the purebred domestic dog.** *Science,* 304:1160–1164, 2004.

Podberscek AL, Paul ES, Serpell JA (Editors). **Companion animals and us: Exploring the relationships between people and pets.** Cambridge University Press, Cambridge, UK, 335 pp., 2000.

Ritvo H. **The emergence of modern pet-keeping.** In: *Animals and People Sharing the World,* AR Rowan, editor, University Press of New England, Hanover, NH, pp. 13–31, 1988.

Rooney NJ, Bradshaw JWS, Robinson IH. **Do dogs respond to play signals given by humans?** *Anim Behav*, 61:715–722, 2001.

Schenkel R. **Submission: its features and functions in the wolf and dog.** *Amer Zoologist*, 7:319–330, 1967.

Serpell JA (Editor). The **domestic dog: Its evolution, behavior, and interactions with people.** Cambridge University Press, Cambridge, UK, 268 pp., 1995.

Tchernov E, Valla FF. **Two new dogs and other Natufian dogs from the Southern Levant.** *J Archaeological Sci*, 24:65–95, 1997.

Turner DC, Bateson P (Editors). **The domestic cat: The biology of its behaviour**, 2nd edition. Cambridge University Press, Cambridge, UK, 244 pp., 2000.

Vigne JD, Guilaine J, Debue K, Haye GP. **Early taming of the cat in Cyprus.** *Science*, 304:5668–5669, 2004.

Vila C, Maldonado JE, Wayne RK. **Phylogenetic relationships, evolution, and genetic diversity of the domestic dog.** *J Heredity*, 90:71–77, 1999.

Wayne RK. **Molecular evolution of the dog family.** *Trends in Genetics*, 9:218–224, 1993.

Wayne RK. **Origin, genetic diversity and genome structure of the domestic dog.** *Bioessays*, 21:247–257, 1999.

Willis MB. **Breeding dogs for desirable traits.** *J Small Anim Pract*, 28:965–983, 1987.

Young, MS. **The evolution of domestic pets and companion animals.** *Vet Clin North Amer*, 15:297–309, 1985.

Zeuner, FE. **A history of domesticated animals.** Harper and Row, New York, NY, 1963.

Footnotes

[1] Coppinger R, Coppinger L. Dogs: *A startling new understanding of canine origin, behavior and evolution.* Scribner, New York, NY, 352 pp. 2001.

[2] Hetts S. *Pet behavior protocols: What to say, what to do, when to refer.* AAHA Press, Lakewood, CO, 1999.

[3] Olson PN, Hall MF, Peterson JK, Johnson GS. Using genetic technologies for promoting canine health and temperament. *Anim Reprod Sci*, 82:225–230, 2004.

[4] Turner DC. Human-cat interactions: relationships with and breed differences between, non-pedigree, Persian, and Siamese cats. In: *Companion animals and us: exploring the relationships between people and pets*, AL Poderscek, ES Paul, JA Serpell, editors, Cambridge University Press, Cambridge, UK, 2000.

CHAPTER 2

Behavior through the Life Cycle: Newborns to Seniors

Although pet owners do not usually acquire dogs and cats until after weaning, a great deal has already occurred behaviorally by the time that most puppies and kittens enter their new homes. Behavior is not a static phenomenon. Dogs and cats develop behaviorally as they mature and continually adapt in response to experiences. Specifically, experiences that occur during early development can have a significant impact on an animal's behavior later in life. In both dogs and cats, behavioral development is traditionally classified into a series of stages, called **sensitive periods** (Sidebar 1). These periods include the neonatal and transitional period, **socialization period**, and juvenile period. As they reach adulthood, dogs and cats continue to learn and to alter their behavior in response to experiences (refuting the adage about "old dogs and new tricks"). Finally, as our best friends enter their senior years, some of the physical changes associated with aging can affect a pet's behavior and temperament.

WHAT ARE SENSITIVE PERIODS OF DEVELOPMENT? — Sidebar 1

The **ethologist** Konrad Lorenz first described sensitive periods in the 1930s. Lorenz observed a dramatic example of imprinting behavior in newly hatched goslings, when the hatchlings rapidly became attached to the first individual they encountered – himself! In **precocial** birds, imprinting behavior is confined to a "critical period" – a short span of time with a definite starting and ending point during which newborns learn to recognize their primary attachment figures. After this period ends, species identification and socialization are less likely to occur.

Subsequent studies of developmental behavior in mammals showed that such dramatic imprinting did not occur in most animals. In contrast to birds, while mammals experienced similar periods of learning and species-identification, the onset and offset of these critical periods was much more gradual. For example, in dogs, the behaviors or preferences acquired during the socialization period are important, but they can still be influenced when the animal is older. As a result, the term "sensitive period" has replaced the term "critical period" when describing developmental behavior of the domestic cat and dog.

Sensitive periods of socialization refer to an age *range* during which a young animal is highly response to socialization and habituation and to opportunities to form social attachments. In the case of domesticated companion animals, these social attachments include members of their own species and their human caretakers. For dogs, the sensitive period of socialization occurs when puppies are between 3 and 12 weeks old. Although the sensitive period of cats is less well defined, kittens are most responsive to socialization when they are between 2 and 9 weeks of age.

THE EARLY WEEKS: NEONATAL AND TRANSITION PERIODS

Both dogs and cats give birth to **altricial** (immature) offspring that are completely dependent upon their mothers for warmth, food, elimination, and protection during the first few weeks of life. Although there are some differences between newborn puppies and kittens, they are at similar stages of development during the first two weeks of life. They have very limited motor ability and are only able to crawl short distances. Their eyes are closed, their ears are not functioning, and they are unable to regulate internal body temperature efficiently. They cannot eliminate without tactile stimulation of the perineal region by the mother dog or cat for the first few weeks of life. It is not surprising, then, that the sensory worlds of puppies and kittens at this age consist primarily of the smell and touch of their mothers and siblings. Indeed, it is through sense of smell that the newborns first learn to recognize their mother and to locate her as the source of warmth and nutrition.

Neonatal Period

Several behavior patterns are unique to the neonatal period. These are adaptive in the sense that they are closely matched to the needs and abilities of newborn puppies and to the mother's ability to respond. Neonatal behaviors fade completely with maturation of the nervous system as neonates approach weaning age. The rooting reflex is triggered by maternal licking and is characterized by a "swimming" motion with the front legs as the back legs push forward toward the warm **stimulus** (Figure 2.1). Rooting helps the newborn to find the mother's teats and prevents puppies and kittens from inadvertently moving away from the nesting site. Suckling behavior is seen as soon as the newborn finds the teat. Suckling is accompanied by kneading of the front limbs and paws, which move against the mammary glands and aid in stimulating milk secretion (Figure 2.2). A final unique behavior seen only in neonates is distress calling. These high-pitched cries are produced by puppies and kittens when they are cold or hungry. The mother responds immediately, feeding the neonates and providing warmth. One vocalization that is unique to cats is the purr (see Chapter 3, pp. 53). Newborn kittens begin to purr within the first few days of life. Content kittens often purr continually while they are suckling or being groomed by the mother cat.

Transition Period

The neonatal period is short, lasting only the first 10 to 14 days of life. It is immediately followed by the transition period, which represents a period of rapid physiological change as the young puppies and kittens begin to "wake up" to their sensory worlds. This period begins when eyes open (12 to 14 days in puppies and 7 to 10 days in kittens). Although the eyes are open by two weeks, puppies and kittens do not see well until they are about four weeks of age. The transition period ends around three weeks when the ear canals open and puppies and kittens consistently react to sounds. First teeth erupt in kittens around 14 to 16 days and in puppies when they are about 20 days old. In both puppies and kittens, the transition from deciduous teeth to adult

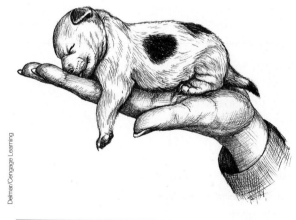

figure 2.1 Rooting reflex in a newborn puppy

figure 2.2 Kneading reflex in a nursing kitten

teeth begins when they are 3½ to 4 months old. Kittens also begin to show rudimentary self-grooming behaviors between the second and third weeks of life.

During the week of transition, neonatal behavior patterns gradually disappear and are replaced by behavior patterns that are more adaptive for older puppies and kittens. Rooting gradually disappears, although many adult cats (and, more rarely, adult dogs) continue to demonstrate kneading behavior when they are content. Anogenital stimulation by the mother is no longer necessary for urination and defecation, and puppies and kittens begin to naturally move outside of their sleeping area for elimination. They are able to stand by about day 17 and begin to walk several days later. As they become more exploratory, associated changes in their distress calls are observed. While neonates distress call only when they are hungry or cold, by the end of the transition period puppies and kittens distress call if they wander too far away from their mother or littermates. During the transition period, learning can occur as puppies and kittens are capable of forming associations between stimuli and responses. However, the rate of learning is slow and the stability of conditioned responses does not reach adult levels until several weeks later.

Practical Applications

Learning capability is limited during the neonatal and transitional periods. However, because neonates are very sensitive to **olfactory** cues and tactile stimulation, early gentle handling by human caretakers is beneficial (Sidebar 2).

THE IMPORTANCE OF EARLY HANDLING — Sidebar 2

In many mammalian species, early daily handling of neonates has been shown to have positive long-term effects on both physical and behavioral development. These effects include accelerated maturation of the nervous system, improved growth rate, and enhanced development of motor skills and the special senses. Studies with newborn kittens have found that kittens who were gently handled on a daily basis for the first five weeks of life were more confident, exploratory, and socially assertive as adolescents than their non-handled counterparts. Similar studies with puppies showed that exposing puppies to frequent handling and mild stressors during the first few weeks of life enhanced puppies' ability to habituate to novel stimuli and produced individuals who were more outgoing and less timid than puppies who had not been handled.

Fox MW, Stelzner D. Behavioural effects of differential early experience in the dog. *Anim Behav*, 14:273–181, 1966.
Fox MW. *The Cat: Its Domestication and Behavior.* Garland STPM Press, New York, NY, 1978.

During the transition period, as puppies and kittens become capable of reacting to olfactory, auditory and visual stimuli, the introduction of toys and other novel objects can help to stimulate investigative behaviors, even though they are not yet able to manipulate these toys. It is also advantageous to expose young puppies and kittens to normal household sounds, smells and sights, daily handling, petting, and gentle brushing.

SENSITIVE PERIOD OF SOCIALIZATION

The sensitive period of socialization (3 to 12 weeks for dogs; 2 to 9 weeks for kittens), represents a time of very rapid behavioral change, coinciding with the maturation of the central nervous system and the ability to perceive and fully respond to environmental stimuli. The term "socialization" refers to the process by which an animal develops species-specific social behaviors, learns to identify primary attachment figures, and forms social relationships. Dogs and cats differ from most other animals in that kittens and puppies can be simultaneously socialized to their own species (conspecifics) and to humans. Adequate socialization and **habituation** to environmental stimuli during this sensitive period can often prevent the development of inappropriate behaviors or deficits later in life. Dogs and cats that are properly socialized to their own species and to human caretakers incorporate both species into their social structure and will tend to direct species-typical behavior patterns, particularly communication signals, to both.

Learning to Be a Dog

After three weeks of age, puppies become increasingly active and curious about their littermates and surroundings. They readily approach and investigate novel stimuli and show little fear or hesitancy to new objects, sounds, or people. After five weeks of age, this investigative response diminishes somewhat, but active approach continues to prevail over nervousness or fear responses until puppies are about eight weeks of age (Figure 2.3). Site attachment develops early in the dog's socialization period. Puppies are able to identify and return to their sleeping and eating areas and, as they are given increased access to other areas, appear to form attachments to particular places.

Puppies play intensively with their littermates during the socialization period. Play provides puppies with the opportunity to practice important social behaviors and communication signals, and teaches valuable lessons about bite inhibition and dominant and subordinate relationships (see Chapter 3, pp. 72–75). The mother's interactions are equally important during this time. Puppies continually solicit care and attention from the mother dog and her responses provide important information to them regarding appropriate behavior with adult dogs. When puppies play too roughly or become too demanding, the mother

figure 2.3 Six-week old puppy

dog disciplines using growls, dominant body postures and physical reprimands such as an inhibited muzzle bite. This important discipline teaches developing puppies to correctly interpret dominant signals from other dogs, to inhibit their bites when playing, and to display appropriate submissive postures to a dog who is displaying dominant signals.

HOW DOES PLAY BETWEEN LITTERMATES TEACH BITE INHIBITION?

Sidebar 3

Scenario: A litter of five-week-old Labrador retriever puppies are playing happily. Two puppies, Marvin and Clyde, begin a rousing chase-and-wrestle game, nipping and mouthing each other as they take turns chasing and being chased. All goes well until Clyde, who is the more confident and bold of the two puppies, gets too excited and bites down on Marvin's ear – hard. One of two possible outcomes occurs:

OUTCOME # 1

Clyde bites Marvin's ear (hard) → Marvin yelps LOUDLY!! (Yip, Yip!)

↓

Marvin runs away, effectively stopping the play.

↓

Clyde is alone, without his playmate.

↓

Clyde learns: Bite hard and play stops (best not to bite hard).

OUTCOME # 2

Clyde bites Marvin's ear (hard) → Marvin yelps LOUDLY!! (Yip, Yip!)
↓
Marvin reacts aggressively, biting Clyde back, HARD
↓
Clyde yelps! (Yip, Yip!) Play stops.
↓
Clyde directs appeasement behaviors (submissive posture, raised paw) to Marvin and play continues
↓
Clyde learns to inhibit his bite to avoid being bitten in return
(and practices appeasement behaviors)

When puppies are between 8 and 10 weeks of age, they become particularly sensitive to new sounds, sights, and experiences, and may react to new experiences with nervousness or fear. This period is commonly referred to as the "fear imprint" stage. Although the age range of this period is consistent, breeds and individual puppies vary significantly in the degree to which they demonstrate fearful responses after eight weeks of age. It is probable that both genetics and early socialization influence the expression of fear imprint in puppies. Regardless, new owners should be aware of this sensitivity and should avoid exposing puppies to startling or potentially frightening stimuli during this time.

Learning to Be a Cat

Kittens also experience a sensitive period of socialization, but its duration is less well defined and is somewhat shorter than the dog's primary socialization period. The kitten's transition period is similar to that of the puppy, occurring during the second to third week of life. Socialization begins near the end of the second week and continues until kittens are 7 to 9 weeks of age. Interactions with the mother cat and with siblings provide important information to kittens about social behavior, predation, and communication. A unique characteristic of kittens is that they are excellent **observational learners**, capable of learning a task simply by watching their mother perform it successfully. Interestingly, the kitten's reliance upon observational learning may be especially important during the sensitive period of socialization when kittens are being handled by human caretakers. Studies have shown that the presence of a mother cat who is calm and friendly towards humans reduces **anxiety** in her kittens and encourages them to interact with their human caretakers.[1]

Puppy Play vs. Kitten Play

The importance of play for normal social development in puppies and kittens cannot be over-emphasized. Puppies and kittens begin to show interest in their littermates and show rudimentary play behaviors when they are three weeks old. Over the next several weeks, play becomes an increasingly large part of their activities. In both species, play teaches young animals to inhibit their bite when interacting with others and provides repeated opportunities to develop and practice communication and social skills, predatory behaviors, and sexual behaviors. However, puppies and kittens have several unique aspects to their play behaviors that illustrate some of the fundamental differences between dogs and cats.

Puppies and kittens both are highly interested in toys and include various types of toys in the games that they play when alone and with each other. Puppies differ from kittens during play in that they spend time chewing and gnawing toys and other objects in their environment. They also begin to compete for possession of toys using chase, keep-away, and tug-of-war games. During these games, the puppies practice communicative body postures that signal confidence (dominance), playfulness, and appeasement (submission) to each other. Finally, puppies (and adult dogs) almost all enjoy shaking, ripping, and tearing apart toys. Violent shaking of the toy followed by "evisceration" – the toy is held between the front feet and ripped apart – appears to be related to the kill portion of the canid predatory sequence. Similarly, extensive periods of chewing and gnawing are part of the normal feeding behavior of most canid species. In contrast, while toys are central to kitten play, possession and keep-away games are not common in cats. Rather, object (toy) play in kittens involves stalking, batting and manipulating toys in mimicry of feline predatory behaviors. While puppies almost exclusively include another puppy in object (resource) play, object play in kittens can be either solitary or social. When they do share a toy with another kitten, both focus on manipulating the toy by batting it back and forth or taking turns pouncing on it. However, kittens do not demonstrate the intense interest in resource guarding and possession that is a typical component of puppy play. This difference can be explained by the cat's history as a solitary hunter that did not need to work cooperatively with others. Conversely, the dog's ancestral history involves group hunting and food sharing, in which possession of resources is an important part of group life and one way in which group cohesion and status are signaled within a social group.

Socialization with Humans

As our companions, puppies and kittens must learn not only how to relate to their own kind, but also to their human caretakers. Puppies are highly responsive to socialization with people when they are between the ages of 5 and 12 weeks, although socialization after 12 weeks is also effective and very important for development. A series of classic studies conducted during the early 1960s with developing puppies showed that a complete lack of human contact between

3 and 14 weeks resulted in prolonged fear and wariness of humans, with most puppies never achieving the ability to trust people or respond to training.[2] Although these studies provide an extreme example of the results of a lack of socialization, they demonstrate the importance of human interaction during the primary socialization period. Ideally, puppies should receive frequent and positive human contact while they are still with their litter, and this contact should be expanded to include many new people and experiences once they are weaned and enter their new home.

Unfortunately, in the real world, not all dogs and cats receive adequate interaction and handling during the sensitive period of socialization. This may occur because of a lack of knowledge on the part of owners, abandonment, or neglect. However, widespread recognition of the importance of puppy socialization has led to the development of puppy kindergarten classes (Sidebar 4). These classes can provide an excellent outlet for a puppy to play with other puppies as well as opportunities for new owners to learn how to take care of and train their new family member. However, puppy socialization should not stop with the sensitive period at 12 weeks, nor with the completion of a puppy class. Puppies vary considerably in their ability

Sidebar 4 | PUPPY SOCIALIZATION CLASSES AND KITTEN KINDERGARTEN

Puppy socialization classes can be found in most communities and provide an opportunity for new owners to introduce their puppy to other puppies and places in a positive and controlled manner. Classes also provide education for new owners regarding behavior, training, and the prevention of behavior problems. Most puppy classes are designed to capitalize on the primary socialization period, so enroll puppies who are between the ages of 8 and 14 weeks.

Properly conducted puppy classes include lessons on house-training, teaching bite inhibition and basic commands, stopping nipping, and redirecting unruly behaviors. Time is set aside for socialization exercises with people and with puppies who are matched according to age and temperament. Some classes also promote habituation by introducing a small agility course or a variety of new objects and walking surfaces.

Although not as common, kitten classes are also offered in some areas, usually through veterinary clinics. Socialization outside of the home is not as important for many cats, but kitten classes can still offer education for owners and opportunities for kittens to become habituated to traveling in the car and to new places.

to respond to socialization after 12 weeks of age, but most continue to habituate to new experiences throughout their juvenile period. Individuals also vary in their need for continued exposure and socialization and in their ability to generalize sociable behaviors to new individuals and situations.[3] For example, some puppies who were well-socialized to other dogs and to people up to 12 weeks may regress to fearful or timid behaviors if socialization is not continued after this age. Therefore, while the age of 3 to 12 weeks is very important for puppy-to-puppy interaction and human handling, it should not be considered a "window" that closes following this age period. Rather, socialization and habituation should continue throughout life for the development of a well-adapted and friendly companion.

For kittens, interaction with humans is very important during the early weeks of life because the kitten's socialization period begins at an earlier age and is of shorter duration than that of the puppy. Daily human handling should begin as early as possible, preferably when kittens are 2 or 3 weeks of age. If no positive human handling takes place before eight weeks, kittens are at increased risk of becoming nervous, fearful of, or unfriendly toward humans as adult cats. Studies have shown that daily handling very early in the sensitive period (4 weeks or younger) and for 30 minutes or more at a time results in adult cats who are more friendly and outgoing toward humans than cats who are handled later in the socialization period or for just a few minutes each day. In older kittens (7 to 14 weeks of age), increasing the amount of time each day that the kitten is handled has the greatest effect on increasing a cat's sociability. There is also evidence that genetics, maternal influences, and environmental factors have a strong influence on a cat's friendliness to humans and other animals, regardless of the amount of early socialization (Sidebar 5).

THAT CAT'S GOT PERSONALITY! Sidebar 5

Although early handling and socialization to humans is critical for kittens to develop into friendly and sociable cats, genetics and maternal influences also influence a cat's ultimate personality. Studies of cats living in homes and surveys of cat owners have found that cats seem to fall into one of two broadly defined personality types. The first are cats who are easygoing, confident and sociable. These cats are usually playful with their family, and friendly and outgoing with familiar and unfamiliar visitors. The second group includes cats who are relatively timid and nervous in nature. These cats prefer a quiet and consistent home environment and often hide when visitors come to the home. While early handling usually improves a cat's degree of friendliness, kittens who are born to nervous parents typically will not become as outgoing or friendly as will kittens born to confident parents. Conversely, kittens of outgoing parents appear to be more capable of developing attachments to humans, even if they are handled later in kittenhood. Interestingly, the temperament of the father cat is highly correlated with the temperament of his offspring, even though the male cat plays no role in raising his kittens. This suggests that these personality traits are strongly influenced by genetics (rather than through observational learning from the mother cat alone). These studies tell us that the temperament and behavior of a kitten's parents are as important as is early handling during the first weeks of life in affecting a kitten's ability to form strong social attachments and to people.

Karsh EB. The effects of early handling on the development of social bonds between cats and people. In: *New Perspectives on Our Lives with Companion Animals*, AH Katcher and AM Beck, editors, University of Pennsylvania Press, Philadelphia, PA, pp. 22–28, 1983.

McCune S. The impact of paternity and early socialization on the development of cats' behaviour to people and novel objects. *Applied Animal Behaviour Science*, 45:109–124, 1995.

Reisner IR, Houpt KA, Hollig NE, Wuimby FW. Friendliness to humans and defensive aggression in cats: the influence of handling and paternity. *Physiology and Behavior*, 55:1119–1124, 1994.

WEANING

Female dogs and cats naturally begin the weaning process when their offspring are 3½ to 4 weeks of age. This is a very gradual process and will be complete by the time the puppies and kittens are 7 to 8 weeks old. As the mother begins to wean, she walks away while her young are nursing, allows nursing for shorter periods, and spends more time away from the litter. This gradual introduction to longer periods of separation teaches puppies and kittens self-confidence and allows them to become progressively independent of their mother's care.

Although puppies and kittens are weaned at about the same age, a clear difference occurs between dogs and cats in behaviors associated with provision of food during weaning. In ancestral dogs and wild canids, the mother dog regurgitates food for the puppies as she begins to wean. Regurgitation is triggered by puppies' greeting and food-soliciting (begging) behaviors. These behaviors include submissive greeting postures, whining, and licking at the corners of the mother's mouth. These behaviors are important social communication skills that will remain part of the dog's behavioral repertoire and will be used during interactions with other dogs (and also with people). Although most female domestic dogs do not regurgitate food for their puppies, the food-soliciting behaviors and submissive postures associated with feeding time are still expressed by puppies and are important for communication between dogs throughout life (Figure 2.4).

figure 2.4 Puppy submissively licking at corner of her mother's mouth (greeting behavior and food-soliciting behavior)

figure 2.5 Mother cat introducing her kitten to prey

In contrast, nutritional weaning of kittens follows a different course. Rather than enhancing social communication and cooperative behaviors, the weaning process directs the attention of kittens *away* from the mother cat and toward solitary predatory behaviors. In free-living cats, the mother cat begins to bring dead prey to her kittens when they are 4 to 5 weeks of age (Figure 2.5). She drops the prey nearby and allows the kittens to manipulate and consume the prey, usually before she allows nursing. This process gradually changes the signal for food availability from the mother cat to the solid food that she provides. As kittens develop, they show increasing interest in the prey that the mother brings and in attempting to manipulate and kill the prey themselves. Most kittens demonstrate fully developed prey-killing skills by the time they are 6 to 7 weeks of age.

In both dogs and cats, the natural weaning by the mother is characterized by a period of *gradual* lessening of the young's attachment to their mother, as opposed to an artificially introduced or abrupt separation. Although puppies and kittens have usually stopped nursing regularly by six weeks of age, they will show intermittent suckling for several more weeks if they are still with their mother. If allowed, most mother dogs and cats continue to spend time with their litter and provide an important source of comfort and discipline until the puppies and kittens are seven weeks or older (Sidebar 6). A primary difference between the two species is that as puppies continue to develop social skills and communicative behaviors related to cooperative hunting and food sharing, kittens are becoming increasingly less social and more independent as they learn solitary hunting skills. Thus, while domestication has resulted in enhancing many of the social behaviors of the cat, it has not created cooperative behaviors such as those needed by a pack-living hunter such as the dog. These differences continue to be evident in the social and feeding behaviors of adult dogs and cats (see Chapter 3, pp. 68–76).

Sidebar 6: WHEN SHOULD PUPPIES AND KITTENS GO TO THEIR NEW HOMES?

- *Weeks 3 to 7:* It is important for puppies and kittens to stay with their mother and littermates during the first part of primary socialization. Emphasis during these weeks is on learning species-specific behaviors and communication signals. Human handling and interactions should also occur during this time, in the form of gentle handling, grooming, and petting, beginning at two weeks (or earlier) and continuing until weaning.

- *Importance of "Natural" Weaning:* Natural weaning typically occurs over a span of four weeks or longer. If allowed to progress at its normal rate, mothers completely wean their puppies and kittens by the time they are 7 to 9 weeks old. In human-controlled settings, it is therefore important to recognize that the abrupt removal of a mother from her young or the premature removal of a puppy or kitten from the litter can have negative long-term consequences.

- *Effects of Early Weaning:* Puppies and kittens who are removed from their litters too early are prevented from learning to interact and communicate normally with other dogs and cats and may identify too strongly with humans. In dogs, this may predispose them to over-attachment to people and associated behavior problems later in life. In kittens, early weaning can affect the development of normal play and predatory behaviors and a kitten's ability to interact with other cats.

- *Welcome Home:* Puppies and kittens should remain with their littermates and the mother cat until they are at least 7 to 8 weeks old. The mother should be allowed to gradually wean her young, through voluntarily decreasing the time that she spends with them each day and by gradually discouraging them from nursing. The best time for puppies and kittens to leave their littermates and go to their new homes is between 7 and 9 weeks of age. In cases in which they remain with their litter longer, daily handling, socialization, and interactions with people is of the utmost importance.

JUVENILE (ADOLESCENT) PERIOD AND ADULTHOOD

The juvenile period extends from the end of the socialization period to sexual maturity. Learning ability is fully developed, and dogs and cats are refining their existing abilities, increasing coordination, and becoming more exploratory. Motor skills become more coordinated and adult-like, and the young animal's attention span gradually increases. Permanent teeth begin to replace the deciduous teeth at about 4 to 5 months of age and are usually fully erupted by 6 months of age. Gradual changes in behavior are seen in response to learning and as a result of previous experiences.

Sexually related behaviors develop with the onset of puberty. Dogs become sexually mature between 6 and 16 months, depending upon breed and adult size. Although most dogs are reproductively mature by the time they are one year old, social behaviors continue to develop and change until dogs are 18 months or older. Behaviors that are closely associated with social maturity include territorial, protective, and dominance aggression. Cats reach sexual maturity between 5 and 9 months of age, although females may show their first estrous cycle when they are as young as 4 months. With the onset of puberty, sexually dimorphic behaviors such as urine **marking**, inter-male aggression, inclination to roam, and sexual mounting begin to occur in males.

Dogs and cats continue to learn and modify their behavior throughout adult life. Although many owners believe that the training of their dog or cat should occur during the first year of life, it is equally important for both their pet's manners and responses to commands, and for continuing quality of life, that training continue throughout adult life. Training sessions can be directed toward preventing unwanted behavior problems, teaching enjoyable games and activities, and reinforcing good manners in the home and when out in public. In addition, socialization throughout life is essential and can take the form of regular excursions to parks and new places, accompanying owners on errands and in some cases vacations, and having opportunities to meet and interact with a variety of people.

SENIOR PETS

As a result of improvements in companion animal health care and nutrition, dogs and cats are living substantially longer in recent decades. Still, even with these changes, dogs and cats have much shorter life spans than their human caretakers, which means that the majority of owners will experience living with an elderly pet at some point in their life. Because of their wide range in adult size, dogs tend to age at different rates. The toy and small breeds live longer and age more slowly than large and giant breeds of dogs. Size is not typically a factor for cats because of the smaller variation in adult size. Additional factors that influence rate of aging include a pet's lifestyle, fitness level, health care and nutrition, and the presence of chronic disease. A common rule of thumb is to consider a dog or cat to be in their senior years once they have entered the final third of their expected lifespan. For example, a spaniel-mix with an expected lifespan of 12 years can be considered a senior after 8 years of age. Conversely, a small or toy breed dog with an expected lifespan of 14 to 15 years is considered elderly when he reaches about 10 years of age. In general, cats are considered to enter their elder years at around 10 years of age. Although such guidelines are helpful, in the end, every dog and cat is an individual, and age-related changes will be affected by genetic, health, nutritional status, and lifetime experiences.

For many owners, living with a healthy older adult dog or cat is a special time, as their pet has established relatively consistent behavior patterns and has a shared history with the owner and other pets in the family. However, just as in people, dogs and cats experience a series of changes with age that can affect the pet's behavior and temperament. Older animals are generally less active, tend to sleep more, play less vigorously, and require less daily exercise than younger dogs and cats. In some dogs and cats, the presence of arthritis or diminished sensitivity of vision or hearing may also contribute to changes in behavior and activity. Older dogs may have trouble maneuvering stairs or jumping into the car, while cats may be less inclined to jump up onto elevated spaces. While many pets "age gracefully" and seem to become more sweet and affectionate as they grow older, others may become less social, more irritable, or less tolerant of changes in their environment or of handling. This may be caused by health changes such as arthritis or by the onset of age-related **cognitive dysfunction**.

Regular veterinary care, proper nutrition, and living in a loving home can help dogs and cats experience healthy older years with minimal behavior changes or problems. Although the pet's overall activity level may decline, daily walks, outings to a favorite park, or supervised retrieving games all help to keep dogs physically and mentally active. For cats, daily games with a favorite toy or supervised excursions outside can keep life interesting and active. Diets designed for older dogs and cats who are less active and experiencing age-related changes in cognitive function and behavior can be helpful in both maintaining weight and in slowing cognitive changes. Avoiding major changes in an older pet's daily routine can be helpful in preventing disorientation or stress. Similarly, continuing with the dog's or cat's established daily rituals such as regular meal times, morning walks, grooming sessions, and play times helps to support the pet's normal behavior patterns and is enjoyable and important to both the pet and to the owner during the late years of the pet's life.

REVIEW QUESTIONS

1. Describe the set of adaptive behavior patterns that are observed in neonatal puppies and kittens.
2. Compare and contrast the period of primary socialization in puppies and kittens.
3. What species-specific behavior patterns do puppies learn from their mother and their littermates during the period of primary socialization?
4. Compare and contrast play behavior in puppies and kittens.
5. What is "natural" weaning? When should puppies and kittens be weaned and adopted into their new homes?

REFERENCES AND FURTHER READING

Bacon WD. **Aversive conditioning in neonatal kittens.** *J Compar Phys Psychol*, 83:306–313, 1973.

Barrett P, Bateson P. **The development of play in kittens.** *Behaviour*, 66:106–120, 1978.

Bateson P, Young M. **Separation from the mother and the development of play in cats.** *Anim Behav*, 29:173–180, 1984.

Beadle M. **The Cat: History, Biology, and Behavior.** Simon and Schuster, New York, NY, 1977.

Beaver BV. **Feline Behavior: A Guide for Veterinarians.** W.B. Saunders Company, Philadelphia, PA, 276 pp., 1992.

Bekoff M. **Social play and play-soliciting by infant canids.** *Amer Zoologist*, 14:323–340, 1974.

Bekoff M. **Play signals as punctuation: The structure of social play in canids.** *Behaviour*, 132:419–429, 1995.

Borchelt PL. **Behavioral development of the puppy.** *In: Nutrition and Behavior in Dogs and Cats*, RS Anderson, editor, Pergamon Press, Oxford, UK, pp. 165–174, 1984.

Bradshaw JWS. **The Behaviour of the Domestic Cat.** *CAB International*, Oxford, UK, 219 pp., 1992.

Caro TM. **Predatory behavior and social play in kittens.** *Behaviour*, 76:1–24, 1981.

Caro TM. **Sex differences in the termination of social play in kittens.** *Anim Behav*, 29:271–279, 1981.

Caro, TM. **Effects of the mother, object play and adult experience on predation in cats.** *Behav Neural Biol*, 29:29–51, 1980.

Duxbury MM, Jackson JA, Line SW, Anderson RK. **Evaluation of association between retention in the home and attendance at puppy socialization classes.** *J Amer Vet Med Assoc*, 223:61–66, 2003.

Elliot O, Scott P. **The development of emotional distress reactions to separation in puppies.** *J Genetic Psychol*, 99:3–22, 1961.

Estep DQ. **The ontogeny of behavior.** *In: Readings in Companion Animal Behavior*, VL Voith and PL Borchelt, editors, Veterinary Learning Systems, Trenton, NJ, pp. 19–31, 1996.

Feldman H. **Maternal care and differences in the use of nests in the domestic cat.** *Anim Behav*, 45:12–23, 1993.

Fox MW. **Behavioral effects of rearing dogs with cats during the "critical period of socialization".** *Behaviour*, 35:273–280, 1969.

Fox MW, Stelzner D. **Behavioural effects of differential early experience in the dog.** *Animal Behav*, 14:273–281, 1966.

Frank D, Minero M, Cannas S, Palestrini C. **Puppy behaviours when left home alone: A pilot study.** *Appl Anim Behav Sci*, 104:61–70, 2006.

Freedman DG, King JA, Elliot O. **Critical periods in the social development of dogs.** *Science*, 133:1016–1017, 1961.

Kolb B, Nonneman AJ. **The development of social responsiveness in kittens.** *Anim Behav*, 23:368–374, 1975.

Lowe SE, Bradshaw JWS. **Ontogeny of individuality in the domestic cat in the home environment.** *Anim Behav*, 61:231–237, 2001.

Markwell PJ, Thorne CJ. **Early behavioral development of dogs.** *J Small Anim Pract*, 28:984–991, 1987.

Martin P, Caro TM. **On the functions of play and its role in behavioural development.** *Adv Study Behav*, 15:59–103, 1985.

McCune S. **The impact of paternity and early socialization on the development of cats' behaviour to people and novel objects.** *Appl Anim Behav Sci*, 45:109–124, 1995.

Miller J. **The domestic cat: Perspective on the nature and diversity of cats.** *J Amer Vet Med Assoc*, 208:498–501, 1996.

Moelk M. **The development of friendly approach behavior in the cat: a study of kitten-mother relations and the cognitive development of the kitten from birth to eight weeks.** *Adv Study Behav*, 10:164–244, 1979.

Nott HMR. **Behavioural development in the dog.** *In: The Waltham Book of Dog and Cat Behaviour*, C Thorne, editor, Pergamon Press, Oxford, UK, pp. 65–78, 1992.

Reisner IR, Houpt KA, Erb HN, Quimby FW. **Friendliness to humans and defensive aggression in cats: The influence of handling and paternity.** *Physiol Behav*, 55:1119–1124, 1994.

Rogerson J. **Your dog: Its development, behaviour, and training.** *Popular Dogs Publishing Company*, London, UK, 174 pp., 1990.

Scott JP, Marston MV. **Critical periods affecting the development of normal and maladjustive social behaviour of puppies.** *J Genetic Psychol*, 77:25–60, 1950.

Scott JP. **Critical periods in behavioral development.** *Science*, 138:949–958, 1962.

Seksel K, Maxurski EJ, Taylor A. **Puppy socialization programs: short and long term behavioral effects.** *Appl Anim Behav Sci*, 62:335–349, 1999.

Serpell JA, Jagoe JA. **Early experience and the development of behaviour.** *In: The Domestic Dog: Its Evolution, Behavior, and Interactions with People*, JA Serpell, editor, Cambridge University Press, Cambridge, UK, pp. 80–102, 1995.

Slabbert JM, Rasa OA. **Observational learning of an acquired maternal behaviour pattern by working dog pups: An alternative training method?** *Appl Anim Behav Sci*, 53:309–316, 1997.

Slabbert JM, Rasa OA. **The effect of early separation from the mother on pups in bonding to humans and pup health.** *J South African Vet Assoc*, 64:4–8, 1993.

Thorne CJ, Mars LS, Markwell PJ. **A behavioural study of the queen and her kittens.** *Anim Tech*, 44:11–17, 1993.

Topal J, Gacsi M, Miklosi A, Virany Z, Kubiny E, Csanyi V. **Attachment to humans? A comparative study on hand-reared wolves and differently socialized dog puppies.** *Anim Behav*, 70:1367–1375, 2005.

West MJ. **Social play in the domestic cat.** *Amer Zoologist*, 14:427–436, 1974.

Footnotes

[1] Rheingold H, Eckermann C. Familiar social and nonsocial stimuli and the kitten's response to a strange environment. *Devel Psychobiol*, 4:71–89, 1971.

[2] Scott JP, Fuller JL. *Genetics and the Social Behavior of the Dog.* University of Chicago Press, Chicago, IL, 1965.

[3] Dehasse J. Sensory, emotional and social development of the young dog. *Bulletin Vet Clin Ethol*, 2:6–29, 1994.

CHAPTER 3

Social Behavior and Communication in Dogs and Cats

Animals who live as enduring pairs or groups of individuals must engage in behaviors that allow the formation and maintenance of social relationships within the group. As discussed in Chapter 1, the types of social relationships that are formed vary in response to the abundance and dispersion of valued resources such as food, shelter, and accessibility to mates. In general, the dog is considered a more social species than the cat in terms of having evolved from an obligatory social species. The domestic cat, on the other hand, evolved from a more solitary animal and only adapted to living in groups as part of the domestication process. However, both dogs and cats who live as companions are members of social groups. At the very least, their social relationships include their human caretakers, and in many families, dogs and cats also maintain stable relationships with other dogs or cats, and other species of companion animals.

In all of these relationships, effective communication is essential. Species-specific communication patterns evolved to enable individuals to identify others of their own species and to interact with each other for social, territorial and reproductive purposes. At its simplest, communication occurs when one animal is attentive to and reacts to the signals that are sent by another animal. This communication is direct and immediate. Examples include two dogs who live in the same household greeting after a short separation, or a cat soliciting and receiving attention from her owner. Alternatively, communication may be achieved indirectly and without a specific target, as in the case of a cat leaving scent marks on a doorframe or a dog howling when isolated. Although the recipient is not present in these cases (or perhaps even known by the sender), the message that is sent has the potential to affect the behavior of others. Our domestic dogs and cats are unique in that the communication signals that they use are directed toward both members of their own species and also toward humans. This chapter examines the ways in which dogs and cats communicate using olfactory, visual, auditory, and tactile signals, and how communication is related to the complex social relationships that individual dogs and cats are capable of developing.

OLFACTORY COMMUNICATION

Odors play an enormous role in many aspects of canine and feline behavior. Both dogs and cats have well-developed olfactory senses that are much more sensitive than our own. The dog's nose contains over 220 million olfactory neurons, while the cat's nose contains about 200 million. By comparison, the human nose includes an unimpressive 5 million! In addition to being highly sensitive to the odors of other animals and of their environment, cats and dogs use their own body odors as important communication signals. Deposited scent has several advantages as a communication tool. Smells remain in the environment for a long period of time and can convey information about the sender's sex and reproductive stage, even after the animal has left the immediate area. Using scent to mark territorial boundaries allows widely spaced individuals within a habitat (cats) or those of another social group (dogs) to recognize both their own living area and the territory of others.

Dogs and cats have several mechanisms that aid in maximizing scenting ability. Most important is the use of sniffing. The sniff is actually a disruption of the normal breathing pattern. This allows the molecules that impart odor to remain within the nasal passages for longer period of time. Sniffing consists of a series of rapid and short inhalations and exhalations. Air is forced into a space above the sub-ethmoidal shelf, a bony structure that functions to trap inhaled air. The "sniffed" air rests within this nasal pocket without being inhaled further into the lungs or being immediately exhaled, as it would during a normal breathing cycle. This allows increased time for the inhaled air and the scent molecules that it carries to interact with scent receptors. For both dogs and cats, sniffing is an important component of greeting behavior and appears to provide information about an individual's sex, age, emotional state, and possibly social status. Interestingly, when cats greet, initial sniffing is usually nose-to-nose, while dogs are more likely to first sniff either the side of the other dog's face or near the **inguinal** (groin) area. Finally, cats and dogs living in pairs or groups may also use scent to produce group-specific odors for identification and to help facilitate group cohesiveness.

Urine Marking

Although the functions of urine marking are not completely understood, this behavior has traditionally been described as serving dual, related purposes – to communicate the identity of an individual and to mark territory. In both dogs and cats, urine marking is evidenced by a characteristic posture. Dogs use a raised-leg urination display when scent-marking, while cats adopt a standing posture, raising their tail and spraying urine directly backwards, usually toward a vertical target (Figure 3.1 and Figure 3.2, respectively). This raised-leg urination posture is most commonly observed in reproductively intact males. However, some female dogs demonstrate a form of raised-leg urination in which they raise one hind leg while squatting. Similarly, urine **spraying** in cats is most commonly observed in adult, intact males, but neutered males and neutered and intact females are also capable of spraying. Intact female cats spray most frequently when they are in estrus, and intact tomcats increase their frequency of spraying when there is an estrus female in close proximity.

When urine is used to mark territory, the pattern of urination involves frequent spraying or squatting at numerous sites, with a small volume of urine deposited at each episode. This differs from normal urination patterns, during which the entire bladder is emptied at a single site. Both dogs and cats intently investigate the urine marks of others. A dog who is out walking with his owner will be attracted to spots where other dogs have urinated and will spend several moments sniffing these areas intently. Dogs scent-mark by over-marking the sniffed area with their own urine before moving on (Sidebar 1). Cats sniff areas where other cats have sprayed or urinated and covered the spot (as in boxes), but unlike dogs, sniffing

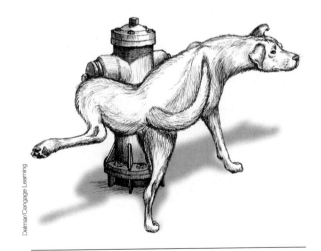

figure 3.1 Raised leg urination posture in male dog

figure 3.2 Urine spraying posture in a cat

URINE MARKING: THE CANINE CALLING CARD — Sidebar 1

Recent studies of urine marking show that dogs can distinguish their own scent from that of others, that urine marking may be used to designate possession of resources, and that scent-marking in females may be more important than previously acknowledged.

In an aptly named experiment, "Tales of Displaced Yellow Snow," University of Colorado ethologist Mark Bekoff asked whether dogs are capable of discriminating between their own urine marks and those of other dogs. Bekoff conducted his study by moving clumps of snow that were saturated with urine to new areas and observing his dog Jethro's reactions. While Jethro consistently sniffed all urine marks, he spent very little time investigating his own urine and was significantly less likely to over-mark his own urine than that of other dogs. Although this work must be repeated with more animals, it was the first study to show that dogs clearly distinguish between their own scent and that of others.

Urine marking may not only be about communicating identity, but may also provide a "label" for valued possessions. A group of researchers studied urine marking patterns in a group of free-ranging dogs living in Katwa, India. Among other findings, they discovered that dogs repeatedly marked around scavenged food sites, even when the sites were not close to their home territory. Surprisingly, the dogs often urine marked over remnant food items after consuming their meal. The purpose of this could be to aid in relocating the spot in the future or as a method of identifying a feeding cache as "taken."

Urine marking has been traditionally viewed as a sexually dimorphic behavior in dogs, used primarily by males to detect estrus females. However, recent studies by researchers at Smith College in Massachusetts have shown that spayed and non-estrus females frequently use urine marking. Females who were four years or older urine marked more often than did younger females, and females being walked in areas away from their home territory urinated more frequently and directed more of their urine to targets than did females walking within their home area. As expected, intact females who were in season marked more often than females who were not in estrus, supporting the role of urine marking in reproduction. However, the use of urine marking in spayed and non-estrus females suggests that this form of communication conveys information that is not necessarily related to sexual status, and that girls have their "calling cards," too!

Bekoff M. Observations of scent-marking and discriminating self for others by a domestic dog (*Canis familiaris*): Tales of displaced yellow snow. *Behav Proc*, 55:75–79, 2001.

Pal SK. Urine marking by free-ranging dogs (*Canis familiaris*) in relation to sex, season, place and posture. *Appl Anim Behav Sci*, 80:45–59, 2003.

Wirant SC, McGuire B. Urinary behavior of female domestic dogs (*Canis familiaris*): Influence of reproductive status, location, and age. *Appl Anim Behav Sci*, 85:335–338, 2004.

Wirant SC, Halvorsen KT, McGuire B. Preliminary observations on the urinary behavior of female Jack Russell Terriers in relation to stage of the estrous cycle, location and age. *Appl Anim Behav Sci*, 106:161–166, 2007.

---

the urine of another cat does not usually cause a cat to over-mark with his or her own urine. Male cats spend a longer period of time investigating than females. Cats also appear to be able to discriminate between the urine of unknown cats, cats from a neighboring group, and cats within their own group.[1]

In addition to sniffing urine marks, cats may react to the odor of another cat's urine by displaying a "Flehmen" or gape response. When gaping, the cat's head is raised and pulled back, the mouth is partially open and the lips are retracted. The cat's tongue moves rhythmically along the roof of the mouth. These movements direct airborne and fluid-borne molecules of scent into the two nasal palatine canals located in the cat's mouth, immediately behind the upper incisors. These ducts connect to the vomeronasal organ located within the hard palate. The vomeronasal organ (also called Jacobsen's organ) contains sensitive olfactory cells that stimulate areas of the cat's brain associated with sexual behavior, feeding behavior, and possibly social behaviors. Although dogs do not demonstrate a classic Flehmen response, some demonstrate "tongue flicking," which appears to function in the same manner. Male dogs typically tongue flick after sniffing or licking a urine spot, especially if the urine was left by a female in estrus. During this behavior, the dog's tongue moves rapidly back and forth against the roof of the mouth, salivation is stimulated and the front teeth often chatter rapidly.

Historically, urine marking has been interpreted as a type of communication meant to deter other individuals of the same species from entering a marked area. However, the "drive off competitors" theory has not been supported with evidence, as the urine deposits left by male dogs do not appear to repel other male dogs.[2] Similarly, studies of free-living cats have found that spray marks in the territory of one group of cats did not discourage cats from other groups from traveling through the area. In fact, most of the tomcats who were studied did not spray on the peripheries of their territories but rather were more likely to spray within their territories in visually conspicuous spots. As stated previously, dogs and cats of both sexes investigate urine marks of others of their species. Males spend more time investigating areas where the urine of females in estrus is found, compared with time spent investigating the urine of non-estrus females. This suggests that urine marking may be a means of providing information about the depositor's identity, reproductive status, and territory, as opposed to acting as a "warning signal" to others to stay out of a designated territory.

Feces Marking

The feces of many carnivore species provide another form of scent marking. Anal sac secretions are discharged during defecation in both dogs and cats. These sacs, located on each side of the anus, empty into ducts that open near the anal orifice and contain secretions from apocrine and sebaceous glands located within the walls of the ducts. The contents of the anal glands are discharged during defecation and contribute **pheromones** to the feces and anal region. Studies have shown that the anal gland secretions of dogs are highly individual in nature, and may provide information to other dogs about age, sex, and identity.[3] Dogs (and, less commonly, cats) also occasionally express their anal glands independently of defecation when they are stressed or frightened. Dogs show an interest in the fecal droppings of other dogs and will often over-mark another dog's feces with urine. However, it is not known how important feces are for marking territory or identifying individuals. Studies of feces as territorial markers in cats have had inconsistent findings and it is not clear if cats use fecal marks to communicate in the same way that they use urine marks.[4]

Allorubbing in Cats

Cats have several types of specialized skin glands that are important for olfactory communication. These include the submandibular gland beneath the chin, perioral glands at the corners of the mouth, temporal glands on the side of the forehead, and sebaceous (caudal) glands along the base of the tail. Another important group of scent glands, the interdigital glands, are located between the pads of the cat's feet. The secretions of the enlarged sebaceous glands located on the face are deposited as scent marks when cats rub on objects at eye level. This behavior is called "**bunting**," or cheek rubbing, and is often directed towards the owner's legs, pieces of furniture or other protruding objects, and even toward other pets, particularly dogs. The targeted person (or dog) or object is rubbed first by the end of the nose, moving down the cat's cheek to the corners of the mouth, and finally ending near the eyes (Figure 3.3). When bunting is intense, the cat's lips draw back and its teeth are exposed, leading some owners to mistakenly interpret this as biting behavior. Because cats often bunt their owners during greeting and following separations, it is possible that bunting functions as a distance-reducing behavior rather than as a form of marking.[5] Allorubbing (rubbing between cats) and sniffing are common in cats who are housemates or belong to the same group, and appears to function in social communication. When cats sniff each other, they tend to concentrate on the head region rather than the flanks and tail, suggesting that the glands of the head provide information about individual identification. Allorubbing is also usually more subtle than bunting, as the cats gently touch along the length of their bodies after a nose-touch greeting.

figure 3.3 Bunting behavior directed toward a dog housemate

Scratching in Cats

Cats scratch rough surfaces to remove the dead outer layers of their claws, which exposes the sharper claw surface underneath. Scratching is also important as a form of marking as it simultaneously deposits secretions of the interdigital glands onto a targeted surface. Most cats prefer to scratch surfaces that have a rough texture, such as wood bark or heavy cloth, and use the same site repeatedly. This leaves a visual mark as well as an olfactory mark. The surface may be either horizontal or vertical, although most cats seem to prefer to be able to stretch upwards while scratching. Free-living cats select scratching sites that are along well-traveled paths and routes, rather than on the periphery of their home range or territory. Similarly, house cats tend to scratch around areas of the house where they spend most of their time. Popular sites include near resting platforms and on doorway trim and furniture. Unlike urine marks, however, cats do not show interest in or investigate the scratched sites of other cats. Scratching as a marking behavior therefore appears to function by providing reassurance and security, and is probably less important as a means of identifying an individual or communicating the boundaries of a cat's territory to other cats.

AUDITORY SIGNALS

While scent marks can convey information over time and after an animal has left the area, vocal signaling has the advantage of being effective over long distances and when vision is impaired. Both dogs and cats are capable of a large range of sounds and use vocal communication frequently. In both species, vocal signals are often contextual, meaning that similar sounds may convey different messages depending upon the situation in which they are used. Common dog vocalizations include grunts, growls, whines, barks, yelps, and howls, while vocalizations of cats are divided into three primary types called murmur patterns, vowel sounds, and strained intensity calls.

Dog Vocalizations

When most people consider dog vocalizations, they immediately think of the bark. The domestic dog is unique among canids in the use of the bark. Although wolves are capable of barking, they bark infrequently and use the bark primarily as an alerting vocalization. Repetitive barking and barking in multiple contexts appears to be unique to the domestic dog. It has been suggested that, during domestication, repetitive barking was desirable because it provided a method of signaling alarm or the approach of intruders. Others suggest that barking is a type of neotenized behavior, representing vestiges of puppy and adolescent vocalizations. Dogs bark in a wide variety of situations, including defense of territory, when announcing the presence of another dog or person, when playing, when isolated, or to solicit attention. Barking is an example of a communicative signal that conveys a warning or change in situation, but is highly variable and context-specific (Sidebar 2).

BARKING UP THE RIGHT TREE — Sidebar 2

Dogs bark in a variety of situations – as an alert, when isolated, during play, and to solicit attention. Recent studies have shown that not only are these vocalizations context-specific, but each type of bark has a set of unique acoustic characteristics that owners are capable of identifying and responding to. Sophia Yin and colleagues at the University of California at Davis recorded and analyzed the barks of dogs in three different contexts: while alerting to the doorbell, when isolated from the owner, and during play. In all three contexts, individual dogs had their own "style" of barking and were easily identified by their barks. When barks within contexts were examined using spectrographic analysis, the researchers found several consistent differences between the three types of barking situations. Dogs alerting to a stranger at the door (disturbance barks) used a harsh, low-pitched bark that did not vary significantly in pitch (i.e. the bark was not modulated). The disturbance bark also continued for a longer duration and was consistently louder when compared with barks that were used during play or in response to isolation. When dogs were playing, their barks were higher in pitch and more tonal, and varied frequently in pitch and loudness. This suggested that dogs use a variety of different types of barks during play, varying the types often. Other researchers have suggested that these varying signals may be important for communicating the non-threatening nature of play behavior. Finally, the barks that dogs used during isolation also varied in frequency and pitch, but tended to be uttered singly and with a longer interval between each bark than the clustered barks used during disturbance or play.

Because barking is an important form of communication between owners and their dogs, the ability of people to identify dogs by their barks and to correctly interpret dogs' vocal communication signals is of interest. Recent studies have shown that people who listened to unknown dogs barking were able to identify dogs barking at strangers (disturbance barks), but were less successful in recognizing isolation barks or in discriminating between two individual (unknown) dogs. However, and perhaps more importantly, human listeners were quite successful at discerning the emotional content of dogs' barks, regardless of whether they knew the dogs and regardless of their own level of experience with dogs. Low-pitched barks with short inter-bark intervals were identified as more aggressive, while high-pitched barks with longer inter-bark intervals were characterized as play barks. Barks that were high-pitched and more tonal tended to be characterized as either fearful or anxious. These results are in agreement with identified vocalization characteristics that signal aggression, appeasement, play, or anxiety in other species. Most importantly, they lend scientific support to what most dog owners are already quite certain of – that they understand what their dogs are barking about!

Molnar C et al. Can humans discriminate between dogs on the base of the acoustic parameters of barks? *Behav Proc*, 73:76–83, 2006.

Pongracz P et al. Human listeners are able to classify dog barks recorded in different situations. *J Camp Psychol*, 119:136–144, 2005.

Yin S, McCowan B. Barking in domestic dogs; context specificity and individual identification. *Anim Behav*, 68:343–355, 2004.

In addition to the bark, dogs have several other types of vocalizations. They whine, howl, grunt, yelp, and growl, and demonstrate enormous variation within each of these vocalization categories. Although it is considered an infantile vocalization in most canid species, adult dogs continue to whine when seeking attention, during displays of submission or when greeting. Whines and whimpers may also be elicited when dogs are fearful or in pain. Howling is one of the most frequent wolf vocalizations but is relatively rare in dogs. It is believed that all dogs are capable of howling, but many do not use this form of vocal communication. Wolves use the howl to seek contact with other pack members when separated or to assemble pack members prior to hunting or travel. Dogs usually howl when they are isolated or in response to an unusual sound in their environment, such as sirens, airplanes flying overhead, or certain types of music. Although the significance of howling is unknown, some think that it may be related to the dog's ability to perceive frequencies of sound that are higher than those perceptible to humans. Puppies and adult dogs grunt when greeting others or as a sign of contentment or relaxation. Conversely, yelps convey severe distress, fear, or pain. Finally, growls usually signal defensive or offensive aggression, but in a modified form also signify playfulness.

Cat Vocalizations

Cats vocalize frequently with their owners, with the classic "meow" being the most easily recognized utterance. In fact, recent studies of cat vocalizations suggest that the meow of the domestic cat may analogous to the bark in dogs. In general, cat vocalizations can be categorized into general groups based upon how the sound is produced. Perhaps the most popular and unusual of the cat's vocalizations is the purr. The purr is a type of murmur pattern, meaning that it is emitted while the cat's mouth is completely closed. Kittens first purr when they are only a few days old, and adults continue to purr throughout life. The purr is also unusual in that it is one of the few vocalizations that is produced during both inhalations and exhalations, rather than only during exhalations, as in most forms of vocalization. Purring is produced in the cat's throat through alternating activation of the intrinsic laryngeal muscles and the diaphragm. This causes a repetitive build-up and release of air pressure in the glottis. It is this turbulence that produces the purr sound. Together, the air movement and muscle vibrations of the chest create the vibrating sensation that can be felt while the cat is purring. Although purring is most commonly associated with pleasure or contentment, cats purr in a variety of circumstances. Kittens purr while they are nursing or being groomed by their mother. Adult cats typically purr when they are in contact with a human caretaker or a familiar housemate, when resting quietly, or when they are rolling and rubbing. While all of these situations are decidedly pleasurable for cats, veterinarians report that some cats purr continuously when they are very ill or even in pain. In most cases, purring is only observed when cats are in the presence of a human or a known cat, so purring may function purely as a form of social communication. Another type of murmur pattern is the trill, also called the chirrup. This vocalization is commonly produced by cats toward their owner or familiar cat housemates upon greeting. Many cat caretakers refer to this call as their cat's "greeting sound." This vocalization is also used in other amicable social encounters, such as to establish contact, when playing, or when asking for food. The volume of this murmur is often very low, and is also heard when a mother cat returns to her litter after a period of absence.

Vowel sounds are made with the mouth open and gradually closing. The most common vowel sound is the characteristic cat "meow." Interestingly, the majority of pet cats use the meow almost exclusively when interacting with humans but do not typically use it when interacting with other cats. Cats' meows are highly variable and context-specific, but all signify some type of amiable social encounter when the cat wishes to establish contact or to request interaction, play, or food. Individual cats are known to develop an entire set of meows for specific situations when interacting with their human caretakers (Sidebar 3). Another less frequently used vowel sound is the female mating call, which is typically uttered during proestrus and estrus, when the female is attempting to attract a mate.

Sidebar 3: THE CAT'S MEOW

Of the various types of sounds that cats utter, the "meow" is unique in that it is directed primarily toward humans and is not a feature of cat-to-cat communication. In fact, the "meow" appears to be an acoustic signal that has been uniquely modified during the cat's domestication. Comparisons of the acoustic characteristics of the meows uttered by domestic cats and those produced by the African wildcat have shown that those produced by domestic cats are distinctly different. A group of human listeners assessed a series of cat meows that had been recorded during several contexts. The listeners knew neither the species of cat (domestic or wild) nor the situation (food soliciting, agonistic, or frustration). Regardless of the context, subjects rated the domestic cat meows to be significantly more pleasant and less urgent than the calls of the wildcats. When the domestic cat meows were evaluated acoustically, they were found to be higher in pitch and frequency, shorter in duration, and more juvenile in tone than the vocalizations of the wildcat.

Interestingly, other studies have shown that domestic cats readily learn to alter both the frequency of their meows and the duration of time that they vocalize in response to interactions with their owners. It is quite possible that selection for cat vocalizations that were more pleasant and "infantilized" occurred during the domestication process. Enhanced flexibility of the meow may have developed as this vocalization became integral for communication during cat-human interactions.

Similar to the bark in dogs, meows that cats direct towards their human caretakers can be classified according to their functionality. When a group of human listeners were asked to assign meaning to meows uttered by cats in five different contexts, they were modestly successful at classifying the sounds correctly. Those study participants who were cat owners or familiar with cats were more successful than those who were not experienced with cats. However, in the absence of other cues such as the cat's body posture and environmental references, most human subjects were unable to derive specific information from the various meows. Most of the subjects readily distinguished between agonistic (conflict) and affiliative meows, but could not correctly assign specific contexts within these broad categories. For example, a subject might know that a cat was meowing in a friendly way but could not distinguish between when she was soliciting food versus when she was seeking attention.

These results suggest that cats use a variety of meow types in a general nonspecific way to attract the attention of their caretakers. It follows that additional cues such as body postures, orientation and activity level are equally important for completing the communication, and for successfully getting the cat's message across.

Nicastro N, Owren MJ. Classification of domestic cat (*Felis catus*) vocalizations by naïve and experienced human listeners. *J Compar Psych*, 117:44–52, 2003.

Nicastro N. Perceptual and acoustic evidence for species-level differences in meow vocalizations by domestic cats (*Felis catus*) and African wildcats (*Felis silvestris lybica*). *J Compar Psych*, 118:287–296, 2004.

Strained intensity sounds are emitted with the mouth held open for the duration of the sound. These include the feline growl, hiss, and pain shriek. All of these vocalizations are used during periods of intense emotional stress. Growls are similar to those made by dogs in that they are low in pitch and are used during aggressive encounters (usually prior to the start of an actual fight). Conversely, cats hiss when they are fearful and defensive. The spit is considered to be a more intense form of the hiss, and is used to deter predators or avert threats. Finally, the male cat's mating call is usually categorized as a strained intensity sound. It is uttered during mating or when competing for estrus females.

VISUAL SIGNALS

Dogs and cats use complex and varied visual signals to communicate with other individuals of their species and with humans. The most important forms of visual communication involve eye contact, facial expressions, and body postures. When evaluating the visual signals of a dog or cat, particular aspects of body posture that should be noticed include the relative placement of the head and the body's topline, placement of the feet, carriage of the tail, and the presence or absence of **piloerection**. In all cases, the context within which the visual signals are given and the identity of the receiver of the signals (i.e. a known versus unknown dog, the owner versus an unknown person) must be considered when interpreting visual signals.

There are some general similarities in the visual signals that are used by dogs and cats, but each species uses body postures and facial expressions that are clearly unique to that species. Once again, we can look to the different evolutionary histories of dogs and cats to help explain some of these differences. The dog evolved from a group-living and cooperatively hunting predator. Within a pack, wolves rely upon clear communication to facilitate cooperation during hunts and when feeding young, defending territory, and protecting the pack. Communication within the social group is also important to strengthen bonds, establish hierarchies, and maintain group cohesiveness. Because of the complex social lives of wolves, their ability to communicate visually with one another is highly developed and complex. The domestic dog has inherited these complex communication patterns and displays similar visual signals. A major difference is that the present-day dog uses visual signals to maintain social relationships with his or her human caretakers as well as with others of his own species and, in some cases, with companion cats living in the same household (see Chapter 1, pp. 8–12).

Conversely, the cat evolved from a solitary species, the African wildcat, an animal that lives and hunts alone. During the wildcat's history, there was no selective pressure for communication signals that would facilitate cooperative hunting or group cohesiveness. Although the domestic cat is more social than his wild ancestor, selective pressure for sociability occurred very recently in the cat's evolutionary history. Moreover, selection during domestication was for social cats who were capable of living peaceably in groups around a central food source. The majority of the cat's history was as a solitary individual who had less need to develop complex communication signals. As a result, many of the cat's communication signals are interpreted as signals designed to minimize contact with other adult cats and to prevent aggressive encounters with strangers. While domestication has resulted in modifications of these signals, many of the cat's inherited visual signals are used either to diffuse a potential fight or to signal defensive or offensive aggression.

Regardless of these differences, visual signals for both dogs and cats can be classified into two broad categories – distance-reducing signals and distance-increasing signals. Distance-reducing signals include those that are used between members of social groups during amiable greetings, play, mutual grooming, and when one animal is showing appeasement or submission to another. Conversely, distance-increasing signals include offensive and defensive threats and fear signals.

Distance-Reducing Signals – Greeting

Most people who live with dogs and cats easily recognize their pet's friendly greeting behaviors. Both dogs and cats use signals that communicate deference or appeasement when greeting another animal or human with whom they wish to interact. A dog who is friendly and confident when greeting has a slightly lowered body posture and the head is held level with the topline of the back. The ears are either forward or shifted slightly down and back. The mouth is usually held open in a "greeting grin" with the lips retracted back horizontally and all of the teeth showing. A confident dog will greet familiar humans and dogs with friendly eye contact or may briefly offer eye contact and then glance away (Figure 3.4). Contrary to popular folklore, eye contact in dogs is not a sign of "dominance" (Sidebar 4). Friendly and non-threatening eye contact is often observed between dogs and between dogs and humans during greeting, play, and social grooming. If the dog is less

figure 3.4 Friendly Rottweiler showing confident greeting signals

figure 3.5 Friendly Labrador Retriever showing appeasing greeting signals

confident or showing deference (appeasement) as she greets, greeting postures change slightly to signal appeasement. The tail is lowered but still wagging, ears are down and shifted backwards, and the dog's body posture is lowered (Figure 3.5). Some dogs may shift their weight back and lift a front paw (an appeasing "intention" gesture). The dog may lick (tongue flick) and her eyes are partially shut with a "squinting" appearance. Dogs who are very excited during greeting and who are showing deference often "wiggle" their hind ends as they greet. A dog demonstrating overt appeasement during greeting will avoid eye contact altogether, but will still approach with her head and tail down. The tail is wagging, and the dog may even have a "groveling" appearance (a look that many owners misconstrue as "guilt").

UNDERSTANDING CANINE EYE CONTACT — Sidebar 4

Eye contact is one of the most important forms of visual communication that dogs use and is also a good example of a signal that must always be interpreted within the context in which it is occurring. When friendly dogs greet one another, some degree of eye contact is initiated. The more confident dog typically initiates eye contact and will maintain contact for a longer period. The second dog demonstrates deference by either averting his gaze or avoiding direct eye contact altogether. Alternatively, one dog may offer a play bow, accompanied by an open-mouth play face and friendly eye contact. This will lead to either play or continued sniffing and friendly interaction.

Conversely, a direct and unwavering stare, when accompanied by a stiff body posture and raised head and tail, is interpreted as an offensive threat.

If the recipient dog does not lower his gaze and show appeasing behaviors, the threat may be escalated through the baring of teeth, piloerection, and growling. Continued escalation without an aversion of gaze by one of the dogs can lead to overt aggression and to a fight.

Similarly, eye contact is an important component of interactions between dogs and humans. Dogs usually interpret a prolonged and direct stare from a person as an overt threat, and most dogs react in the same manner that they would with another dog to this form of eye contact. Some show deference and appeasing behaviors, while other respond aggressively with threat postures. However, dogs who have established a secure and loving relationship with their human caretakers regularly use and accept friendly eye contact to communicate. Examples include dogs who are soliciting attention, playing, or begging for food. There is also recent evidence that dogs are capable of correctly interpreting human gaze to direct their attention to an object of interest in the environment and that dogs understand eye contact as a form of communication signaling attentional state. Similarly, pairs or groups of dogs who live together and have a stable social group show friendly and non-threatening eye contact to each other when greeting, playing, and grooming each other, and use gaze to direct the attention of others in their group. In peaceful social groups, dogs rarely use prolonged direct stares when interacting. When this does occur it is usually only seen during disputes over a desired resource such as a treasured toy, food, or possibly the caretaker's attention.

Fukuzawa M, Mills DS, Cooper JJ. More than just a word: Non-semantic command variables affect obedience in the domestic dog (*Canis familiaris*). *Appl Anim Behav Sci*, 91:129–141, 2005.

Miklosi A, Soproni K, Topal J, Csanyi V. Comprehension of human communicative signs in pet dogs (*Canis familiaris*). *J Compar Psychol*, 115:122–126, 2001.

Schwab C, Huber L. Obey or not obey: Dogs (*Canis familiaris*) behave differently in response to attentional states of their owners. *J Compar Psychol*, 120:169–175, 2006.

All of these postures are normal distance-reducing signals that convey an interest in interacting and greeting and communicate a non-threatening intent.

Non-threatening greeting behaviors between two dogs includes similar body postures plus sniffing of the face followed by the side or groin (inguinal) region (Figure 3.6 and Figure 3.7). The dogs' body postures are relaxed, their ears are relaxed or back, their mouths are closed and lips are pulled back, their eyes or either slightly squinted or open wide by avoiding direct stares and their tails are gently wagging. Although one dog may be more confident than the other (in this case, the dog on the left), this does not signal a threat in the absence of other signs. After sniffing faces and then groin regions, the greeting either ends peacefully or one dog invites the other to play.

Cats also show characteristic visual signals when greeting their owners or another friendly cat in their social group. Some authors refer to this set of behaviors in the cat as "affiliative" postures. When a cat is approaching a human caretaker or a known and friendly member of her social group to greet, the tail is held vertically and the body posture is raised slightly by extending the back legs (Figure 3.8). When greeting a person, the cat will first touch the hand or leg very softly with her nose, followed by head rubbing. Many cats then rub along the person's legs from head to tail. When meeting another cat, the nose-touch occurs nose-to-nose and is followed either by sniffing of the face and mild rubbing, sitting next to each other to groom, or by simply walking away (Figure 3.9).

figure 3.6 Two friendly (non-threatening) dogs greeting for the first time; face sniffing

figure 3.7 Continuing to greet; side presentation and sniffing

figure 3.8 Greeting signals in a friendly cat

figure 3.9 Two familiar cats greeting ("nose touch")

figure 3.10 Cat rolling to solicit petting

Some cats also roll onto their backs after greeting their owners. This is a species-specific behavior that is normally demonstrated by female cats immediately after mating and appears to be reappropriated as an invitation to play or a solicitation for petting directed toward people but not to cats (Figure 3.10).

A comparison of the greeting postures of dogs and cats reflect an important difference between dogs and cats. Specifically, a dog who is not entirely confident and is showing appeasement still desires to greet (in fact, this type of greeting is typical of many companion dogs). Conversely, the cat does not express such a wide range in types of greeting postures. A cat will generally either greet in friendly and confident manner, as shown, or will not approach at all (i.e. will demonstrate fear or aggression). Rolling in cats is considered to be a learned behavior that solicits interaction and petting, and should not be confused with signs of appeasement or submission. Because cats do not engage in the complex dominant/submissive interactions or social relationships that are described in dogs, the greeting behavior of cats does not show the large degrees of variation that are observed in dogs. A greeting cat who becomes less confident simply will not greet, whereas a dog who is more deferential will greet just as will a dog who is more confident.

Distance-Reducing Signals – Play Solicitation

Puppies and kittens all play frequently, but play is also an important part of the lives of many adult dogs and cats. Most dogs enjoy playing with their owners and with other dogs, provided they are well-socialized. Similarly, cats who live in multiple-cat homes often have certain "play buddies" with whom they play a variety of games, and many also enjoy playing with the humans in their lives. Both dogs and cats have a set of highly ritualized behavioral cues that communicate to another individual the intention of play. However, once again, these signals appear to be more highly developed in the dog than in the cat. Clear signals for play are important because many of the behaviors that occur during play mimic those typically used during aggressive encounters and predation. Therefore, a signal that communicates "we are playing now – this is not real" is important to prevent misinterpretations and fights.

Perhaps the most well-recognized play posture in dogs is the play bow. This body posture is displayed when dogs invite another dog or a person to play. The dog lowers his forequarters slightly while leaving the back legs extended and the rump raised (Figure 3.11). When inviting another dog to play, a sideways approach is often used, as this type of approach is interpreted as non-threatening. The inviting dog may raise a front paw in an intention movement and may paw at the face of the other dog (or at the leg of the human playmate). The mouth is usually open wide, the lips are retracted back horizontally, and the tongue is lolling out. The ears are forward and, if the dog is confident, he will offer friendly eye contact. Indeed, many dogs attempt to look directly into their owner's eyes when offering a play bow, perhaps as an attention-seeking gesture (Figure 3.12). If the slight forward movement and paw does not have the desired result, the dog may then progress to a full play bow. This is often followed by either a pounce forward

figure 3.11 Canine play bow as an invitation for play

figure 3.12 Play bow in dog showing direct eye contact to owner

or running away in an invitation to chase. Dogs (and humans) interpret a play bow as an invitation to play and will often respond with a similar bow, or will immediately begin a chase to instigate a "catch-me" game. Both puppies and adult dogs use play solicitation postures to signal friendly intentions. Interestingly, recent studies of how dogs and their caretakers interact have found that one of the most common body postures that humans use to entice their dogs to play is a simulation of the canid play bow.[6] Most dogs interpreted this posture as invitation to approach and to make contact with the person for play or petting.

Cats demonstrate several types of play. The two most common are object play and social play. Kittens also show exploratory play, which involves investigating, climbing, and jumping on novel items within the environment. Exploratory play in cats tends to decrease in frequency as cats mature. Object play may or may not involve another cat or human caretaker, while social play always involves interactions with other animals or people. An important difference between dog and cat play is that cats are more likely than dogs to engage in solitary play. Favorite toys are often batted around and pounced upon, simulating the cat's normal predatory sequence. Similarly, hanging toys that move erratically are attractive to many cats. When soliciting play with their owner or other cats, however, cats (like dogs) use several ritualized visual signals that appear to convey the intention to play rather than to fight. Kittens (and, less frequently, adult cats) will partially open the mouth (the cat version of the play face) at the start of interactions. This is often followed by the "belly-up" position, with the kitten or cat lying with the belly exposed and all four feet in the air. The cat paws with the front feet toward the target play partner, keeping the claws retracted (Figure 3.13). Belly-up is the most common play solicitation posture that cats offer to their human owners. A second posture that cats frequently demonstrate with other cats and occasionally with humans is the "side step." This posture is characterized by a lateral body position with a slightly arched back and a slightly raised tail. The cat then either approaches from the side or hops sideways in an invitation to play. This posture is most commonly seen when a cat is inviting another cat for a chase game.

figure 3.13 Feline invitation to play

Distance-Increasing Signals – Offensive Threat

A dog who is alert and attending to a stimulus will direct his ears forward and up, pointed in the direction of the stimulus. If the dog remains confident, the ears remain forward and up and the eyes are directed to the stimulus. *Offensive threat* refers to a confident dog who is displaying threat signals when in control of a valued resource or when responding to a challenge to a defended territory or social status. The dog will stand tall on his toes, raise his head, lean forward, and elevate his tail. As an interaction becomes more agonistic, piloerection will develop along the dog's shoulders and back, and facial expression will change. The dog's lips will be retracted vertically into a snarl, sometimes called a "C" lip position, and the dog may growl deeply (Figure 3.14). The dog continues to lean forward and hind feet are broadly spread. The tail is often wagging stiffly (flagging) at high frequency (Sidebar 5). If the offensive threat display is directed toward another dog, the threatening dog may attempt to place his forepaws over the shoulders of the second dog. If the second dog shows deference (reduced body posture, avoiding eye contact), the interaction may end peaceably. However, if the second dog responds with direct eye contact or growls, a fight may ensue. When a dog showing offensive threat is guarding a valued resource such as a bone, toy, or territory, the dog will "freeze" over the object and direct eye contact toward the intruding person or dog. If the intruder continues to approach, an aggressive bite will usually occur as the dog defends against the other's access to the resource.

Similar to the dog, a cat who is confident and on the offensive adopts a body posture that makes the cat's body appear larger in size. Cats showing offensive threat have piloerection over the back, and will extend their hind legs to stand at full height. The ears will be drawn partially back and diverted

figure 3.14 Offensive threat posture in a dog

| Sidebar 5 | **TALES OF THE TAIL: WHAT WAGGING TELLS US** |

Although a wagging tail is often considered to be the universal sign of a friendly dog, in reality dogs wag their tails in a variety of circumstances and when experiencing a wide range of emotional states. In short, tail wagging is a highly context-specific behavior that conveys excitability or a high degree of stimulation (arousal). Friendly and calm dogs greet with a relaxed and loosely wagging tail. Highly excited dogs and dogs exhibiting deferential behaviors also have an exuberantly wagging tail, often described as the "entire rump-moving" tail wag. By comparison, an anxious dog will lower her tail and wag it comparatively slowly as she greets. In all of these situations, a wagging tail tells us that we are indeed meeting a friendly dog.

However, the tail can also tell a different story. A dog who is holding her tail very high and stiff as she wags it at high frequency is not usually conveying friendly intentions. This posture can signal offensive threat and the possibility of aggression. This is most commonly seen when two unfamiliar dogs are greeting or when a dog is defending territory. The best approach to assessing the meaning of a tail wag is to always look at the entire dog. Although the tail can tell us a lot, it is never the entire story. The dog's body posture, facial expression and use of eye contact, along with the tail's position and movement, will provide the most complete information about the dog's emotional state. Watch the entire dog, not just the tail!

to the side. The cat gazes directly at his opponent, and the pupils of the eyes are constricted. The tail is held down and pointing toward the ground (not tucked), and the tip of the tail may be flagging slowly (Figure 3.15). A cat whose territory has been invaded by another cat will often first freeze in place and alert to the new cat. If the cat advances or does not leave the area, the resident cat progresses to offensive threat. In many cases, posturing alone causes the invading cat to exit the area, thereby preventing an aggressive attack. However, if the intruding cat remains, a fight may result. With the exception of territory, cats are less likely than dogs to guard resources such as toys, food bowls, resting places, or access to their owners. This difference between cats and dogs can probably be explained by both the cat's more solitary history (i.e. they have less of a need to form strong and enduring attachments to others in a social group) and to the related difference in eating patterns (cats evolved as solitary hunters who consumed small prey that was quickly consumed, thereby having little need to protect or guard a food cache; see feeding behavior, pp. 71–72).

figure 3.15 Offensive threat posture in a cat

Distance-Increasing Signals – Fear

Animals who are fearful are said to display distance-increasing signals because their primary motivation is to avoid or escape from the fearful stimulus. If escape is prevented through the use of a cage, leash, or by cornering the animal, defensive aggression may occur. However, not all dogs and cats who are fearful will become aggressive if they are unable to escape. Some react to extreme fear and the lack of an escape route by freezing in place. Regardless, it is important to distinguish between body postures that signify fear and body postures that indicate that the dog or cat is becoming defensively aggressive. Similarly, it is of utmost importance in dogs to distinguish between signals that convey appeasement and those that convey fear. (Sidebar 6)

IS IT APPEASEMENT, SUBMISSION, OR FEAR? | Sidebar 6

When assessing communication signals in dogs, it is crucial to distinguish between appeasement behaviors, submission, and fear. The most obvious difference between appeasing behaviors and fear is that a fearful dog wishes to avoid interaction if at all possible and so will either move away from the cause (flight), become immobile (freeze), or become defensively aggressive (fight). In direct contrast, appeasement behaviors are social behaviors that a dog expresses during interactions with others. Appeasement and submission differ from fear in that they typify a dog who *desires* interaction and is communicating that she is not a threat. Appeasement behaviors are typically seen when a dog greets other dogs or humans or when a deferential dog approaches another dog who has possession of a desired resource. Appeasement communicates the intent for peaceable interaction and conflict avoidance. Submission is said to occur when an actual conflict does occur and the appeasing dog averts aggression through submissive postures.

Because some of the visual signals seen with fear and appeasement are the same, it is important to consider all of the visual cues that a dog is showing, as well the stimulus to which the dog is reacting and the context of the behavior. Fearful and submissive dogs lower their body posture, pivot their ears back and down, and usually avoid eye contact. However, a dog who is fearful will not be attempting to interact, has dilated pupils, wide open (rounded) eyes, a lowered and motionless tail, and may have raised hackles (piloerection). A dog who is showing appeasing behaviors will approach with a lowered body posture, a lowered and wagging tail, squinty eyes, and a greeting grin. If a conflict has occurred and the dog shows submission, he will freeze in place or lie down and expose his inguinal area, keeping his face averted. (A fearful dog who lies down and "freezes" will *not* expose his belly, and will try to make himself look as small as possible.)

Last, it is important to always consider the context of the interaction. Is this a situation in which you would expect this dog to be friendly and social, or is it perhaps a situation that might cause fear? This is especially important for animal professionals who are required to handle dogs for medical or behavior reasons. The outcome of attempting to restrain a fearful dog can be decidedly different from the outcome of attempting to restrain a submitting dog – it is important to be able to recognize the difference!

The general signs of fear in dogs include a lowered body posture, tucked tail, wide and rounded eyes, dilated pupils, closed mouth with retracted lips, and piloerection (Figure 3.16). Although the dog's ears may be directed forward when the dog initially notices the stimulus, fearful dogs usually keep their ears pivoted back and down. The piloerection in a fearful dog is concentrated primarily over the shoulders and over the hips, rather than as a continuous line along the entire length of the back, as is seen in an offensive threat posture. Because the dog's primary desire is to escape, his body posture is often directed backwards rather than forwards and he may make an effort to escape or will hide behind his owner. Dogs who are nervous and conflicted – in that they would like to approach but are fearful – often emit a characteristic "woof" sound in alarm, followed by rapid eye movements sideways as they search for an escape route. A dog who is non-reactive (i.e. will not become aggressive) and extremely fearful will often lie down, with the head lowered and turned away and gaze averted. Finally, a dog who is moderately fearful and very anxious will often pant excessively or may show displacement behaviors such as yawning or lip licking. Conversely, a dog who is more reactive and perceives himself to be cornered and unable to escape may become defensively aggressive (see below).

figure 3.16 Fearful dog

figure 3.17 Fearful cat

Similar to dogs, cats who are fearful reduce their body size. If the cat is in a cage, she will crouch towards the back, lower her head, and tuck her front feet underneath her (Figure 3.17). Her pupils will be dilated and her ears are either directed toward the frightening stimulus or pivoted backwards. Cats who are frightened and suddenly pin their ears back should be approached cautiously as pinned ears indicate that the cat may become defensively aggressive. It the cat is not confined or otherwise prevented from fleeing, an observation of fear is quite fleeting, because most fearful cats choose to escape. A common difference between dogs and cats demonstrating fear is that dogs are more likely to feel conflicted if the fear-inducing stimulus is a person or friendly dog, and so are more likely to attempt to remain in visual contact and to interact than are cats.

Distance-Increasing Signals – Defensive Threat

Dogs and cats exhibit defensive threat postures when they are fearful and desire escape, but perceive that there is no escape route available. A dog showing defensive threat will pin his ears back, but the muzzle is now retracted in a snarl and the dog is gazing toward the stimulus. Some dogs show a characteristic "whale eye" during this type of threat posture, with the head averted slightly as the dog still attempts to watch the stimulus that is causing fear. This posture causes the white of the eye to become easily visible. Some describe the fearful aggressive dog's body posture as depicting a state of ambivalence, since the dog is both insecure and defensive. In other words, the dog shows the lowered body posture of fear, but also will snarl and possibly snap at the threatening person or dog (Figure 3.18).

figure 3.18 Defensive aggressive body posture in a dog

figure 3.19 Defensive aggressive body posture in a cat

Similar to dogs, cats on the defensive send somewhat conflicting signals. While a fearful cat reduces his body size, a defensive cat will often present laterally (sideways) towards the aggressor with an arched lateral display and piloerection. This sideways display is thought to be an attempt to maximize the cat's visual impact and, as such, is an attempt to bluff. However, the cat's facial expressions belie this and reflect the cat's true state of mind. A defensive cat's ears are flattened down and back, the whiskers are drawn back, the corners of the mouth are pulled back to bare the teeth, and the pupils of the eyes are dilated. The tail is either erect or is concave with the tip pointed down. If the cat is hiding or is confined (such as in a cage), the lateral display is not shown but the cat's pinned ears, hiss, and open mouth will signify as change from passive fear to defensive aggression (Figure 3.19). During fights between cats, the defensive cat will flip over onto his back and, keeping his ears pinned back, will use his back legs and claws to attack the more confident aggressor.

PREDATORY AND FEEDING BEHAVIORS OF DOGS AND CATS

Dogs and cats are both members of the order Carnivora which, as the name implies, includes many species that hunt other animals for food. However, while the cat is considered to be a "true" or obligate carnivore, the dog is more omnivorous, capable of consuming and deriving nutrition from a variety of animal and plant foods. In addition, the dog evolved from a species that hunted cooperatively in groups, using coordinated efforts to prey upon large mammals. In contrast, the cat evolved from a solitary wildcat that hunted small mammals such as mice and voles. Domestication has led to significant changes in the daily eating patterns of dogs and cats, but both species have retained vestiges of the predatory and feeding behaviors of their wild ancestors.

Predation Sequence and Feeding Behavior of Dogs

The dog's wild relative, the wolf, obtains much of its food supply by hunting co-operatively with other wolves in its social group. Cooperative hunting behaviors allow the wolf to prey on large mammals that would otherwise be unavailable to a wolf hunting alone. As a result, most wolf subspecies tend to be intermittent eaters, gorging themselves immediately after a kill and then not eating again for an extended period of time. Competition between members of the pack at the site of a kill leads to the rapid consumption of food and the social facilitation of eating behaviors. Social facilitation refers to increased rate of eating or the amount that is consumed when an animal is eating with in the presence of conspecifics. Wolves and other wild canids also exhibit food-hoarding behaviors. Small prey or the remainder of a kill is buried when food is plentiful and later dug up and eaten when food is not readily available. Finally, most wild canids, including wolves, will consume carrion and will scavenge for a wide variety of food items. Feral dogs have been shown to subsist almost entirely by scavenging, which may reflect adaptations that occurred during the early stages of domestication as wolves exploited human waste sites (see Chapter 1, pp. 5–8).

Predatory Behavior: Like other behavior patterns, predatory behavior has been modified in the domestic dog through domestication and selective breeding. In most breeds, the intensity of the predatory response has also been significantly reduced. For example, while most dogs retain a strong orient-and-chase response, the final "kill bite" is not expressed. In some breeds, certain aspects of predation have been selected for and enhanced. For example, many hunting breeds excel at the detection of prey, but the stalk, chase, and bite portion of predation has been truncated from the sequence. In contrast, the stalking and chasing of herding breeds represents another segment of predatory behavior that has been modified through selective breeding. Terrier breeds have a strong predatory response that is directed toward small mammals (rodents) and usually does include the final killing bite (but not the dissecting or consuming bite). In all dogs predatory behavior is not directly associated with hunger, but rather with the presence and movement of prey.

Dogs who chase small animals, cars, children, or bicyclists may be demonstrating either predatory behavior (i.e. stalk and chase) or territorial behavior. If the dog is silently stalking the target, and continues to chase and attempt to attack and bite the target regardless of location, this suggests predation. Conversely, many dogs demonstrate territorial aggression by chasing runners, bicyclists, and cars until they are past the dog's established territory. While these dogs may nip and even bite, this signifies territorial aggression, *not* predatory behavior (see Chapter 11, pp. 293–295 for a complete discussion of territorial aggression). While in both cases the aggression can be severe, a dog who shows stalking behavior with the intent of turning the target into prey is potentially far more dangerous, especially if this behavior is directed towards children or small pets.

Feeding Behavior: Similar to their wild cousins, many dogs tend to eat rapidly. Although there is limited evidence for this, it has been suggested that certain breeds of dog are more likely to consume their food rapidly when compared with other breeds.[7] This tendency can be a problem because it may predispose a dog to choke or swallow large amounts of air. In multiple-dog homes, social facilitation can influence rate of eating and may encourage food bowl-guarding behaviors in some individuals. Feeding dogs separately to remove the competitive aspect of mealtime often normalizes the rate of eating and reduces tension. Training dogs to eat only from their own bowls and teaching good mealtime manners can also be used to prevent food competition problems from developing (see Chapter 11, pp. 290–293).

Vestiges of the wolf's food-hoarding behaviors are also seen in domestic dogs. It is not uncommon for dogs to bury bones in yards or, much to the owner's chagrin, to hide coveted food items in furniture or under beds. However, unlike their wild ancestors, many domestic dogs forget about these hidden caches and rarely return to dig them up. Related activities that are common to dogs (but found to be undesirable by most owners) are scavenging and coprophagy (stool-eating). Many dogs readily consume carrion, feces, plant material, and insects. Although these behaviors are considered normal, they present a health and sanitation risk, and should be prevented. Keeping the yard picked up, using supervision, and teaching dogs a reliable "leave it" command are the best approaches to controlling scavenging and stool-eating behaviors in dogs (see Chapter 9, pp. 218–219 for a complete discussion of teaching "leave it").

Predation Sequence and Feeding Behavior of Cats

It is common to think of the domestic cat as a descendant of the wild felids that prey on large grazing animals. However, owners must remember that the primary ancestor of the cat is the small African wildcat, *Felis libyca.* This desert-dwelling cat's primary prey is small rodents about the size of field mice. Therefore, the immediate ancestor of the cat is not an intermittent feeder like the larger wildcats, but rather is an animal that feeds frequently throughout the day by hunting many small rodents. Like the majority of wild felids, the African wildcat is a solitary animal, living and hunting alone for much of its life and interacting with others of its species only during mating season. This solitary nature has resulted in an animal that eats slowly and is generally uninhibited by the presence of other animals.

Predatory Behavior: Unlike dogs, predatory behaviors of cats have remained essentially unchanged during domestication. This may reflect the shorter period of time since domestication as well as the fact that many cats continue to live as outdoor cats on farms and in barns, in environments that continue to exert selective pressure for predation. When hunting, cats locate prey either while moving slowly through a hunting area or while sitting quietly and waiting. They initially detect prey in their surroundings using primarily acoustic cues (sound). The cat has exceptional ability to discriminate between different high-pitched sounds (such as those made by a mouse or mole) and to locate the source of sounds. In addition to being able to hear the high-pitched vocalizations of mice, cats also respond to the sound of mice scratching and moving. Once the cat's attention has been alerted by the sound of prey, visual cues become important for hunting and capture. The cat slowly crouches low to the ground, staring at the source of the sound or movement. The cat's head is stretched forward and the ears are erect and pointing toward the area of interest. The entire body is pressed flat to the ground with forepaws drawn back beneath the shoulders. The cat remains frozen in this position until the prey animal appears. Most cats show mounting tension at the sight of prey by shifting their weight back and forth, flagging the end of the tail, and paddling the back feet. The cat then pounces on the prey. Mice are usually killed with a bite to the nape of the neck, while smaller prey (such as large insects or baby birds) are killed by pouncing with the paws.

After the prey has been killed, most cats do not immediately consume the killed animal but rather continue to toy with their prey by batting it around. This tendency to "play" with prey is thought to be a displacement activity caused by the conflict between hunting behavior and a fear of being injured. (These post-predatory behaviors are a common form of object play in most cats.) After this manipulation, many cats carry their prey to a secluded area to eat. However, not all hunting cats actually end up consuming their prey. Cats who are well-fed will continue to hunt, regardless of how recently they have eaten. Although well-fed cats do not typically spend as much time hunting as do free-living cats, predatory behavior is not completely eliminated by providing ample food. Given the opportunity, well-fed cats continue to hunt but consume their prey less consistently. Many cats simply "play" with the prey animal, while others bring the mouse back to their owners, often depositing it in the house or at their caretaker's feet (a behavior that some cat lovers find both puzzling and somewhat upsetting).

Feeding Behavior: Most domestic cats consume their food slowly and do not respond to other cats by increasing either rate of eating or the amount of food that is consumed. In multiple-cat homes, cats often eat peaceably from the same bowls either together or at different times of the day. When problems do occur, they are often very subtle, with one or more cats intimidating a less assertive cat and not allowing access to the food bowl. For this reason, multiple feeding stations in different areas of the home should always be provided in multiple-cat homes. If cats are fed free-choice, most will nibble at the food throughout the day, as opposed to consuming a large amount of food at one time. Several studies of eating behavior in cats have shown that if food is available free-choice, most cats eat frequently and randomly throughout a 24-hour period.[8, 9] However, meal feeding is often preferred by owners because it represents a time of pleasurable interactions with their cats, characterized by daily and familiar feeding routines of communication, petting, and handling. In addition, the predisposition of many housecats to overweight conditions makes meal feeding necessary in order to control intake and weight gain.

SOCIAL RELATIONSHIPS OF DOGS

It is useful to consider that many dogs (like humans) are members of several different social groups. The first, and probably most important to owners, is comprised of the relationships that dogs have with their human caretakers. These relationships may include a single human owner or a large group of individuals, ranging in age from a toddler to a grandparent. In addition, many dogs live with one or more other dogs. The relationships that they have with the other dogs in their household comprise a second (and equally important) social group. Finally, just as human relationships are varied and complex, so too are those of our dogs. The importance of considering the context within which interactions occur cannot be overstated. Much of what dogs are concerned with involves access to and possession of valued resources. Such resources include food bowls, toys, favorite resting places, access to doorways and the outdoors, and attention from owners. An individual dog's reaction to the removal of resources or to being prevented from attaining a resource is influenced by how much the dog values that particular resource, the dog's relationship to the dog or person with whom they are competing, and the dog's past learning experiences regarding appeasement and threat behaviors.

When describing canine social behavior, the terms *dominance* and *subordinance* are sometimes appropriate when describing an interaction between two individuals in which one dog gains or retains control of a desired resource and the second dog demonstrates either deference (appeasement behaviors) to avoid a conflict or submissive behaviors in response to an aggressive threat. However, it is less clear that the terms *dominant* and *submissive* should be used to describe *individuals* (Sidebar 7). For example, Figure 3.20 and Figure 3.21 illustrate these two scenarios. In Figure 3.20, the dog on the right is in possession of a favorite chew toy. The visual signals that he shows include freezing

THE DOMINANCE MODEL: WHAT WORKS AND WHAT DOESN'T

Sidebar 7

The concept of dominance was first developed in the 1950s as a theory to describe the social relationships observed in certain bird species. Over time, this model of social behavior became popularized to reflect linear "pecking orders" in which dominant animals retained control over all resources and actively defended their status against others. This model was then assumed (with some evidence) to describe the social behavior of the wolf, and then extrapolated (with little evidence) to the social behavior of the dog.

First, the facts: Dominant behavior is defined as occurring when one individual actively competes with another for access to or control over a desired resource. The individual (dog) who maintains control of the resource in question is said to be demonstrating dominant behavior or showing dominant signals, and the dog who relinquishes the resource or declines to compete for it is showing appeasement or subordinate behavior (signals). These interactions are typically resolved through the use of ritualized and highly stereotypic communication patterns that rarely involve overt aggression. When used to describe the expressed behaviors between two individuals in a specific situation, the dominance model can be appropriate as a "short-cut" method of describing the results of that interaction.

Now, some problems: Dominance (and its counterpart, submission) does not describe a dog's personality or temperament (although it is often misguidedly used in this way, especially in the popular press). Nor is dominant behavior synonymous with aggressive behavior (another common mistake). There are serious problems with these portrayals. The first, and most important, is that "dominance" is not a personality trait. A dog cannot be "dominant" in isolation because dominant signals are (by definition) always relative to the behaviors shown by another individual. While there are some dogs who seem to "care" about resources more than others and who display more offensive threat signals than deference signals, it is more accurate to describe such a dog as "behaving dominantly" rather than as a "dominant dog." Second, a portrayal of dominance as aggression implies that conflicts over resources between dogs often end in fighting. As described previously, while all dogs are capable of fighting and aggression, the highly ritualized posturing that dogs use to resolve conflicts typically function to prevent aggression, not cause it. In fact, it is generally more common for dogs to aggressively bite when their signals of fear or deference have been ignored or misinterpreted, rather than during an offensive (dominant) threat.

Steinker A. Terminology think tank: Social dominance theory as it relates to dogs. *J Vet Behav*, 2:137–140, 2007.

figure 3.20 Dog showing appeasement signals to another dog in possession of a valued resource (bone)

figure 3.21 Dog showing offensive threat posture when in possession of a desired resource (bone); Responding dog showing submissive signals

over the toy, forward-directed ears, and a direct stare at the approaching dog. He keeps his head positioned above the bone and one paw on top of the bone. The approaching dog reacts to these signals with appeasement. He looks away (avoids eye contact), lowers his body posture slightly, and turns away from his housemate and his bone. The appeasing dog's ears are rotated partially back but not pinned. His lips are pulled back and his mouth is almost closed, and he may be tongue-flicking. These appeasing signals send the message that his intent is to avoid conflict (threat aversion) and that he will not challenge the other dog for possession of the bone. The interaction ends peacefully.

In Figure 3.21, the dog in possession of the bone shows escalated guarding behavior as an offensive threat. He is standing over the toy, his body posture is raised and forward, his lips are pulled forward in a snarl, and he is giving direct eye contact in a hard stare to the approaching dog. The other dog responds to these threat signals submissively. He has lowered his body posture significantly, is looking away, and has raised a paw. These submissive signals should have the effect of "turning off" aggression in the dog who is showing an offensive threat.

The scenarios illustrated in both of these figures depict the visual signals used during a possessive conflict between two dogs. It is important to note that in homes with several dogs living peaceably together, this type of interaction is often decidedly more subtle. For example, one dog might simply freeze over his bone as another dog approaches. The approaching dog reads these signals, realizes that his housemate is in possession of a desired resource, and so immediately averts his gaze (threat aversion signal) and simply walks away.

While the previously described conflict reflects dominant and submissive signaling, these behaviors should not be erroneously interpreted as reflecting "dominant" or "submissive" dogs. First, not all dogs demonstrate a desire to compete for resources, attain "dominant" status within a group, or have the potential to show offensive threats or aggression during conflicts. The process of domestication has resulted in a tempering of the need for dominance ranking in dogs. The selection for neotenized features and temperament characteristics such as enhanced deference and trainability has resulted in a species in which many individuals are simply "born subordinate." As a result, many dogs never display signs of offensive threat towards their owners or toward other dogs, even upon reaching social maturity or when challenged for possession of a desired resource. For many dogs, such challenges are not an issue because they never develop a strong need to compete for social status or for desired resources. Some of these traits are believed to be influenced by breed. For example, some breeds, such as beagles and fox hounds, were developed to work in very large packs and as a result show little need for social hierarchies. Similarly, some hunting breeds such as the retrievers and spaniels were required to tolerate the presence of other adult dogs who were not in their social group during group hunting excursions. For these breeds, increased sociability and reduced dog-reactivity as adults were probably selected for early in the breed's creation. In contrast, some of the working breeds have been developed to guard and protect, and so may naturally demonstrate more interest in rank order and have a lower threshold for offensive threats and aggression. However, although some generalities can be made in accordance with the original function of a breed, each dog must be viewed as an individual whose propensity to display offensive threat, deferential, and submissive signals is also dependent upon learning, living situation, and the context of each interaction with other dogs and humans.

SOCIAL RELATIONSHIPS OF CATS

Similar to companion dogs, many house cats maintain multiple social relationships. Today, multiple-cat homes are as common as single-cat homes, and many cats share their owners and their living space with a resident dog as well. Adult house cats interact with their human caretakers very similarly to the way in which they interact with other cats within their social group. It also appears that some cats interact with their owners in a more infantile way, treating humans somewhat as surrogate mothers. Bonded pairs of cats and group-living cats typically show loosely structured cooperative or commensal relationships. However, the concepts of dominance and submission that can be helpful in explaining some aspects of the dog's social behavior do not suitably describe the social behavior of cats. The relationships that cats develop with humans and with other cats are not based upon dominant-subordinate dichotomies. Cats neither respond to nor understand attempts by humans to impose such a relationship. Indeed, most owners know that attempts to "dominate" cats will result in either fear or aggression from the cat, not "submission." In many ways, the theoretical model of commensalism better explains the social behavior of the cat, and may apply equally well to the relationships that cats develop with humans. **Commensalism** refers to relationships in which individual (or species) benefit from associations and interactions with others on a temporary or permanent basis.

The strength of companion animal relationships is typically measured by examining the type and frequency of contact between owners and their pets. Because cats do not commonly follow their owners about the house as do many dogs, daily human-cat interactions are typically shorter in frequency and duration than human-dog interactions. Most cats initiate contact with humans as a form of greeting, at feeding time, as a request to go outside or to play, or as a request for petting and cuddling. The behavior patterns that cats show in all of these situations are the same that they use when greeting or initiating play or allogrooming with other cats. Some of the behaviors that adult cats show to their human caretakers may represent neotenized behaviors or behaviors that were in the cat's normal repertoire but are now being used in a new context. For example, many cats greet their owners with their tail held vertically and hind limbs extended slightly. After a brief touch of the nose, the cat will then head and/or flank rub on the owner's legs. Some cats will alternatively rub on objects and furniture that are in close proximity to the owner. Kittens show these same patterns of behavior when greeting their mother or an adult cat in their social group, and adult cats living together in social groups often greet with a gentle nose touch followed by briefly rubbing against each other. Other species-specific behavior patterns that cats demonstrate during interactions with humans include purring, kneading with the paws, and body postures and vocalizations associated with the initiation of play. Cats also use a wide range of meows and purrs when interacting with humans, and most caretakers are very adept at interpreting their cat's intent from these.

REVIEW QUESTIONS

1. Compare and contrast urine marking as a form of communicative behavior in dogs and cats.
2. Describe the ways in which cats use different types of "rubbing" behaviors to communicate.
3. List the types of feline vocalizations and provide an example of each.
4. How have the evolutionary histories of dogs and cats influenced the respective visual communication signals that evolved in each species?
5. Compare and contrast greeting behavior in dogs and cats.

REFERENCES AND FURTHER READING

Beaver BV. **Feline Behavior: A Guide for Veterinarians.** W. B. Saunders Company, Philadelphia, PA, 276 pp., 1992.

Beaver BV. **Distance-increasing postures of dogs.** *Vet Med: Small Anim Clinician*, 77:1023–1024, 1982.

Beaver BV. **Friendly communications by the dog.** *Vet Med: Small Anim Clinician*, 76:647–649, 1981.

Bekoff M. **Scent-marking by free ranging domestic dogs: Olfactory and visual components.** *Biol Behav*, 4:123–139, 1979.

Biben M. **Predation and predatory play behaviour of domestic cats.** *Anim Behav*, 27:81–94, 1979.

Byrne RW. **Animal communication: What makes a dog able to understand its master?** *Curr Biol*, 13:R347–R348, 2003.

Bradshaw JWS. **The Behaviour of the Domestic Cat.** CAB International, Oxford, UK, 219 pp. 1992.

Bradshaw JWS. **The evolutionary basis for the feeding behavior of domestic dogs (*Canis familiaris*) and cats (*Felis catus*).** *J Nutr*, 136:1927S–1931S, 2006.

Brown KA, Buchwald JS, Johnson JR, Mikolich, DJ. **Vocalization in the cat and kitten.** *Develop Psychobiol*, 11:559–570, 1978.

Cooper JJ, Ashton C, Bishop S, West R. **Clever hounds: Social cognition in the domestic dog (*Canis familiaris*).** *Appl Anim Behav Sci*, 81:229–244, 2003.

Coppinger RP, Feinstein M. **Why dogs bark.** *Smithsonian Mag*, January: 119–129, 1991.

Edwards C, Heiblum M, Tejeda A, Galindo F. **Experimental evaluation of attachment behaviors in owned cats.** *J Vet Behav*, 2:119–125, 2007.

Feldman H. **Methods of scent marking in the domestic cat.** *Can J Zoology*, 72:1093–1099, 1994.

Feuerstein N, Terkel J. **Interrelationships of dogs (*Canis familiaris*) and cats (*Felis catus* L.) living under the same roof.** *Appl Anim Behav Sci,* 113:150–165, 2008.

Frazer-Sissom DE, Rice DA, Peters G. **How cats purr.** *J Zoology*, 223:67–78, 1991.

Frommolt KH, Gebler A. **Directionality of dog vocalizations.** *J Accous Soc Am*, 116:561–565, 2004.

Fukuzawa M, Mills DS, Cooper JJ. **More than just a word: Non-semantic command variables affect obedience in the domestic dog (*Canis familiaris*).** *Appl Anim Behav Sci*, 91:129–141, 2005.

Hare B, Tomasello M. **Human-like social skills in dogs?** *Trends in Cognitive Sci*, 9:439–444, 2005.

Houpt KA. **Ingestive behavior: The control of feeding in cats and dogs.** In: *Readings in Companion Animal Behavior*, VL Voith and PL Borchelt, editors, Veterinary Learning Systems, Trenton, NJ, pp. 198–206, 1996.

Jongman EC. **Adaptation of domestic cats to confinement.** *J Vet Behav*, 2:193–196, 2007.

Kienzle E, Bergler R. **Human-animal relationship of owners of normal and overweight cats.** *J Nutr*, 136:1947S–1950S, 2006.

Kleiman DG, Eisenberg JF. **Comparisons of canid and felid social systems from an evolutionary perspective.** *Anim Behav*, 21:637–659, 1973.

Leyhausen, P. **Cat behavior: The predatory and social behavior of domestic and wild cats.** Garland STPM Press, New York, NY, 1979.

Lowe SE, Bradshaw JWS. **Ontogeny of indivuality in the domestic cat in the home environment.** *Anim Behav*, 61:231–237, 2001.

McConnell PB. **Acoustic structure and receiver response in domestic dogs,** *Canis familiaris*. *Anim Behav*, 39:897–904, 1990.

Mertens PA. **The concept of dominance and the treatment of aggression in multidog homes: A comment on Kerkhove's commentary.** *J Appl Anim Welfare Sci*, 7:287–292, 2004.

Miklosi A, Soproni K, Topal J, Csanyi V. **Comprehension of human communicative signs in pet dogs (*Canis familiaris*).** *J Compar Psychol*, 115:122–126, 2001.

Molnar C, Pongracs P, Doka A, Miklosi A. **Can humans discriminate between dogs on the base of the acoustic parameters of barks?** *Behav Processes*, 73:76–83, 2006.

Natynczuk S, Bradshaw JWS, Macdonald DW. **Chemical constituents of the anal sacs of domestic dogs.** *Biochem System Ecol*, 17:83–8, 1989.

Nott HMR. **Social behaviour of the dog.** In: *The Waltham Book of Dog and Cat Behaviour*, C Thorne, editor, Pergamon Press, Oxford, UK, pp. 97–114, 1992.

Pal SK. **Urine marking by free-ranging dogs (*Canis familiaris*) in relation to sex, season, place and posture.** *Appl Anim Behav Sci*, 80:45–59, 2003.

Pongracz P, Miklosi A, Kbinyi E. **Social learning in dogs: The effect of a human demonstrator on the performance of dogs in a detour task.** *Anim Behav*, 62:1109–1117, 2002.

Pongracs P, Molnar C, Miklosi A. **Acoustic parameters of dog barks carry emotional information for humans.** *Appl Anim Behav Sci*, 100:228–240, 2006.

Riede T, Fitch T. **Vocal tract length and acoustics of vocalization in the domestic dog (*Canis familiaris*).** *J Exp Biol*, 202:2859–2867, 1999.

Rooney NJ, Bradshaw JWS. **Breed and sex differences in the behavioural attributes of specialist search dogs – a questionnaire survey of trainers and handlers.** *Appl Anim Behav Sci*, 86:123–135, 2004.

Schwab C, Huber L. **Obey or not obey: Dogs (*Canis familiaris*) behave differently in response to attentional states of their owners.** *J Compar Psychol*, 120:169–175, 2006.

Scott JP. **The evolution of social behaviour in dogs and wolves.** *Amer Zoologist*, 7:373–381, 1967.

Steinker A. **Terminology think tank: Social dominance theory as it relates to dogs.** *J Vet Behav*, 2:137–140, 2007.

Turner DC, Bateson P (Editors). **The Domestic Cat: The Biology of its Behaviour**, 2nd edition. Cambridge University Press, Cambridge, UK, 244 pp., 2000.

Yeon SC. **The vocal communication of canines.** *J Vet Behav*, 2:141–144, 2007.

Yin S. **Dominance versus leadership in dog training.** *Compend Contin Educ Pract Vet*, July, pp. 414–417, 2007.

Yin S, McCowan B. **Barking in domestic dogs; context specificity and individual identification.** *Anim Behav*, 68:343–355, 2004.

Footnotes

[1] Passanisi WC, MacDonald DW. Group discrimination on the basis of urine in a farm cat colony. In: *Chemical Signals in Vertebrates*, 5th edition, DW MacDonald, D Muller-Schwarze, SE Natynczuk, editors, Oxford University Press, Oxford, UK, 1990.

[2] Bekoff M. Scent-marking by free-ranging domestic dogs: olfactory and visual components. *Biological Behav*, 4:123–139, 1979.

[3] Bradshaw JWS, Natynczuk SE, MacDonald DW. Potential applications of anal sac volatiles from domestic dogs. In: *Chemical Signals in Vertebrates*, 5th edition, DW MacDonald, D Muller-Schwarze, SE Natynczuk, editors, Oxford University Press, Oxford, UK, pp. 640–644, 1990.

[4] MacDonald DW, Apps PJ, Carr GM, Kerby G. Social dynamics, nursing coalitions and infanticide among farm cats (*Felis catus*). *Adv Ethol*, 28:1–64, 1987.

[5] Beaver B. *Feline behavior: A guide for veterinarians*. W. B. Saunders Company, Philadelphia, PA, 1992.

[6] Rooney NJ, Bradshaw JWS, Robinson IH. Do dogs respond to play signals given by humans? *Anim Behav*, 61:715–722, 2001.

[7] Bradshaw JWS. The evolutionary basis for the feeding behavior of domestic dogs (*Canis familiaris*) and cats (*Felis catus*). *J Nutr*, 136:1927S–1931S, 2006.

[8] Kanarek RB. Availability and caloric density of diet as determinants of meal patterns in cats, *Physio Behav*, 15:611–618, 1975.

[9] Bradshaw JW, Cook SE. Patterns of pet cat behaviour at feeding occasions. *Appl Anim Behav Sci*, 47:61–74, 1996.

SECTION 2

How Dogs and Cats Learn – Training Techniques and Practical Applications

CHAPTER 4

How Dogs and Cats Learn: Principles of Learning Theory

The behavior of dogs and cats changes throughout life and is influenced by both their environment (experiences) and by genetic makeup. The simplest definition of *learning* involves an enduring change in behavior that occurs in response to stimuli in the environment. (In a rather circular series of definitions, a stimulus is defined as anything that causes a response.) Dogs and cats, like humans, are capable of several types of learning. The major forms that we observe in our companion animals and that we use to teach dogs and cats new behaviors are classical conditioning, operant conditioning, and social learning. Although owners are not often aware of it, habituation and its counterpart, sensitization, also result in changes in behavior as our two best friends learn about their worlds.

CLASSICAL (PAVLOVIAN) CONDITIONING

Classical conditioning is a form of **associative learning** that occurs when an animal responds to relationships between two or more stimuli. Many people are familiar with the most frequently cited example of stimulus-stimulus conditioning – Pavlov's salivating dogs and the ringing bell. The basic elements of this type of learning are a meaningless stimulus (initially called a "neutral" stimulus) that elicits no response and a meaningful (unconditioned) stimulus that elicits a response without any prior conditioning. The consistent pairing of the two stimuli, with the neutral stimulus always immediately preceding the unconditioned stimulus, results in a change in the meaning of the neutral stimulus. The animal learns that the neutral stimulus, while meaningless at first, consistently *predicts* the unconditioned stimulus. Classical conditioning has occurred when the subject begins to show the same or a very similar response to the neutral stimulus that was initially elicited only by the unconditioned stimulus. This is called a stimulus-stimulus association and forms the basic principle of classical conditioning (Sidebar 1).

Returning to Pavlov's dogs: Pavlov was studying dogs' salivary and gastrointestinal responses to the presentation of food. When hungry dogs were presented with food (the unconditioned stimulus), they began to salivate (an unconditioned

CLASSICAL CONDITIONING | Sidebar 1

Negative Example: Fear Response
Cat Visiting a Veterinary Clinic

Appearance of Veterinarian ⟶ Restraint/Pain ⟶ Fear/Anxiety
(neutral stimulus)   (unconditioned stimulus)   (unconditioned response)

After Several Visits

Appearance of Veterinarian ⟶ ⟶ ⟶ ⟶ Fear/Anxiety
(conditioned stimulus)   (conditioned response)

Positive Example: Excitement/Happiness Response
Owner Returning Home

Sound of Car ⟶ Owner Appears ⟶ Happiness/Excitement
(neutral stimulus)   (unconditioned stimulus)   (unconditioned response)

After Several Repetitions

Sound of Car ⟶ ⟶ ⟶ ⟶ Happiness/Excitement
(conditioned stimulus)   (conditioned response)

response). The neutral stimulus was the opening of the door as the assistant entered the room each day with food (or, later, the ringing of a bell immediately before food was presented). As the dogs learned that the opening of the door or the ringing of a bell reliably predicted the presentation of food, they began to salivate in response to the door opening (or bell ringing), prior to seeing or smelling food. The sequence presented in this example is important because classical conditioning is most likely to occur when the two stimuli (neutral and unconditioned) are closely paired in time and when the pairing is consistent and reliable (i.e. one stimulus does not occur without the other). These two conditions are referred to as **contiguity** (the stimuli are closely paired in time and place) and **contingency** (the neutral stimulus is reliably followed by the unconditioned stimulus).

In addition to stimuli contiguity and contingency, the value that a particular unconditioned stimulus has for the dog or the cat affects the animal's tendency to pay attention to predicting stimuli. For example, a dog is more likely to become classically conditioned to events that matter to him than to events that are of no concern. While an owner picking up a shampoo bottle can quickly become classically conditioned to predict bath time (and result in an absent dog!), the same dog is much less likely to associate the owner picking up a pen to write a letter with anything pleasant or unpleasant. This is because writing a letter is an irrelevant event in the life of a dog. Bath time, on the other hand, is often a very undesirable event for most dogs and something they would prefer to avoid. For this reason, attending to stimuli that consistently predict the likelihood of a bath are quickly learned.

Finally, classical conditioning most frequently occurs with behaviors that are respondent in nature. Respondent behaviors are those that are relatively innate and so are under partial or complete involuntary control. The most common respondent behaviors with which we are familiar in dogs and cats (and which are often the subject of classically conditioned behavior problems) include those behaviors expressed when the dog or cat is fearful, anxious, excited, or happy. A few practical examples demonstrate just how prevalent classical conditioning is in the lives of our dogs and cats (Box 4.1 and Box 4.2).

Box 4.1 FRANKIE AND HIS LEASH

Like most dogs, Frankie enjoys going for walks with his owner, Stan. Their typical routine involves Stan returning home from work each day at 5:30 PM. After greeting Frankie and Ralph (the cat), Stan feeds Ralph, puts the paper and mail on the kitchen table, changes his shoes into walking shoes, picks up the lead, and calls Frankie. He then attaches Frankie's collar and lead, and off they go for their daily walk to the park (which of course, Frankie and Stan both thoroughly enjoy!). Most dog owners, at this point in this example, have correctly assumed that, at some point during this daily sequence, Frankie shows increasing levels of excitement. Depending upon the dog, and upon Stan's daily reliability and consistency, this may occur very early or rather late in the sequence. Regardless, if Frankie begins to show excitement and joy at any stage of the preparations for his walk, he is demonstrating the results of classical learning. Let's examine what exactly is occurring in this sequence:

```
                (predicts)
PICKS UP LEASH ─────────▶ GOES TO DOOR ─────────▶ WALK !! (EXCITEMENT) !!!!
(leash = neutral stimulus)              (opportunity to walk = unconditioned stimulus)

                         (conditioning)
LEASH PREDICTS "WALK" ─────────────▶ LEASH ALONE ─────────▶ EXCITEMENT!!
```

Box 4.2 FLUFFY AND HER DINNER

Fluffy's owner Fran provides two meals to Fluffy each day. In the morning, Fluffy gets a small bowl of dry cat food and, in the evening, she is given dry food with two added teaspoons of canned cat food. At 5:00 every evening, Fran asks Fluffy if she would like dinner and then walks to the refrigerator to remove Fluffy's can of food. Fluffy follows Fran and begins to meow repeatedly as she dances back and forth in excited anticipation of her meal. What stimulus-stimulus association has Fluffy learned?

The unconditioned (meaningful) stimulus (US) is the presence of Fluffy's evening meal. Nothing needs to be learned for Fluffy to enjoy her food when it is presented to her. The neutral stimulus (NS) may be Fran asking "Do you want to eat?" going to the refrigerator, and/or opening the refrigerator door. Initially, such a question will not evoke a response because Fluffy has not yet associated it with the presentation of her food. However, if Fran is both reliable and consistent (contingency and contiguity), Fluffy learns that this spoken phrase *predicts* Fran's walking to the refrigerator and opening the door, which in turn predicts the presentation of food. After multiple repetitions, the question "Do you want to eat?" is now a conditioned stimulus (CS) as it now evokes the same responses as the actual presence of Fluffy's food – meows and excitement!

```
            (predicts)              (predicts)
QUESTION ─────────▶ REFRIGERATOR ─────────▶ DINNER!! (EXCITEMENT!!)
(question/refrigerator = neutral stimuli)    (food = unconditioned stimulus)

                        (conditioning)
QUESTION PREDICTS FOOD ─────────▶ QUESTION ALONE EXCITEMENT!!
```

Classical Conditioning in Training Classical conditioning is useful in dog and cat training in two rather different situations. The first involves the deliberate pairing of neutral and unconditioned stimuli to promote positive and desirable responses. For example, many young dogs resist brushing and grooming because the feel of the brush is unfamiliar. The pleasurable emotions that are associated with the presentation of food treats (unconditioned stimulus) can be used to initially condition the presence of the brush as a pleasant stimulus, signifying something to look forward to rather than something to avoid. The brush (a neutral stimulus) is presented, followed immediately by food treats, which are given to the dog. The brush (and the food treats) are then removed and the training session is ended. This sequence is repeated several times until the dog's reaction to seeing the brush is positive and excited (i.e. he anticipates food treats). At this point, the brush has been transformed from a neutral stimulus to a conditioned stimulus that predicts something pleasurable. (Note: This type of training differs from counterconditioning, in which an attempt is made to *change* the association of a previously classically conditioned stimulus. Using the same example: If the dog had previously developed a fear or dislike of the brush, counterconditioning could be used to gradually, in a step-wise manner, change a negative association into a positive association. See Chapter 5, pp. 129–130 for a complete discussion and examples).

When using classical conditioning to promote desirable responses in dogs and cats, the timing with which the neutral stimulus is presented is an important factor. The neutral stimulus must be presented immediately before and, if possible, overlapping with, the unconditioned stimulus. This facilitates the development of a strong and enduring connection between the two stimuli and the transformation of the neutral stimulus into a conditioned stimulus. Conversely, if the conditioned stimulus is presented after the unconditioned stimulus, learning proceeds slowly, if at all. Using the previous example, it is a common mistake for owners to present the food treats and *then* present the brush. Given this sequence, there is no reason for the dog to pay attention to the brush if she is already enjoying food treats. For this reason, classical conditioning rarely occurs if the sequence is mistakenly reversed.

The second way in which classical conditioning is used in training involves teaching specific words or hand gestures as behavioral cues (commonly called "commands"). One example with dogs is teaching a dog to eliminate on command when taken outdoors. In this case, unconditioned stimuli include internal signs and external events which trigger eliminative behavior. External events that can be used in conditioning include events such as eating, waking up after a nap, or sniffing an area that was previously used for elimination. Neutral stimuli include taking the dog outside to a specific spot and the command, "hurry up" or "go potty" (or "make popcorn," for that matter; the words themselves do not matter to the dog at all). The verbal command should be consistent and should be repeated several times as the dog is circling or sniffing an area of the yard that is used for elimination. If the new verbal **cue** is consistently paired with elimination, it can be used to encourage dogs to eliminate in specific areas or at particular times during the day. Conversely, if the command is given after the dog is already in the process of elimination or intermittently, classical conditioning will proceed much less efficiently, if at all. A second example is the introduction of the command "Fluffy, come" when teaching a cat to come when called. One approach to teaching this behavior is to pair an unconditioned stimulus (presence of food) and unconditioned response (cat approaching to consume the food) with a neutral stimulus (the words "Fluffy, come"). If the verbal cue is presented consistently and immediately prior to the presentation of food, classical conditioning will result in the cat associating the two words with an opportunity to eat, responding by offering the response, coming toward the owner. As we will see later, this example demonstrates the way in which classical conditioning and operant conditioning are linked and overlap in many learning situations.

OPERANT CONDITIONING (INSTRUMENTAL LEARNING)

Learning that occurs as a result of the consequences of behavior is called operant conditioning or **instrumental learning**. This terminology originates from the concept that animals are always "operating on" their environment, and subsequently alter their behavior in response to consequences. Specifically, animals tend to repeat behaviors that have had desirable consequences. We say that these behaviors are **reinforced**. Conversely, animals tend avoid repeating behaviors that have aversive consequences. We say that these behaviors have been **punished**. While classical conditioning is concerned with establishing relationships between stimuli, operant conditioning involves primarily response-consequence relationships. For many behaviors that concern owners of companion dogs and cats, both types of learning are taking place.

Four types of consequences are of interest with operant conditioning: **positive reinforcement, negative reinforcement, positive punishment**, and **negative punishment**. *Positive reinforcement* occurs when a behavior results in the delivery of a pleasant or desirable stimulus. For example, a dog quickly learns that sitting next to her owner and nudging her owner's hand with her nose results in petting and attention. The behaviors of sitting close and "nose nudging" are therefore positively reinforced. *Negative reinforcement* occurs when a behavior results in the prevention (avoidance) or termination (escape) of an unpleasant stimulus. A dog is lying in the sun and, as he starts to feel too warm, he rises and moves to a shady spot. The unpleasant feeling of being overheated is ended (or avoided) by moving. In this case, we say that the behavior of moving to a shady spot has been negatively reinforced. The frequency of moving to a shady spot when overheated will *increase* in the future as a result of being negatively reinforced. *Positive punishment* occurs when a behavior produces an aversive consequence. A cat jumps onto a kitchen counter and activates a motion-sensitive canister that emits a harmless but unpleasant blast of air. The cat jumps off the counter to avoid the air blast. In this example, jumping onto the kitchen counter has been positively punished. (We could also say that jumping off the counter has been negatively reinforced because getting off the counter ended the **aversive stimulus**.) Finally, behaviors that result in the prevention or termination of a pleasant stimulus are being *negatively punished*. A young puppy is playing with his owner and starts to nip. The owner stops play (removes the positive interaction) and the puppy stops nipping. In this scenario, nipping has been negatively punished. If the owner consistently plays with the puppy only if the puppy plays gently, then playing gently is being positively reinforced.

Sidebar 2: THE USE AND ABUSE OF POSITIVE PUNISHMENT

Historically, dog training has emphasized compulsion as an approach to controlling (or to use popular parlance, "correcting") a dog's undesirable behaviors. The prevailing paradigm was that physical corrections and coercion are necessary to control a dog's misbehavior. Although some of the aversive stimuli that were recommended were relatively innocuous and did not pose a risk to a dog's physical or emotional health, others, such as alpha rolls, choking with a slip collar, kneeing a dog in the chest, or hitting the pet were abusive and dangerous, and should never be condoned. In addition to its potential for abuse, relying upon positive punishment as a principal approach to changing a pet's behavior has other important limitations. These include:

- Because pets are often highly motivated to engage in the undesired behavior (and because it has often been unintentionally reinforced by the owner), they often habituate to mild aversive stimuli, forcing an escalation to increasingly harmful punishers.
- It follows that in order to be severe enough to suppress the unwanted behavior, most punishments also cause fear, avoidance, and, in some cases, aggression.
- Punishment does not address underlying causes or motivations for the unwanted behavior, nor can it provide clear information regarding alternate behaviors that are available to the animal to avoid punishment in the future.
- Because it is very difficult to deliver punishment effectively and remotely, its repeated use damages the relationship between the pet and owner.

Positive punishment is effective in a few specific situations. These include the use of remote punishers to keep pets away from certain objects or areas and using a conditioned punisher such as "no" or "wrong" to immediately stop an unwanted behavior. However, stopping the behavior at the time it is elicited should not be confused with training the dog or cat not to offer the behavior again in the future. For this, removing reinforcers, changing the pet's motivation, and reinforcing alternate and incompatible behaviors is effective. In cases in which positive punishment is used to suppress behavior, it can be effective (and safe) only if the following four conditions are met:

- The punishment must be administered promptly, at the moment that the undesired behavior is occurring; never after the behavior has occurred.
- It must be consistent; applied every time that the pet offers the unwanted behavior.
- The punishment must be aversive enough to cause a startle response and interrupt the behavior, but *not* so aversive as to cause fear.

- When punishment is delivered by the owner, an alternative behavior should be made available that can be reinforced. The purpose is to overcome the problem of punishment not being as good of a source of information as is reinforcement.

American Veterinary Society of Animal Behavior. Guidelines on the use of punishment for dealing with behavior problems in animals. AVSAB, 2008.

Reinforcement vs. Punishment

These four consequence quadrants are shown graphically in Table 4.1. Most behaviors can be taught using any one or combination of the four types of consequences. However, the four consequences are associated with different emotional responses which must be carefully considered when designing a training or behavior modification program. Specifically, the use of aversive stimuli is associated with causing fear and anxiety, and the use of negative punishment (removal of a desired stimuli; taking away something that the subject wants) typically causes some degree of frustration. For these reasons (and others discussed in Chapter 5), the four types of consequences should not be viewed as equally effective or valuable for use in practical dog and cat training. Simply eliciting a behavioral response is not the desired outcome of most trainers and pet owners. Aversive stimuli have negative emotional effects because noxious events result in avoidance or escape behaviors. The emotions that are associated with avoidance include fear, anxiety, and possibly defensive aggression. Therefore, while the intended change in behavior may be quickly accomplished through the use of aversive stimuli, reliance upon positive punishment and negative reinforcement also cause fear, anxiety, and avoidance behaviors (Sidebar 2). Therefore, in most training situations, the risk of causing such negative reactions and emotions prohibits the use of two of the four possible consequences: negative reinforcement and positive punishment. An example to illustrate this point is teaching a dog to remain in

table 4.1 Consequences in Operant Conditioning

TYPE OF STIMULUS	BEHAVIORAL CHANGE	
	INCREASE FREQUENCY	DECREASE FREQUENCY
DESIRABLE	Positive Reinforcement (add stimulus)	Negative Punishment (remove stimulus)
AVERSIVE	Negative Reinforcement (remove stimulus)	Positive Punishment (add stimulus)

a sitting position for a short period of time (called a sit-stay). Each of the four possible consequences that would increase a dog's tendency to sit-stay in the future are described:

1. Positive reinforcement: The dog is sitting. The trainer provides quiet praise, gentle petting, and food treats to the dog during the time that the dog remains in the desired position. In this example, the behavior of remaining in a sit position is being positively reinforced; increasing in frequency. (Figure 4.1)

2. Negative reinforcement: The dog is sitting. The trainer stands quietly, saying nothing, and waits until the dog moves out of position or attempts to lie down. Collar jerks or harsh reprimands are given until the dog stops attempting to move and remains sitting. In this example, the sit-stay is being negatively reinforced; increasing in frequency. (Figure 4.2)

3. Positive punishment: The dog is sitting. The trainer waits until the dog moves out of position or attempts to lie down. Collar jerks or harsh reprimands are given until the dog stops attempting to move and remains sitting. Moving out of position is being positively punished; decreasing in frequency.

4. Negative punishment: The dog is sitting. The trainer provides quiet praise, gentle petting, and food treats to the dog during the time that the dog remains in the desired position. If the dog attempts to move, the trainer immediately stops all interactions, praise, and food treats. In this example, moving out of position is being negatively punished; decreasing in frequency.

This example illustrates the connection between reinforcement and punishment in practical training situations (see below). When one behavior is being positively punished, an alternate behavior is simultaneously being negatively reinforced. Positive reinforcement and negative punishment are similarly connected (see below). In this sit-stay example, the combined used of positive reinforcement and some degree of negative punishment can effectively teach sit/stay and results in a dog who is relaxed, attentive and calm. Conversely, the repeated use of negative reinforcement and positive punishment results in a dog who stays in position but is also stressed and anxious. This occurs because any behavior that does not comply with a "sit-stay" results in aversive consequences. A sit-stay might be achieved, but at the expense of also instilling anxiety or fear as the dog learns that the only way to avoid aversive consequences is to freeze in place. The repeated use of negative punishment (withholding treats/praise/petting) can also cause stress and frustration, because removing a desirable stimulus repeatedly also punishes a behavior repeatedly. Frustration occurs when a dog is not able to offer a behavior that earns the desired stimulus and so has little opportunity to earn

figure 4.1 Training Sit-Stay with Positive Reinforcement

figure 4.2 Training Sit-Stay with Negative Reinforcement (and Positive Punishment)

positive reinforcement. This is a common mistake seen with new trainers who are expecting too much from a dog, do not design the training session to allow opportunities for positive reinforcement, or are not using **shaping** effectively (see Chapter 5 for details).

The Link between Positive Reinforcement and Negative Punishment: The examples that are provided above demonstrate the intricate connection between positive reinforcement and negative punishment. The behavior that earns the desirable consequences (sit-stay results in treats, praise, and petting) is linked with a behavior that causes the absence or withdrawal of the desirable consequences (moving from a sit-stay causes treats/praise/petting to stop). More subtly, when a dog offers the behavior that is desired, other behaviors are essentially "off-limits" in the sense that opportunities for positive reinforcement for alternative behaviors (such as moving out of place) are not available. The use of negative punishment must always be used judiciously because withholding desirable consequences (treats, petting) from all behaviors other than the target behavior is only effective if the dog or cat is capable of offering the desired behavior at a frequency that facilitates learning. Otherwise, negative punishment of undesirable behaviors (moving, lying down) is being used without sufficient opportunities for positive reinforcement of the desired behavior (sit-stay) and the animal may become stressed or frustrated. Therefore, training situations should be designed to allow ample opportunities for the dog or cat to "earn" positive reinforcement and at the same time minimize "mistakes" that have the consequences of negative punishment. This approach allows the dog or cat to maximize positive consequences and minimize negative consequences, even when those consequences appear relatively mild to the trainer. These considerations and more examples are provided in Chapters 5 through 7.

The Link between Negative Reinforcement and Positive Punishment: Similarly, negative reinforcement and positive punishment are intricately linked. In the case of teaching a sit-stay, the behavior that elicits the aversive stimulus (moving from a sit-stay) is punished (decreases in frequency) while the behavior that allows the dog to escape or avoid the aversive stimulus (sit-stay) is negatively reinforced. Although it is primarily a matter of desired outcomes, in some situations, positive punishment alone is used without being paired with negative reinforcement. In these situations positive punishment is used to stop an undesirable behavior, without attempting to increase the frequency of another behavior. For example, positive punishment is often used to stop dogs from raiding garbage cans. A noxious-smelling substance is applied to the outer surface of the waste bin. When the dog smells or tastes the area, the behavior of disturbing the trash is positively punished by an aversive stimulus (the noxious-tasting substance). The dog is less likely to offer the behavior of raiding the garbage can in the future (i.e. garbage-can raiding has been punished; decreasing in frequency). In this example, the trainer is not concerned with what specific behavior is increased as long as the dog does not raid the garbage. The dog may do any number of things after leaving the garbage alone – walk away, lie down, or leave the room.

Types of Reinforcing Stimuli for Dogs and Cats

It is important for trainers to always remember that reinforcement is defined by its effect on the subject, not by its intent. The value that a particular stimulus has for an individual directly affects its power as a reinforcer or punisher. This means that, by definition, the subject (dog or cat) determines whether a particular stimulus is a desirable or aversive consequence. For example, most dogs enjoy petting and social interaction, and this type of interaction can be a strong positive reinforcer in a variety of training situations (Figure 4.3). However, a dog who is timid or who is unusually sensitive to touch may not find the experience of petting to be either pleasant or desirable (Figure 4.4). For this dog, petting would not be a desirable consequence that could be used to positively reinforce targeted behaviors. Similarly, reinforcers are often affected by circumstances and the training environment. While food treats are highly useful as a stimulus to reinforce behaviors in many settings, a dog who is fearful will often refuse to eat. This instantly makes the presentation of food as a consequence to that dog in that setting irrelevant and thus not reinforcing. Again, although circular in definition, a reinforcer is something that reinforces a behavior (causes it to increase in frequency in the future). If a behavior is not increasing in frequency in response to the consequence, then the consequence is not a reinforcer for that animal in that situation.

In dog and cat training, the trainer provides reinforcement for targeted behaviors by presenting pre-selected consequences (reinforcers) to the dog or cat. For example, gently petting a cat who approaches will positively reinforce (strengthen) approaching behavior. The cat is likely to increase her frequency of approaching in the future in order to obtain the consequence of enjoyable petting (Figure 4.5). Conversely, squirting a cat with a spray of water as he jumps up onto a kitchen counter uses an aversive stimulus (sprayed water) to positively punish counter

figure 4.3 Petting as a positive reinforcer for sit-stay

figure 4.4 Timid dog avoiding petting during sit-stay

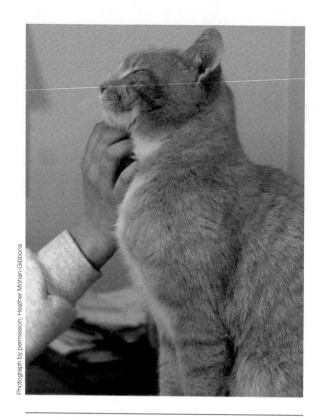

figure 4.5 Cat responding positively to petting and interaction

jumping. Stimuli that serve as highly effective positive reinforcers for dogs include social interaction (petting, praise, petting), opportunities to play a favorite game, and various types of food treats or tidbits. Similarly, most cats react to petting, small food treats, and an opportunity to play with a favorite toy as positive reinforcers. Aversive stimuli that are used as positive punishers or negative reinforcers include any event that causes pain, discomfort, or anxiety, and which the dog or cat prefers to avoid or escape.

Primary and Secondary (Conditioned) Reinforcers

Desirable and aversive stimuli that are used to reinforce or punish behaviors are further classified into primary (unconditioned) and secondary (conditioned) reinforcers (Table 4.2). **Primary reinforcers** are stimuli that have a biological basis and therefore need little or no prior conditioning (learning) to be effective. These are stimuli that an animal inherently enjoys and wants more of, or inherently dislikes and wishes to avoid. Positive (desirable) primary reinforcers for most well-socialized companion dogs and cats include food, opportunities for social interaction, and opportunities for exercise or play. Primary aversive stimuli

table 4.2 Types of Reinforcers for Dogs and Cats

CATEGORY	TYPE OF STIMULUS	
	PRIMARY (Unconditioned)	SECONDARY (Conditioned)
POSITIVE REINFORCEMENT	Presentation of: Food treats Exercise (play) Social interaction	Presentation of: Bridging word/click Toys Praise/petting*
NEGATIVE REINFORCEMENT	Removal of: Pain (collar jerk, shock) Discomfort Fear/anxiety	Removal of: Harsh voice Threatening gesture "*No!*" and "*Bad dog!*"

* Some words of praise and some types of petting/handling may be conditioned rather than primary positive reinforcers for some dogs and cats (see text for explanations)

include any event that causes discomfort, pain, anxiety, or fear, and which the dog or cat naturally wishes to avoid.

Secondary reinforcers are stimuli that are either completely neutral or that have weak positive or negative reinforcing properties. These stimuli are paired with a primary reinforcer and, once conditioned, take on the reinforcing power of the primary reinforcer. For example, telling a puppy "good girl" followed immediately by a treat will establish a predictive (classical) relationship between the words "good girl" and the presentation of a food treat. Once this relationship has been established, the secondary reinforcer can occasionally be used without the primary reinforcer because it has acquired similar reinforcing properties. In addition to word phrases, a **clicker** (cricket) is commonly and very successfully used as a secondary reinforcer by dog and cat trainers today (more about clicker training in Chapter 5).

Value of Reinforcing Stimuli

A final factor to consider with reinforcing stimuli is the value that a particular stimulus has for the dog or the cat. Because every animal is an individual, a hierarchy of desirable and aversive stimuli can often be constructed. Understanding which stimuli are highly valued and which are moderately valued can be very helpful when choosing appropriate consequences to use as reinforcers or punishers in training. For example, a trainer may identify several types of food treats that her dog enjoys. She learns from observing her dog that hot dog pieces are highly valued while soft-moist dog treats are enjoyed but not as intensely, and that dry kibble has the least strength as a primary positive reinforcer. The same dog may also enjoy several types of play (fetch a squeaky toy, fetch a tennis ball, fetch a toy stick) and several types of social interaction or games (tummy rub, head petting, ear rubs). These hierarchies become important when teaching dogs and cats new behaviors because the trainer can select a set of reinforcers to match behaviors on which the trainer is focusing within a training session (see Chapter 5 for more details).

HABITUATION AND SENSITIZATION

Habituation is considered to be a type of non-associative learning because it involves the reduction of a response through repeated exposure to a stimulus that has neither positive nor negative consequences. Habituation occurs as young animals are learning which stimuli in their environment are relevant and which are irrelevant. In general, relevant stimuli are those that have the potential to either harm or benefit the animal, while irrelevant stimuli are those that are of no consequence to the animal and so can be "screened out" from his or her perceptual world. A common example of habituation is a new puppy being startled by the sound of the dishwasher and then gradually, after several exposures, no longer reacting to the sound. Similarly, a new kitten may initially be startled by the movement of the floor sweeper, but slowly becomes habituated to the appliance after repeated exposures demonstrate that the sweeper is a harmless stimulus.

Habituation can be short-term or long-term. Short-term habituation occurs when an animal is exposed to a novel stimulus repeatedly, but for a relatively short period. For example, an owner may vacuum for 20 minutes, moving the vacuum cleaner to several rooms during that time. If the noise is not excessive and the vacuum cleaner is managed in a non-threatening way, a new puppy in the home quickly develops short-term habituation, ignoring the noise after several minutes. However, if the machine is reintroduced after several days, the puppy may once again show a startle or fear response. This phenomenon is called spontaneous recovery. Long-term habituation occurs when many repetitions of the stimulus no longer evoke a response. After the vacuum cleaner has been used over a period of weeks, the puppy no longer reacts.

During development, habituation is important for dogs and cats to learn to recognize a specific set of non-threatening stimuli in their environments. In addition, habituation also affects an individual's long-term ability to react to novel stimuli. Dogs and cats who are habituated to many stimuli are more likely to show stimulus generalization, which is the rapid habituation to a new stimulus that is similar to one that is familiar. For example, a dog who has been habituated to the sound of cars on the street where he walks with his owner each day will usually demonstrate stimulus generalization to the sound of larger or smaller cars when out walking in a new area. Conversely, a lack of habituation, especially in mature animals, can be a very serious problem. Adult dogs and cats who experienced limited exposure to a variety of stimuli during their first year of life are more inhibited when exposed to new experiences, are more apt to react fearfully, and may not be capable of readily habituating to new stimuli.

Sensitization is the opposite of habituation. Sensitization occurs when repeated exposure to a stimulus results in an increase in responsiveness, rather than a decrease. In dogs and cats, sensitization usually involves either fearful (flight), or aggressive (fight) responses. For example, a new kitten who sees a dog for the first time may approach the dog with curiosity. If the dog suddenly barks and rushes toward the kitten, the kitten becomes frightened and runs away. If the kitten is severely frightened by this experience, she may become sensitized to dogs, and future meetings with any dog will elicit a fearful reaction in the kitten, regardless of the dog's behavior. Many kittens would also generalize this response to other similar stimuli. In this case, the kitten may become fearful of all dogs, and continue to be fearful as an adult cat. The effects of sensitization are typically less stimulus-specific than habituation. This means that any small change in a stimulus can lead to sensitization, even if the cat was previously habituated to the stimulus. For example, the kitten who was habituated to the vacuum cleaner may suddenly become sensitized to the machine if it malfunctions and makes a much louder noise. Similarly, a cat who had been completely habituated to dogs becomes sensitized to all dogs if a new dog barks or attempts to chase.

SOCIAL LEARNING

Social learning is a broad term that refers to the tendency of an individual to attend to the behavior of others in his or her social group and to change behavior in response to the behavior of others. Social learning is most commonly observed in group-living species in which communication between individuals and attention to the behavior of others is an important contributor to survival. In addition, social learning is important for acquiring behaviors from littermates or parents, regardless of sociability of the species. Within the broad category of social learning, several sub-classifications exist. These include allelomimetic behavior, social facilitation, local enhancement, and observational learning.

Types of Social Learning Allelomimetic behaviors are group-coordinated behaviors that reflect the tendency of animals within a social group to follow the example of others. These behaviors are most commonly observed in puppies and kittens with their littermates, and in groups of dogs and cats who live together. Examples in young animals include sleeping and nursing together, and investigating new objects as a group. Adult dogs and cats who live together and share similar daily experiences also tend to sleep, eat and play at the same time. These all are examples of allelomimetic behavior.

The second type of social learning is social facilitation. This is related to allelomimetic behavior in that it too involves the behavior of a group. However, while allelomimetic behavior is simply a group of animals engaging in the same behavior, social facilitation affects an individual's motivational state or the intensity with which he engages in the group-facilitated behavior. The most common example of social facilitation in companion animals is meal-feeding behavior in dogs. The majority of dogs are strongly affected by the presence of other dogs during meal times. Most dogs change their eating habits by over-consuming or eating more rapidly when in the presence of another dog, compared with when they are alone. There is also recent evidence that food preferences in dogs can be learned from other dogs (Sidebar 3).

Sidebar 3 — FOOD PREFERENCE LEARNING IN DOGS

The feeding behavior of the domestic dog has long been known to have a strong social component. Dogs consume more food and eat more rapidly when in the presence of other dogs and food guarding behaviors may be learned from other dogs within a social group. In addition, a recent study conducted by Gwen Lupfer-Johnson and Julie Ross at the University of Alaska suggests that social learning may also be involved in how dogs learn individual food preferences. In this pilot study, the researchers studied 12 pairs of adult dogs in which one dog acted as a demonstrator and the other acted as an observer. Demonstrator dogs were fed dog food that was flavored with either dried basil or dried thyme leaves, out of sight of their paired observer. After consuming the meal, the demonstrator dog was reunited with the observer dog and the two dogs were allowed to interact for 10 minutes. During this time, observer dogs all intently investigated and smelled around the face and mouth of the demonstrator dogs. Following the interaction period, the observer dog was taken to another room and offered equal amounts of each type of flavored food. Food preference was assessed by measuring the proportion of the demonstrator food (basil or thyme) that the observer dog consumed relative to the total amount of food consumed.

Results showed that dogs consumed more of the food that their respective paired demonstrator had been given than of the alternative flavor. Dogs who were paired with demonstrator dogs fed basil-flavored dog food consumed significantly more basil-flavored food than did dog who were paired with demonstrator dogs fed thyme. These results suggest that domestic dogs may learn food preferences by observing the food preferences of other dogs, without having direct access to the food itself.

The evolutionary significance of learning what foods to consume by observing others is similar to that suggested for the Norway rat, a species that learns food selection from others in its social group. The social transmission of food preferences would be beneficial as dogs were evolving to live close to human settlements and began to consume a more varied diet as they scavenged trash sites. Taking social cues from other dogs would provide information about the types of foods that were currently available, edible and safe. Additional studies may tell us whether dogs who are well acquainted with each other (socially bonded pairs) communicate food preferences more efficiently than do those who do not live together, as has been shown in rats, and whether dogs are capable of also learning food preferences by observing human caretakers.

Lupfer-Johnson G, Ross J. Dogs acquire food preferences from interacting with recently fed conspecifics. *Behav Proc*, 74:104–106, 2007.

Local enhancement refers to the tendency of an individual to attend to and react in the same manner to environmental cues as others in their social group. Because the individual attends to a stimulus to which he sees another animal reacting, the resulting behavioral response is similar. Common examples include dogs who learn to roll in noxious-smelling substances or to engage in **coprophagia** (eating feces) when in the presence of a housemate who first smells the area (or feces) and then begins to roll. The second dog typically learns this behavior by first responding to the other dog's sniffing and investigations, and then engaging in the same behavior. An example in cats is the kitten who learns to climb up a set of stairs by following an adult cat. While some behaviorists identify this type of behavior as observational learning, others maintain that local enhancement is less complex than observational learning because the inexperienced individual has the opportunity to be exposed to the same set of stimuli as the demonstrator animal.

The final (and most complex) type of social learning is observational learning. Observational learning occurs when a subject learns to perform a behavior by simply observing another animal perform or learn to perform the behavior. An important distinction between observational learning and other forms of social learning is that the animal who is demonstrating the behavior in question is physically separated from the animal who is observing the behavior. This constraint eliminates the opportunity for the observing animal to attend to or react to the same stimuli as the demonstrator animal. Therefore, when defined in this manner, observational learning is exhibited when an individual learns without having had the opportunity to react to a relevant stimulus or to be reinforced through trial and error. Both dogs and cats are capable of learning to perform a task simply by watching another animal complete the task. A series of recent studies with companion dogs have shown that not only are dogs capable of observational learning from one another, but that this type of learning also occurs between dogs and their owners (Sidebar 4). In the cat, one of the most highly social periods of life is during kittenhood, when the kitten is learning from her mother and from her littermates. This is a period during which observational learning is most important for this species. For example, kittens quickly learn to perform an arbitrary task, such as pressing a lever for food, if they first watch an adult cat successfully perform the task. Conversely, if they are just allowed to use trial-and-error (operant) learning, they learn very slowly or do not learn the behavior at all. Interestingly, kittens learn the new task fastest if the demonstrator cat is their mother than if it is a known adult cat who is not their mother.

Sidebar 4 | OBSERVATIONAL LEARNING IN DOGS

Similar to kittens, puppies are capable of observational learning when they watch their mothers or littermates perform a task. An early study found that pre-weaned puppies could learn to grasp a ribbon and pull a packet of food into their pen after watching another puppy learn the task through trial and error. In another study, puppies who were allowed to observe their mother as she searched for and found hidden drug caches performed significantly better at learning scent detection tasks when they were six months old than puppies who had not observed their mothers working.

Perhaps even more interesting is recent evidence that dogs not only learn from observing other dogs, but are very adept at learning by observing people! In a series of studies conducted by Peter Pongrácz and colleagues, a "detour task" test was used to investigate the ability of dogs to learn to maneuver around an obstacle through observation. Dogs were placed within a V-shaped, transparent fence that had food or a toy placed behind the apex. Dogs in one group were allowed to determine how to move out of the fence to obtain the food through trial and error (operantly), while a second group first observed a human successfully maneuver the obstacle prior to attempting to solve the task. Dogs who had first observed a person walking the detour learned the task significantly faster than dogs who had not, with many correctly solving the detour problem on their very first attempt! Further studies showed that a dog's age and breed did not affect rate of observational learning, but that verbal communication from the demonstrator human significantly enhanced learning.

In another study, the ability of dogs to learn how to extract a ball from a box by observing their owners push a handle on the box was tested. Dogs who watched their owner push the handle to release a ball learned to touch and push the handle, while those who did not observe were less likely to touch or maneuver the handle and removed the ball through other means. Together, these results show that humans are efficient demonstrators to dogs, suggesting that social learning is well developed and highly flexible in the domestic dog. These results may have important implications for human-dog interactions as well as for the development of new training techniques.

Adler LL, Adler HE. Ontogeny of observational learning in the dog (*Canis familiaris*). *Devel Psych*, 10:267–280, 1977.

Kubinyi E, Topal J, Miklósi I, Csányi V. Dogs (*Canis familiaris*) learn from their owners via observation in a manipulation task. *J Comp Psychol*, 117:156–165, 2003.

Pongrácz P, Miklósi A, Kubinyi E, et al. Social learning in dogs; the effect of a human demonstrator on the performance of dogs in a detour task. *Anim Behav*, 62:1109–1117, 2001.

Pongrácz P, Miklósi A, Timar-Geng K, Csányi V. Verbal attention getting as a key factor in social learning between dog (*Canis familiaris*) and human. *J Comp Psychol*, 118:375–383, 2004.

Pongrácz P, Miklósi A, Vida V, Csányi V. The pet dogs' ability for learning from a human demonstrator in a detour task is independent from the breed and age. *App Anim Behav Sci*, 90:309–323, 2005.

Slabbert JM, Rasa OAE. Observational learning of an acquired maternal behaviour pattern by working dog pups: An alternative training method? *App Anim Behav Sci*, 53:309–316, 1997.

REVIEW QUESTIONS

1. Provide one practical example for each of classical conditioning that might occur with a dog and a cat.
2. What are the four types of consequences of operant conditioning?
3. List three desirable (pleasant) stimuli that are effective as positive reinforcers with dogs and cats.
4. Provide an example of habituation and an example of sensitization with a dog or a cat.
5. What risks are associated with reliance upon negative reinforcement and positive punishment in dog or cat training?

REFERENCES AND FURTHER READING

Adler LL, Adler HE. **Ontogeny of observational learning in the dog (*Canis familiaris*).** *Devel Psych*, 10:267–280, 1977.

Borchelt PL, Voith VL. **Punishment. In:** *Readings in Companion Animal Behavior,* VL Voith and PL Borchelt, editors, Veterinary Learning Systems, Trenton, NJ, pp. 72–80, 1996.

Burch MR, Bailey JS. **How Dogs Learn.** Howell Book House, New York, NY, 188 pp., 1999.

Donaldson J. **Culture Clash: A Revolutionary New Way of Understanding the Relationship Between Humans and Domestic Dogs.** James and Kenneth Publishers, Oakland, CA, 221 pp., 1996.

Dumas C, Page DD. **Strategy planning in dogs (Canis familiaris) in a progressive elimination task.** *Behav Proc*, 73:22–28, 2006.

Hare B, Brown M, Williamson C. **The domestication of social cognition in dogs.** *Science*, 298:1634–1636, 2002.

Johnson P. **Twisted Whiskers: Solving Your Cat's Behaviour Problems.** The Crossing Press, Freedom, CA, 154 pp., 1994.

Kubinyi E, Topal J, Miklósi I, Csányi V. **Dogs (*Canis familiaris*) learn from their owners via observation in a manipulation task.** *J Comp Psychol*, 117:156–165, 2003.

Ley J, Coleman GJ, Holmes R, Hemsworth PH. **Assessing fear of novel and startling stimuli in domestic dogs.** *Appl Anim Behav Sci*, 104:71–84, 2007.

Lindsay SR. **Handbook of Applied Dog Behavior and Training,** *Volume 1: Adaptation and Learning*. Iowa State University Press, Ames, IA, 410 pp., 2000.

McKinley S, Young R. **The efficacy of the model-rival method when compared with operant conditioning for training domestic dogs to perform retrieval-selection task.** *Appl Anim Behav Sci*, 81:357–365, 2003.

McKinley J, Sambrook TD. **Use of human-given cues by domestic dogs (*Canis familiaris*) and horses (*Equus caballus*).** *Anim Cogn*, 3:13–22, 2000.

Miklósi A, Soproni K. **A comparative analysis of animals' understanding of the human pointing gesture.** *Anim Cogn*, 9:81–93, 2006.

O'Farrell V. **Dog's Best Friend: How Not to Be a Problem Owner.** Methuen Press, London, UK, 1994.

Pearce WD, Cheney CD. **Behavior Analysis and Learning,** 3rd edition. Lawrence Erlbaum Associates, Inc, Mahwah, NJ, 388 pp., 2004.

Pongrácz P, Miklósi A, Kubinyi E, et al. **Social learning in dogs; the effect of a human demonstrator on the performance of dogs in a detour task.** *Anim Behav*, 62:1109–1117, 2001.

Pongrácz P, Miklósi A, Timar-Geng K, Csányi V. **Verbal attention getting as a key factor in social learning between dog (*Canis familiaris*) and human.** *J Comp Psychol*, 118:375–383, 2004.

Pongrácz P, Miklósi A, Vida V, Csányi V. **The pet dogs' ability for learning from a human demonstrator in a detour task is independent from the breed and age.** *App Anim Behav Sci*, 90:309–323, 2005.

Pryor K. **Don't Shoot the Dog.** Bantam Books, New York, NY, 187 pp., 1984.

Pryor K. **Karen Pryor on Behavior.** Sunshine Books, North Bend, WA, 405 pp., 1995.

Reid P. **Excel-Erated Learning: Explaining How Dogs Learn and How Best to Teach Them.** James and Kenneth Publishing, Oakland, CA, 172 pp., 1996.

Rogerson J. **Your Dog: Its Development, Behaviour, and Training.** Popular Dogs Publishing Company, London, UK, 174 pp., 1990.

Schilder MBH, van der Borg JAM. **Training dogs with help of the shock collar: Short and long term behavioral effects.** *App Anim Behav Sci*, 85:319–334, 2004.

Schwab C, Ludwig H. **Obey or not obey? Dogs (*Canis familiaris*) behavior differently in response to attentional states of their owners.** *J Compar Psychol*, 120:169–175, 2006.

Slabbert JM, Rasa OAE. **Observational learning of an acquired maternal behaviour pattern by working dog pups: An alternative training method?** *Appl Anim Behav Sci*, 53:309–316, 1997.

Spreat S, Spreat SR. **Learning principles.** *Vet Clin North Amer: Small Anim Pract*, 12:593–606, 1982.

Voith VL, Wright JC, Danneman PJ. **Is there a relationship between canine behavior problems and spoiling activities, anthropomorphism, and obedience training?** *Appl Anim Behav Sci*, 34:263–272, 1992.

Voith VL, Borchelt PL (Editors). **Readings in Companion Animal Behavior.** Veterinary Learning Systems, Trenton, NJ, 276 pp., 1996.

Yin S. **Classical conditioning: Learning by association.** *Compend Contin Vet, Small Anim Pract*, 472–476, June 2006.

Yin S. **Dominance versus leadership in dog training.** *Compend Contin Educ Pract Vet*, 414–417, 2007.

Zentall TR. **Imitation: Definitions, evidence and mechanisms.** *Anim Cogn*, 9:335–353, 2006.

CHAPTER 5

Practical Applications: Training and Behavior Modification Techniques

Training and behavior modification techniques provide interventions that allow pet owners and companion animal professionals to guide a dog or cat's learning toward desired outcomes. Training methods manipulate stimuli and reinforcers and modify motivation to promote desired behaviors and eliminate or prevent undesired behaviors. In most training situations, classical and operant conditioning are the primary forms of learning that trainers manipulate to teach new behaviors and solve behavior problems in their dog or cat.

THE ROLES OF CLASSICAL AND OPERANT CONDITIONING

Although classical and operant conditioning are often discussed as if they are separate and distinct ways in which animals learn, in actuality the two types of learning are intricately linked. As an animal learns about connections between stimuli (for example, the presence of the shampoo bottle predicting the unpleasantness of a bath), he also may offer a behavior such as leaving the room and hiding under the owner's bed. In this example, the relationship between the shampoo bottle and an impending bath is classically learned, while leaving the room to hide under the bed is operantly learned. Hiding under the bed is negatively reinforced in that the dog avoids the bath, at least for that moment! Similarly, when we teach dogs and cats to respond to spoken commands ("sit," "come," "roll over"), these verbal cues are initially neutral stimuli (meaningless to the animal) that become meaningful when they are repeatedly associated with the pleasantness of a treat or other form of desired stimuli that is used as a reinforcer (Sidebar 1). The connection between operant and classical conditioning becomes an important consideration when a trainer is deciding when to introduce the command for a particular behavior, called "putting the behavior on cue."

THE RELATIONSHIP BETWEEN OPERANT AND CLASSICAL CONDITIONING

Sidebar 1

Both classical and operant conditioning are taking place during most training exercises. The example below uses teaching a dog to sit on command. If this is taught using a conditioned reinforcer such as a clicker, the association between "click" and "treat" is classically learned. The consequence for offering "sit" is a click-treat (positive reinforcement). Sitting increases in frequency, signifying operant learning. Finally, when the command "sit" is added to the exercise, the relationship between hearing the verbal word "sit" and the eventual presentation of the treat is classical once again.

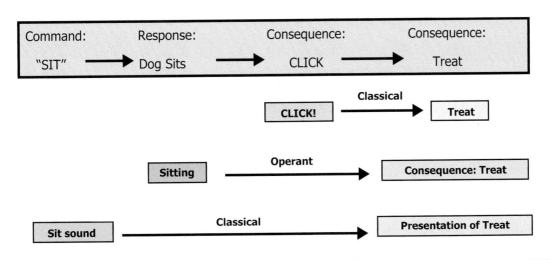

POSITIVE VS. NEGATIVE (AVERSIVE) CONTROL OF BEHAVIOR

An Historical Perspective

Most of the early dog training programs that were developed in the United States were for the purpose of training dogs for the military. These programs were initially developed during the mid-1900s and were based primarily upon the use of various aversive stimuli to negatively reinforce desirable behaviors or positively punish unwanted behaviors. Following the first two world wars, these methods became the foundation for pet dog training for more than 30 subsequent years. Methods that emphasized the use of aversive stimuli produced rapid results, which were needed for military dogs, and so were accepted as being the most efficient and appropriate approach to dog training at that time. Although these early methods were using operant learning, the term "escape/avoidance training" was coined because it aptly described the learning process in a program that focused primarily on the aversive control of behavior.

When escape/avoidance is used to teach new behaviors, the dog or cat first learns to escape an aversive consequence by changing his or her behavior. An example is teaching a dog to walk on a loose lead using negative reinforcement in the form of collar corrections (jerks on a training collar). The aversive stimulus (jerks on the collar) is applied as a consequence whenever the dog pulls forward out of position (forges) or hangs back out of position (lags). During the learning phase, the dog could *escape* the discomfort of these jerks by changing her behavior (not pulling on the lead). As the dog learns the contingencies associated with the negative reinforcement (consequences of pulling ahead, lagging behind, lying down are all collar jerks), she begins to *avoid* the negative stimulus altogether by continually offering the desired behavior – walking in position on a loose lead.

More about Aversive Stimuli

Escape/avoidance conditioning relies upon an animal's natural desire to avoid pain and discomfort. It must be recognized that if a stimulus is not sufficiently aversive, the animal will not change his behavior to avoid it and learning will not take place. Therefore, by its very definition, the escape/avoidance training approach (i.e. aversive control of behavior) involves causing some level of discomfort or pain to the animal. Basic emotions that are associated with pain and discomfort are fear and anxiety. Although these can be minimized by using the mildest intensity of an aversive that is necessary to cause a change in behavior, in most practical training situations it is difficult to accurately estimate a level of intensity that is "enough, but not too much." Some trainers suggest using an increasing hierarchy of intensity, starting at the lowest possible level and working up to an intensity that results in avoidance behaviors. However, this approach can lead to habituation at each level of intensity, necessitating the use of increasingly severe or inhumane aversive stimuli. In addition, even when mild aversive stimuli are used, the introduction of stress or fear ultimately will interfere with an animal's ability to learn new behaviors (Sidebar 2).

A second problem with reliance upon negative reinforcement or positive punishment is that the exact nature of each animal's response to an aversive stimulus is highly variable. Although some dogs and cats will move away from an aversive stimulus if there is an escape route available, others may freeze in place, panic, overreact, or become aggressive. As a result, the risk is that the response of the animal is not always what was intended by the trainer. This is confounded by the fact that applying an aversive stimulus (either to negatively reinforce or positively punish a behavior) only provides an animal with information about *what not to do*, but does not provide information about *what to do*. Essentially, the animal is

"THE QUIZ" — Sidebar 2

A Parable about Aversive vs. Pleasurable Stimuli

Let us digress from dogs and training for a moment, and consider how human subjects learn. Imagine that you have enrolled in a college math course. You have been attending lectures for several weeks and, although the material is quite difficult, you feel that the instructor has been explaining the concepts very clearly and that you have been able to learn a great deal from his lectures. You have completed and handed in three homework assignments and received over 90 percent on each one. When you enter the lecture hall for the weekly lecture, your instructor informs the class of 50 students that he will be administering a surprise quiz. He has two options for taking the quiz; one in Classroom A and the other in Classroom B. The instructor has randomly assigned students to each room. He tells you that the questions on the quiz are exactly the same in each room, but the consequences for answering quiz questions are slightly different.

Classroom A: The chairs in this room have been electronically wired so that a mild but quite uncomfortable electric shock can be sent to each chair by the instructor. The quiz instructions are as follows: The instructor will put an equation to be solved up on the overhead screen. When a student has correctly solved the problem, he or she should raise a hand and provide the answer aloud. If the student answers correctly, he or she is dismissed and may have the remainder of the day off. If the student provides an incorrect answer, he or she will receive an electric shock and must remain in the class and try again. All students must answer one question correctly to complete the quiz and leave the room.

Classroom B: There is an enormous bowl of fancy chocolates sitting on the front desk. The quiz instructions are as follows: The instructor will put an equation to be solved up on the overhead screen. When a student has correctly solved the problem, he or she should raise a hand and provide the answer aloud. If the student answers correctly, he or she will receive a chocolate. The student may then either remain to try for another correct answer or can be dismissed with the remainder of the day off. If the student provides an incorrect answer, he or she is simply ignored and will need to try again. All students must answer one question correctly to complete the quiz and leave the room.

Would Classroom A or Classroom B be better facilitate learning?

Which room would you rather be in?

(*Dogs may not learn math, but they do respond to aversive and desirable stimuli in the same manner as do math students*)

---

forced to learn through the process of elimination. Negative reinforcement relies on the subject's ability to select the desired behavior that will allow escape from the negative stimulus. Because a variety of behaviors are often equally successful in avoiding the stimulus (for example, running away or showing aggression), the behavior that is elicited each time the correction is applied may not be the behavior that the trainer was expecting to elicit.

Finally, because stress or fear is often introduced with the use of negative reinforcement and punishment, the use of escape/avoidance training as a humane approach to training is questionable. In addition to the potential for intentional or unintentional abuse, negative stimuli that are associated in any way with the animal's human caretaker have the potential for causing fear and damaging the pet's relationship with his or her owner. For example, a cat who is squirted with water from a water bottle quickly associates the squirted water with the presence of the person. Therefore, the cat may learn to stay off a counter when her owner is in the room because she is avoiding getting close to the owner, not necessarily because she has learned that staying off the counter is the desired behavior. (The cat will also learn the alternative situation; when the owner is not present, neither is the water bottle, and jumping on the counter has no negative consequences.) For this reason, using punishment that is associated with the caretaker is generally ineffective and may even be counterproductive. Severe aversive stimuli may result in a generalized fear and avoidance of the owner altogether as the animal learns that a behavior that will allow her to avoid negative stimuli is to simply avoid her owner.

Maximizing Pleasant and Minimizing Aversive Stimuli

New knowledge about learning in dogs and cats, coupled with a shift toward more humane and caring attitudes toward non-human animals in general, have led to an increase in the use of positive reinforcement in training and behavior modification programs. Specifically, positive reinforcement and the judicial use of negative punishment or **extinction** (see below) have been found to be more effective in modifying behavior than negative reinforcement and positive punishment. The advantages are numerous. First, relying on positive reinforcement is more effective and efficient because a positive consequence can provide very specific information to the dog or cat about the exact behavior that is affected. Using the sit/stay example from Chapter 4, food treats and praise are provided only while the dog is sitting, and are simply withheld if the dog moves or changes his position. Most dogs rapidly associate the targeted behavior (sitting) with the positive consequences (praise, petting, food treats). These consequences convey specific information to the dog regarding the exact behavior that is being positively reinforced. This scenario can be compared with the use of aversive consequences that inform the dog about incorrect behaviors such as moving out of position or lying down, but provide no information about the behavior that is desired (sit-stay). In this same manner, unlike negative reinforcement, positive reinforcement lends itself well to successive approximation, which facilitates teaching complex behaviors and exercises that require a great deal of control (see pp. 119–122).

Second, an emphasis upon pleasurable stimuli and positive reinforcement during training promotes subjective emotional states such as pleasure, contentment, and enjoyment in the animal. This can be directly compared with the varying levels of stress, fear, or distress that are associated with reliance upon aversive stimuli during training. Because learning new tasks is not associated with aversive stimuli, the dog or cat is relaxed and comfortable during training and is more likely to attempt new behaviors. In other words, a dog who has learned that "guessing wrong" simply goes unrewarded and is not punished is much more likely to try new behaviors than is a dog who has a history of receiving aversive consequences for any incorrect behaviors that she attempted. For example, an alternative approach to squirting a cat with water to punish her for jumping onto counters is to provide an acceptable elevated ledge in the home that the cat enjoys. Whenever the cat approaches or sits on the ledge, she is given cat treats or attention. Because this approach both provides for the cat's natural enjoyment of elevated resting spots and reinforces a desired behavior, the cat usually changes her resting spot from counters to the new ledge. This learning can be augmented by blocking the cat's access to the counter tops while training her to use her new spot.

TRAINING PREPARATIONS

Whether you are training your new puppy to have good manners, teaching a kitten to use a scratching post, or solving a barking problem in an adult dog, you are modifying behavior and your pet is learning. Dog and cat training can be divided into two broad categories: teaching new and desirable behaviors (e.g. manners training) and preventing or eliminating undesirable behaviors (e.g. solving behavior problems). In an ideal situation, the trainer is teaching "good manners" in his or her pet *before* undesirable behaviors and bad habits have developed. Regardless, animals like people, are complex beings who exhibit a wide range of behaviors that are influenced by age, the situation or context, and prior experiences. Therefore, it is important to develop a training plan that considers the following factors.

Identify Objectives and Goals

The trainer should develop a set of clear objectives and goals for training. When the objectives are to prevent problem behaviors and teach desirable behaviors, the trainer should list the specific behaviors that he wishes to teach, in their order of importance. For example, the owner of a two-year-old Border Collie may be interested in teaching his dog agility exercises, but the dog has not yet had any basic obedience training. In this case, the initial objective is to enroll in a basic obedience class to teach the dog to walk on a loose lead, sit and lie down on command, stay in position, and come when called. When basic manners have been introduced, a second objective can be added – introducing some of the basic agility exercises.

Similarly, if the trainer is interested in preventing behavior problems, objectives should identify the unwanted behaviors and alternate, incompatible behaviors that are desired. For example, the owner of a new and very playful kitten understands the importance of teaching her kitten to play gently. Her objectives may be to teach the kitten to play with appropriate toys and to not use her claws or teeth when playing with her human family members.

Finally, a training plan is especially important when the objectives are to solve existing behavior problems. In this case, the first step is to complete a behavioral history profile that fully describes the problem behaviors and identifies all potential underlying causes. (General behavior profiles that can be used for initial screening with dogs and cats are provided in Appendix 3 and Appendix 4, respectively). For example, dogs and cats develop house-soiling problems for a variety of reasons. A behavior profile for house soiling should identify all of these possible causes to allow proper diagnosis and treatment. Once the profile has been completed and reviewed, a set of treatment goals is developed. These goals must be attainable and reasonable and should consider the severity of the problem, its underlying cause, and the owner's commitment to the pet and dedication to solving or managing the problem behaviors.

Select a Training Program or Behavior Modification Technique

Teaching a dog or cat new behaviors requires the use of effective reinforcers (See pp. 112, will refer to "Selecting Primary Reniforcers") and a plan for introducing, shaping, and maintaining new behaviors. There should be a clear set of the training steps and changes in behavior that reflect learning. Flexibility is also important during training, as the trainer must be able to modify the training plan in response to the animal's rate of learning. When solving behavior problems, an appropriate behavior modification program should be identified, along with any management approaches that may aid in reducing or solving the problem behavior.

Develop a Reasonable and Achievable Training Schedule

Once goals have been identified and an approach to changing behavior has been selected, it is important to create a realistic training schedule that includes short- and long-term goals. These goals will be influenced by the owner's level of commitment, time available, and ability to train and modify his or her pet's behavior. In many cases, 100-percent success is not a reasonable goal for either the pet or the owner. For example, a dog who jumps up excitedly to greet may be doing so because she is isolated for much of the work day and is very exuberant when her owner returns home. A reasonable solution would include providing the dog with more exercise and interaction and teaching her to sit for greeting when meeting new people. The owner may increase the dog's exercise but is inconsistent with teaching the dog to sit for greeting. As a result, the behavior may become more manageable but the dog will still occasionally jump up to greet. Educating owners regarding reasonable goals and their own responsibilities in achieving success can enhance appreciation for the behavior changes that are achieved and prevent "blaming the dog" for those that are not.

Evaluate Progress

Just as small steps should be included in the training or treatment plan, progress can also be evaluated in a step-wise manner. Using the example above, if the owner is aware that her dog jumps up simply because she is being friendly and exuberant (rather than having an idea of "jumping as misbehavior"), progress can be assessed with that motivation in mind. If the dog becomes reliable at greeting new people in a sit position once she has been out on her walk for 10 minutes, this is an excellent indication of progress. Conversely, she may continue to have difficulty staying off when she has been isolated or has not received her daily exercise and interactions. Over time, the owner can focus on the more challenging goals, while recognizing that significant progress has been made.

SELECTING PRIMARY REINFORCERS

Once a plan has been developed, the trainer must identify and select effective reinforcing stimuli. Chapter 4 describes the types of reinforcing stimuli (desirable or aversive), their relative values (high, moderate, low), and their inherent reinforcing properties (primary vs. conditioned stimuli). For the reasons discussed in here and in Chapter 4, desirable stimuli (commonly referred to as positive reinforcers) are most effective for use in the majority of training situations. Primary positive (desirable) reinforcers are stimuli that the dog or cat enjoys and will respond to with no prior conditioning or exposure.

For most dogs and many cats, food treats are a potent primary reinforcer. Individuals do have very specific preferences, however. While one dog may show a very strong response to small bits of cheese, another may prefer soft-moist dog treats. Similarly, one cat may respond best to a small piece of tuna while another may prefer a bite of canned cat food. In addition to food treats, most dogs and cats respond to the opportunity for positive social interactions, especially if those opportunities involve a member of the pet's primary social group (human caregiver). Playing and petting can be primary reinforcers for many dogs and cats. Verbal praise is thought to be a reinforcer that has weak primary reinforcing properties which can be strengthened through association with a primary stimulus such as food or play. While there is some dispute over the degree to which socialization plays a role in conditioning this type of reinforcer in dogs and cats, it is clear that social interaction is not a completely neutral stimulus to the majority of pets. Similar to food treats, individuals vary in the types of interaction and the types of play that they enjoy. A terrier may respond with great excitement to a squeaker toy as a primary positive reinforcer, while a retriever may prefer a tennis ball. Some cats enjoy small fuzzy balls that they can bat around while others get very excited about the opportunity to play with a laser light beam. Once several primary reinforcers have been identified for an animal, it is often possible to rank these in terms of their level of value to the animal and use them judiciously in training.

USING A CONDITIONED REINFORCER

A **conditioned (secondary) reinforcer** is a neutral stimulus that becomes associated with a primary reinforcer through classical conditioning (see Chapter 4, pp. 94–95). Over time, this stimulus acquires the same reinforcing capabilities as the primary (or unconditioned) reinforcer. A commonly used example in dog and cat training is verbal praise, specifically a single word such as "good." If the pet owner tells her dog "good boy" for sitting and then immediately offers a food treat (primary reinforcer), the phrase "good boy" becomes classically conditioned as a predictor of the primary reinforcer (the food treat). When the phrase "good dog" has become a conditioned reinforcer, it will serve as a positive reinforcer, even when not paired with food. (It is important, however, to still consistently pair the conditioned reinforcer "good dog" with the primary reinforcer to maintain its reinforcing properties.) Although it can be argued that praise alone has at least moderate primary reinforcing properties for dogs and cats, pairing a specific word or phrase with a primary food reinforcer still conditions the word to take on secondary reinforcer properties.

A benefit of a conditioned reinforcer, as well as its primary function, is that it allows the trainer to precisely target a desired behavior at the exact moment it is occurring. In many situations, providing a food treat (primary reinforcer) to the animal at the exact time that the animal is offering the targeted behavior is difficult or even impossible. The most common situation is when the dog or cat is not physically close to the trainer. Attempting to provide a food treat to reinforce behaviors when the animal is separated from the trainer by more than about 12 inches often results in the dog or cat moving toward the trainer, and subsequently engaging in a new behavior. Therefore, having a word (or click) that effectively communicates to the animal "Yes! It is that behavior that is occurring at exactly this moment that will be reinforced!" is an invaluable training aid.

Clicker as a Conditioned Reinforcer

A very popular conditioned reinforcer that is used in dog and cat training is the clicker. The click sound made by a small metal or plastic device (commonly called a "cricket") is a neutral stimulus (unlike verbal praise, which may have some inherently reinforcing properties). As a bridging stimulus, the clicker produces a sound that is both unique and of very short duration, which are important characteristics for effective conditioned reinforcers. Unlike the trainer's voice or words, which can vary significantly in tone and pitch, a clicker does not vary and so does not have the same potential for confusion. As described previously, the click-treat connection is established through classically conditioning the sound of the click with the presentation of the primary stimulus, a food treat. As with all classically conditioned responses, the click must immediately *precede* the provision of the treat. Conditioning is apparent when the dog or cat reacts with a startle or alerting response to the click sound, indicating that the click now *predicts* the arrival of the primary reinforcer (the treat). Once conditioned, the trainer then begins to vary the temporal (time) relationship between the click and the treat. This teaches the dog or cat to expect a treat within a second to up to several seconds after hearing the click. This training is imperative for the use of the clicker in any type of exercise that the animal is required to engage in while physically away from the trainer.

GETTING STARTED

Because the sound that a clicker makes is neutral and meaningless to the animal, it must be conditioned as a reliable predictor of a primary reinforcer (treat). This is easily achieved by clicking and immediately following the click with a treat. Dog trainers typically use "treat pouches" attached to their belt or a pocket that hold a quantity of small treats that can be consumed quickly by the dog. For most dogs, several repetitions of click-treat (with the treat always following the click; never the opposite), leads to a visible "startle response" to hearing the click. When the dog clearly begins to look for a treat upon hearing the click sound, the trainer has successfully trained the click as a conditioned reinforcer. An effective approach with cats is to place a teaspoon of canned food or other moist treat on the end of a spoon. The cat is allowed one bite of the food following each click. Although a startle response is not always clearly evident in cats, attention to the owner or searching behaviors for the spoon after hearing the click indicate that the stimulus-stimulus association has been learned.

The next step is to select a very simple behavior that the dog or cat is already offering to attempt to manipulate operantly. In dogs, a behavior that is often very readily learned is "eye contact." Most dogs will spontaneously look into their owners' eyes at some point during a training session. The trainer simply waits for his dog to glance into his eyes, and immediately click-treats (hereafter; CT). After several repetitions, the dog will increase the frequency of offering eye contact, presenting an increasing number of opportunities for reinforcement. In cats, friendly approach is a good behavior to begin with. When the trainer sees his cat approaching for petting or to ask for dinner, the behavior of approaching is reinforced with CT. To facilitate learning, even a step towards the trainer should receive a CT. Both of these simple behaviors allow the trainer to become accustomed to accurately isolate (target) behaviors with a clicker, and provide the dog or cat with repeated opportunities to learn the classical association between the behavior, the sound of the clicker, and the primary reinforcer. As dogs and cats (and their human caretakers) become more proficient, more complex behaviors can be introduced and shaped. There is some recent evidence suggesting that providing a primary reinforcer alone may be as efficient when introducing simple operant tasks to dogs (Sidebar 3). However, the ubiquitous use and success of conditioned reinforcers and specifically of clicker training in professional dog training provides strong evidence in support of this approach to training (see Appendix 1 for a list of recommended books).

TIMING AND SCHEDULES OF REINFORCEMENT

The timing with which stimuli and reinforcers are presented to the animal is an important consideration when teaching new behaviors. To be effective, positive and negative reinforcement must occur within *one second or less* of the behavior that

CLICKER TRAINING VS. FOOD ALONE AS A PRIMARY REINFORCER

Sidebar 3

A recently published study examined the ability of a group of purebred Basenjis to learn a simple operant task (touching an orange traffic cone with their nose) when trained with either a conditioned reinforcer (click-treat) or with a primary reinforcer alone (treat). The number of repetitions and amount of time needed for proficiency were recorded during a training phase (continuous reinforcement) and during a strengthening phase (switching to an intermittent schedule of reinforcement). After the dogs had become proficient at "nose-touch," the number of repetitions and duration (seconds) needed for extinction of the behavior were measured in both groups. The CT-trained group continued to receive a click for touching, but no food treat followed the click, while the group trained with food alone received no reinforcement at all during the extinction phase.

Results showed that although the mean number of repetitions and the length of time required by dogs trained with CT to learn "nose-touch" were lower than values for the group of dogs trained with the presentation of food alone, these differences were not statistically significant. Interestingly, when the dogs were switched to an intermittent schedule of reinforcement, the number of repetitions and duration to achieve a reliable response did not differ at all between the two groups. When the final phase of extinction was introduced, the dogs who had been trained using a clicker as a conditioned reinforcer continued to offer "nose-touch" for a significantly longer time and for a larger number of repetitions than did dogs who were trained using a primary reinforcer alone.

The researchers concluded that this study found no difference between dogs who were clicker-trained and dogs who were trained using a primary reinforcer alone in their ability to learn a new operant behavior, but that clicker-trained dogs appeared to be more resistant to extinction of the behavior in the absence of a primary reinforcer once the behavior had been learned. However, the data supported a trend in favor of enhanced learning in the clicker-trained dogs, suggesting that additional studies would be helpful. Additionally, the type of task that was trained allowed immediate offering of a primary reinforcer (the treat was placed into a bowl located immediately next to the traffic cone, within one second of the targeted behavior's occurrence). Because one of the reported benefits of using a conditioned reinforcer (clicker) is to mark behaviors that are occurring at a distance or which do not lend themselves to immediate reinforcement, a study comparing CT to primary reinforcement alone in this type of task is needed.

Smith SM, Davis ES. Clicker increases resistance to extinction but does not decrease training time of a simple operant task in domestic dogs (*Canis familiaris*). *Appl Anim Behav Sci,* 110:318–329, 2008.

is intended to be affected. Learning occurs most rapidly if there is some degree of overlap between the behavior and its consequence. Conversely, if reinforcement follows the behavior by more than one second, the animal is often already engaging in another behavior, and it is this behavior that is inadvertently reinforced. In fact, the underlying basis for many behavior problems that seem to persist regardless of training can often be traced back to the trainer's use of inaccurately timed reinforcement (Box 5.1).

Box 5.1 WHY CAN'T CASSIE SIT-STAY?

Cassie is an eight-month-old Australian shepherd. Her owner, Sally, is teaching Cassie to sit and to stay in position. Strong reinforcing stimuli for Cassie include soft-moist treats, pieces of cheese, petting, and soft petting on her chest. Sally trained Cassie to sit on command when she was four months old and just recently started to teach Cassie to stay in place (command = "stay") until she is released (release word = "okay"). Cassie wears a flat buckle collar and a six-foot nylon lead for her training.

Sally begins her training session by telling Cassie to sit and verbally reinforcing the sit ("good girl, Cassie"). Because Sally knows that it is important to shape this behavior, she begins her training with very short intervals of time. She waits 5 seconds and releases Cassie with her release command: "okay." Cassie jumps up, and Sally praises her exuberantly and gives her a cheese treat. Sally repeats this sequence, waiting 15 seconds before releasing Cassie. Each time that Sally releases Cassie with "okay" she immediately gives a cheese treat.

After several training sessions, Sally begins to notice something odd. Cassie seems to avoid the "sit" command, often jumping up exuberantly on Sally as soon as Sally asks her to sit. Alternatively, if Cassie does offer sit, she has started to crouch, as if in a play bow, shortly after being asked to stay. As soon as Sally releases her with "okay" (and, now, sometimes before she is released – Sally fears Cassie is regressing!), Cassie jumps up and down excitedly and happily.

What Is Happening? Why Is Cassie Not Learning?

Sally has made a common (but serious) mistake in her application of positive reinforcement. Instead of reinforcing her targeted behavior (sit quietly and stay in a sitting position), she is inadvertently but consistently reinforcing "jump up exuberantly!" Remember, if positive reinforcement (petting, praise, food treats) is provided while *Cassie is sitting quietly* in position, this is the behavior that will be reinforced (i.e. increases in frequency). However, if Sally waits too long, and reinforces Cassie upon the release word, then Cassie's behavior at this moment movement is reinforced, *not* the sit-stay!

The Solution?

Sally continues with her training, slowly shaping a sit-stay behavior. However, instead of providing a reinforcer *after* the sit-stay is complete, she now provides positive things to Cassie during the sit-stay. While Cassie is sitting, Sally quietly praises her, gives her cheese treats, and gently pets her chest. When Cassie is released, Sally simply says "okay" and it is the end of the behavior (no treat for the release!). Cassie learns quickly that the behavior that is "working" now is sitting and staying, while jumping up afterwards earns no reward.

When the trainer reinforces some responses and not others, this is referred to as **differential reinforcement**. All operant training is based upon differential reinforcement. For example, when a trainer waits for eye contact, he is *not* reinforcing other behaviors that the dog may be offering. Behaviors such as looking away, yawning, pawing, sighing, or lying down are simply ignored. As soon as the dog offers the smallest glance up towards the trainer's eyes, that behavior is reinforced with CT. Differential reinforcement is also used during shaping when a trainer begins to shift the criteria for reinforcement (see pp. 119–122). Learning occurs most rapidly if every correct response that meets the trainer's criteria is positively reinforced and every response that does not meet the criteria is ignored.

Reinforcing every correct response is a type of reinforcement schedule called a **continuous reinforcement** schedule. A continuous reinforcement schedule should be used when new behaviors are being introduced and continued until the dog or cat is reliably offering the behavior in response to the selected cue. For example, the dog reliably and consistently sits when asked to "sit" or the cat reliably and consistently comes to her name when called. Each time the desired behavior is offered, a positive reinforcer is provided. The trainer should consider that each time the dog or cat offers the desired behavior, presenting positive reinforcement in the form of a CT, verbal praise, or petting provides specific information that says "Yes! That's it! That is the behavior that I wish to see increase!" to the dog or cat.

An **intermittent reinforcement** schedule occurs when reinforcement is provided occasionally as the consequence to the desired behavior. A useful analogy of this concept is gambling behavior in humans. A person who is addicted to the slot machines of Las Vegas is responding to an intermittent schedule of reinforcement. "Slot-machine playing" behavior is strongly maintained because it is occasionally and unpredictably reinforced by the arrival of money. Similarly, a dog who has been trained to come when called and gradually switched from continuous to intermittent reinforcement will continue to offer this behavior in anticipation of possibly receiving reinforcement. Several types of schedules of intermittent reinforcement can be used. These include fixed ratio, fixed interval, variable ratio, and variable interval (Table 5.1). Of these four types of schedules, variable ratio schedules are most often used in dog and cat training when teaching any type of active behavior. Variable interval schedules are helpful when teaching a dog or cat to stay in place (sit-stay or down-stay) for a period of time.

In general, intermittent reinforcement schedules produce stronger responses of learned behavior than continuous reinforcement schedules. In addition, because the total amount of reinforcement that is used in a training session is less, satiety is delayed, allowing for longer training sessions. However, it is still

table 5.1 Reinforcement Schedules

TYPE	DEFINITION
Continuous	Reinforcement is provided for every correct response
Variable Interval	Reinforcement is provided at irregular time intervals (For example: A food treat is given to a dog during a sit-stay exercise at 30 seconds, 40 seconds, 90 seconds, 200 seconds, and 240 seconds)
Fixed Interval	Reinforcement is provided at regular time intervals (For example: A food treat is given to a dog during a sit-stay exercise at 30, 60, 90, 120, 150, 180, 210, and 240 seconds)
Variable Ratio	Reinforcement is provided after an irregular number of correct responses (For example: A food treat is given to a dog for coming when called after the first, second, fifth, seventh, and twelfth correct response)
Fixed Ratio	Reinforcement is provided after a regular number of correct responses (For example: A food treat is given to a dog for coming when called after the first, fourth, seventh, tenth, and thirteenth correct response)

always best to err on the side of caution when considering changing from a continuous to an intermittent schedule. Switching to an intermittent schedule of reinforcement prematurely can cause confusion if the dog or cat does not yet fully understand the desired behavior that is being trained. Similarly, when an intermittent schedule of reinforcement is introduced, the ratio of reinforced to not-reinforced repetitions of the behavior should still be very high (i.e. reinforcement should still occur frequently) and the ratio can be gradually decreased (or interval of time increased).

Although intermittent schedules of reinforcement have been shown to have advantages in laboratory settings and under very controlled conditions, in practical dog and cat training most behaviors can be easily maintained using a continuous schedule of reinforcement. Because trainers always have their voices and hands available, and because many dogs and cats respond to a variety of forms of positive reinforcement, there is often no need to switch to an intermittent schedule of reinforcement. However, if a particular behavior cannot be consistently reinforced in practical situations, taking the time to introduce some form of intermittent reinforcement will help to make the behavior more resistant to extinction. A second situation in which intermittent reinforcement is helpful is when the desired behavior is occurring at a very high frequency. One practical example is the use of positive reinforcement to increase "loose lead" while out walking with the dog. Rather than reinforce every step that the dog takes with a loose lead, the trainer can use an intermittent schedule and reinforce proper heel position after a variable number of steps.

Interestingly, many problem behaviors in dogs and cats are unintentionally maintained by owners who are using an intermittent schedule of reinforcement. For example, the dog who barks for attention during dinnertime has often been fed tidbits at variable times, reinforcing (and often strengthening) this behavior. Similarly, the cat who repeatedly bats pens off of the desk until her owner stops working on her behavior book to pet her has learned through intermittent reinforcement schedule to be persistent in her pen-playing behavior because at some point in this daily game she will be reinforced!

SUCCESSIVE APPROXIMATION (SHAPING)

Because operant conditioning relies upon the consequences of behavior for learning, the dog or cat must first be able to *offer* the behavior so that it can be reinforced. However, many of the behaviors that trainers are interested in reinforcing either occur at very low frequencies or are not a part of the animal's behavioral repertoire in their complete form. An example is waiting for a dog to spontaneously sit and stay in place for a period of time. Although most dogs do sit now and then, they do not sit in place without moving (a sit-stay) for more than several seconds. For this reason, operant conditioning techniques almost always use successive approximation, also called shaping. Shaping involves starting with a small part or an approximation of a desired response. The animal should be already offering this little "piece of the behavior" so that the trainer can differentially reinforce it. As the animal begins to increase the frequency with which he offers the "piece" (approximation), the trainer then shifts the criteria to reinforce closer approximations, moving gradually towards the final and desired response. As the dog or cat is successful at each level of response, the criteria for reinforcement are shifted slightly toward the final behavior, and previous forms of the behavior are no longer reinforced.

Perhaps the best way to describe successive approximation is through an example. In the case of teaching a sit-stay, a first-level response would be simply to sit, without requiring the dog to stay in position. When the dog reliably sits on command, the criterion is shifted and the trainer now reinforces sit followed by a two-second stay in place. This criterion (time) is gradually shifted (increased) as the dog is successful, until he is capable of staying for several minutes. A second criterion for the stay is distance away from the trainer. This can be shaped using the same procedure, but should be taught separately from shaping time intervals. In other words, the trainer concentrates on increasing time while staying close to the dog. Once the dog is reliably staying, distance can then be shaped. When the trainer begins to shape distance, she will move further and further away from the dog, but only for very short periods of time. Once both time and distance have been shaped, the two criteria can then be combined and increasing intervals of time with longer distances can be simultaneously shaped.

If failure occurs at any level, the trainer simply drops back down to a lower response level until the dog or cat is again proficient. One of the most common mistakes made by novice trainers is to expect too much too soon, reflected by shifting criteria at a rate that causes the dog or cat to fail. Proceeding gradually ensures that the subject is proficient at each stage before attempting a more difficult response and prevents repeated mistakes (and lost opportunities for positive reinforcement). Trainers must also always be willing to drop back to an earlier stage if the animal experiences difficulties and to creatively develop various forms of intermediate (successive) steps to facilitate learning. When used correctly, successive approximation is a powerful tool because very complex behavior patterns can be achieved and many problem behaviors can be solved, without the need of negative reinforcement or punishment (see Chapters 6 and 7).

Prompting vs. Free-Shaping

Because operant conditioning is a consequence-based learning process, training can only proceed when the animal offers a behavior that meets the criterion set by the trainer in order to earn reinforcement. A training criterion refers to the behavior or response that the trainer has identified as desirable and which will be reinforced in a given training session. With shaping, criteria will shift as the animal begins to offer behaviors that approach the final desired behaviors. Prompting (also called "**luring**") is a training method that is used to induce a dog or cat to offer all or part of a desired behavior so that the opportunity for reinforcement can occur. Prompting is most effective when teaching behaviors that are occurring at a very low frequency or when the subject is very distracted. For example, a young dog may not be inclined to offer even the first part of lying down (moving front feet out in front of her) simply because she is an active young dog and is very distracted by other dogs in the training class. The trainer can prompt the first portion of the "lying down" behavior by slowly moving a hand containing a food treat from the puppy's nose towards his front toes (the "nose-to-toes" technique), and reinforcing for any movement towards a down position (Figure 5.1). Food treats held in hands are a very effective training tool because most dogs and cats readily orient to the treat and the hand, and can then be induced to move into a desired position by simply following the hand. Holding or tossing a favorite toy is another effective lure for some training situations.

Free-shaping refers to simply waiting for the dog or cat to "offer" a desired behavior (or, more practically, a criterion for part of a desired behavior), and then immediately reinforcing that behavior with "good!" or CT. Trainers often refer to free-shaping as "capturing" behaviors because, when performed correctly, it involves targeting naturally occurring behaviors and then gradually shaping these to approach the new behavior that is desired. An example of free-shaping is teaching a dog to shake hands. Many dogs naturally use their paws to request attention or during play. If the trainer wishes to capture this behavior with the intent of teaching "shake hands" she waits

figure 5.1 Prompting (luring) "down" using a food treat and hand motion downward (nose-to-toes)

figure 5.2 Free-shaping "shake hands"

until the dog offers even a slight paw lift and immediately reinforces it ("good" or CT). Once the dog begins to increase the frequency of "lift paw," the trainer then shifts the criterion slightly and reinforces a higher lift of the paw touching a hand (Figure 5.2). Although trainers often debate the pros and cons of prompting vs. free-shaping, in actuality, both approaches to teaching new behaviors are useful (Sidebar 4).

PROMPTING (LURING) VS. FREE-SHAPING — Sidebar 4

Prompting

- *Reduces frustration and increases rate of learning:* Using a lure to "jump-start" a behavior allows the dog or cat to be successful very quickly. Luring provides the animal with an opportunity to rapidly learn from consequences, thus reducing the frustration that can occur with free-shaping and enhancing the rate of learning.

- *Risks: Prompting a behavior contributes to poor stimulus discrimination:* This is the major disadvantage of prompting new behaviors. The movement of the hand holding the food or the presence of the lure or prompt (toy or target) quickly becomes associated with the behavior. In fact, if the trainer is not careful, the prompt itself becomes the cue to offer the behavior.

- *Solution:* The prompt (lure) MUST be removed as early in the learning process as possible. For example, if food in the hand is used to prompt lying down, the food must be removed from the hand after the dog has offered several repetitions and has not learned to "wait for the food" to offer the behavior.

Free-shaping

- *Increases creativity and willingness to try new behaviors:* Dogs and cats who learn that offering new behaviors may earn them a desired reinforcer become more willing to offer new (and often silly!) behaviors during training sessions. This can lead to rapid learning and the shaping of new and unusual behaviors.

- *Risks: Free-shaping can be frustrating to the dog/cat if unusual or complex behaviors are targeted:* Because the subject (dog or cat) does not know what behavior is desired, free-shaping can lead to frustration if the animal continually tries new behaviors that are not approximations of the behavior that the trainer is attempting to shape.

- *Solution:* When the subject is not being successful and is at risk for becoming frustrated, the trainer must can either lower the criteria for reinforcement (i.e. select an approximate behavior that the animal is offering and reinforce that), or introduce a lure to "jump-start" the learning process.

BEHAVIOR CHAINS AND THE PREMACK PRINCIPLE

Teaching a new behavior to a dog or cat often involves a sequence of two or more relatively simple behaviors that occur in a specific order. These sequences are called **behavior chains** and are taught by first breaking the chain of behaviors into small individual components and training each component separately. When the subject is proficient at each piece of the behavior, they are then linked together, starting with just two behaviors, then adding more as the dog or cat learns to perform each sequence reliably. Many behavior chains are best taught using a technique called "back-chaining." In back-chaining, the very last behavior in the sequence is taught first, thus establishing a strong reinforcement history (i.e. the dog or cat has had the opportunity to practice the last behavior most often). When all of the steps have been taught and the trainer is ready to link them together, the chaining proceeds from the end of the chain towards the beginning. The first two behaviors to link together are the final behavior and the second-to-last behavior. The trainer then works backwards toward the start of the entire chain (Box 5.2).

Although back-chaining seems counterintuitive, it functions to "drive" the behavior chain because the dog is always working towards the final behavior that has been most strongly reinforced. As the dog or cat works her way through the chain, the behaviors near the latter part of the chain are more familiar. Interestingly, the opportunity to engage in a familiar (and enjoyable) behavior can serve as a reinforcer in and of itself! This phenomenon is called the "Premack Principle," named after the behaviorist who first described it, David Premack. A helpful analogy in human experience is the feeling that many people have during a very long car trip to a much anticipated vacation site. The early and middle hours of the trip can become quite boring and fatiguing, but during the last few hours the traveler

Box 5.2 USING BACK-CHAINING TO TEACH RUCKUS TO PUT HIS TOYS AWAY

George has a two-year-old Golden Retriever named Ruckus. Ruckus loves to retrieve and has many and assorted toys that he plays with and retrieves. George has provided a large wicker basket for these toys and would like to teach Ruckus to retrieve his toys around the house and place each one in the basket. Here are some tips that George can use when back-chaining this rather complex trick:

STEPS:
1) *Break the behavior into its component parts, starting from the last behavior in the chain:*
    (a) Coming when called from the basket.
    (b) Dropping the toy into the basket.
    (c) Running to the basket with the toy.
    (d) Retrieving (fetching) the toy.
    (e) Marking (indicating) the toy.
    (f) Staying close to trainer (sit/down) and toy until asked to fetch.
2) *Teach (and strongly reinforce) the <u>last</u> behaviors first:*
    (a) Dropping the toy into the basket (second-to-last) and coming back to George from the basket (last behavior).
    (b) Ruckus is in a sit/stay next to the basket and is reinforced for coming when called to George (away from the basket).
    (c) Ruckus is standing next to the basket and is given a toy to hold. He is reinforced for dropping to toy into the basket (and leaving it there).
    (d) Each behavior in the sequence is trained separately (using shaping), but these final two should receive the most attention and Ruckus should earn the greatest or most intense reinforcement for performing these steps.
3) *Assembling the steps in the chain:*
    (a) Consecutive steps are linked, starting with the final two or three steps.
    (b) George gives Ruckus a toy as he stands with his head over the basket; he asks Ruckus to drop the toy into the basket and then calls him to come (strong reinforcement!).
    (c) Steps are then added working backwards towards the beginning of the trick (run to the basket with toy is added next; then pick up toy; then sit at side), until – ta-da! – Ruckus puts his toys into a basket!

often feels energized as she anticipates the reward of vacation. In this case, the positive reinforcement for a long task (driving for many hours) is the arrival at a pleasant location. Similarly, a dog who is performing a behavior chain that has a well-reinforced final behavior will show increased motivation near the end of the chain if the last behavior is anticipated as having a very positive "pay-off" (strong reinforcement history).

From the pet's point of view, if an entire series of behaviors is required for the performance of a new exercise or trick, there must be a strong motivation to perform the entire sequence and not quit early after one or two of the steps in the sequence. By making the final step of the series the "best" part, we teach the dog or cat to anticipate the final portion (or, as we know it, the completion) of the trick. The final steps are said to then drive the entire sequence. Behavior chains can also be taught using forward-chaining – simply linking each behavior in sequence of their occurrence. However, this can lead to a decrease in performance as the dog or cat works his way through the series, because the first few steps of the sequence were those that were most strongly reinforced rather than the final steps. Therefore, the animal's anticipation of reinforcement would focus on the first portions of the sequence rather than the last, leading to a tendency to quit early without completing the entire trick.

Back-chaining behavior chains invoke the Premack Principle, which simply states that the opportunity to engage in a stronger (more desired) response can serve to reinforce a weaker response. If the final behavior is the strongest, the opportunity to engage in each subsequent behavior serves to reinforce the previous behavior as the animal anticipates more reinforcement as he goes through the chain. The Premack principle can also be used when selecting effective reinforcers for training. For example, if an owner trained his dog to retrieve, and retrieving is a highly valued activity for that particular dog, the owner can then use the "fetch" behavior to reinforce behaviors that are less strong (or that the dog is less naturally motivated to perform), such as waiting at the owner's side until he is released. Used properly, the dog will quickly learn that he will have the opportunity to retrieve the toy only if he waits in a sit position until released.

Fading Cues and Attaining Stimulus Control

When to Put a Behavior "On Cue": Cues are the verbal commands or hand signals that a trainer uses to elicit the targeted behavior. Examples of verbal cues include "sit," "down," "come," and "get it." Examples of hand signals include a motion downward (down), or an arm swing into the body (come). Operant trainers typically add the cue for new behaviors *after* the dog or cat has learned the new behavior or at least part of the new behavior. This may appear "backwards" but it is actually more effective because it prevents incorrect responses and reduces confusion in the dog or cat. For example, once a dog is consistently lying down for a food treat, the trainer introduces the verbal command "down," followed by any prompts that are used, reinforcing the dog for the down behavior. This facilitates fading the hand cue or lure to a hand signal as the verbal command becomes classically conditioned and evokes the response without the hand cue.

Fading Cues: As animals learn, they are responding to one or more of the subtle nonverbal cues that a trainer provides. This occurs in addition to (and, in some cases, in place of) responding to the verbal cue that is used to elicit the behavior.

An example is a trainer who is using the command "shake" immediately prior to the dog offering her paw. However, he suddenly notices that the dog begins to offer his paw for shaking when he simply bends towards the dog, or reaches out with his own hand, without necessarily waiting for the verbal "shake" command. These movements have all served as additional (though inadvertent) cues that have been repeatedly paired with the "shake" command. From the dog's point of view, these movements have the same meaning as the verbal command itself (classical conditioning again!). Trainers must be aware of these extraneous cues (and should attempt to minimize them) as they can interfere with stimulus control once the behavior has been completely shaped and put on cue.

Once a dog or cat has learned to perform a particular trick "on cue" and is beginning to reliable in his responses, the trainer should identify all of the cues that the dog or cat is responding to and should determine whether any of the cues require fading. Evaluating cues can be as simple as deciding whether to use a voice command, a hand signal, or both as the cue to elicit a behavior. For example, some dog trainers teach dogs to respond to both a hand signal and a voice command to lie down or sit, while others prefer to use primarily voice commands. In addition to the type of cue, the trainer must also fade any prompting properties that the cue may have. Using the "down" command as an example: If the command is initially taught using the hand prompt described previously, it is often helpful to fade this cue to a weakened or shortened version, for example a brief movement of the hand downward. For many behaviors, a weakened version of the lure is often an appropriate hand signal to use as the command for the behavior. If the trainer wishes to change the cue that is used to elicit the behavior, this can be accomplished with classical conditioning. The new (neutral) cue is given immediately prior to the established cue, each time the behavior is requested. Over time, the new cue becomes classically conditioned to elicit the behavior without being paired with the established cue.

Stimulus (Cue) Control: A stimulus is anything that causes a behavioral response. In the case of practical dog and cat training, cues are stimuli that elicit trained behaviors. Because these stimuli do not naturally elicit the desired response and have to be taught through association and shaping, they are called conditioned stimuli. When the dog offers the behavior (lying down) only in response to the conditioned stimulus (the verbal command "down"), the behavior is said to be under stimulus control. However, most dogs and cats tend to offer behaviors when they are not asked for or in response to a different cue. This is called a lack of stimulus control. Anticipating a command is a common example of a lack of stimulus control. For example, a dog who offers "shake" prior to the verbal cue is offering a correct behavior, but he is not offering it in response to the correct cue. Another example of a lack of stimulus control is the dog who happily rolls over in response to his owner's command, "shake." Although the dog is offering a trained behavior, he is not offering the correct trick for the stimulus (cue) that the trainer has given.

Stimulus control is a part of the normal training process that must be consciously taught. If it is not, most dogs and cats will simply choose behaviors that they most enjoy (or that are easiest for them to do or have the strongest reinforcement history), and offer them repeatedly in response to various cues. Others will continue to happily offer their favorite trick to anyone who even remotely appears to be getting ready to offer them the cue. While this is not a terrible problem to have in the grand scheme of dog and cat ownership, trainers are most successful if they consciously teach stimulus control once they have a behavior on cue. Helpful tips for attaining stimulus control are outlined in Sidebar 5.

Sidebar 5 — TECHNIQUES FOR ATTAINING STIMULUS CONTROL

When several tricks or training exercises have been taught, stimulus control problems can arise when the dog or cat offers a particular (*favorite*) trick in response to several different (and unrelated) cues, or offering a behavior spontaneously in the absence of the cue. Using the trick of "shake hands" as the example, the four rules of attaining stimulus control are

- *ALWAYS reinforce "the desired behavior for the correct cue":* Each time that the trainer says "shake hands" and the dog responds correctly (offers paw to shake), the trainer offers a positive reinforcement. This is important when training for stimulus control because the classical pairing of the verbal cue ("shake hands") with an opportunity to earn a treat (dog shakes and earns treat) must be consistent.

- *NEVER reinforce "spontaneous offers of the desired behavior":* If the dog offers to shake but the trainer has not asked for the behavior with a cue, the behavior is ignored (no reinforcement). Some dogs will persist (this is called an extinction burst), in an attempt to earn a reinforcer. However, stimulus control cannot be attained if spontaneous offerings are reinforced!

- *NEVER reinforce "the desired behavior for an incorrect cue":* It is sometimes tempting for trainers to reinforce a new behavior when the dog offers it in response to a different command; for example, the dog offers to shake hands when asked to sit. A trainer may reason that any offering of the new behavior should be reinforced in order to strengthen the behavior. However, not reinforcing the incorrect response is a more effective approach to gaining stimulus control.

- *NEVER reinforce "an incorrect behavior for the correct cue":* This is the opposite of the previous mistake – the dog offers a sit behavior when asked to shake hands. No reinforcement sends the message, "Sorry, that is not correct; that is not what was asked for." Release the dog, try again, and reinforce when he gets it right!

A Few Additional Tips:

- *Change the reinforcer:* When a new behavior is taught and being "put on cue," it is sometimes helpful to switch to a high-level treat as a reinforcer for that behavior while using lower-value reinforcers for behaviors that the dog knows well and performs reliably. This functions to focus the dog's attention on that particular behavior in a training session because it is the one that is "paying off" the most.

- *Modify the cue:* If a dog repeatedly confuses the cues of a trick or repeatedly offers the wrong response for a particular cue, this may indicate a need to re-teach that cue or modify the magnitude or intensity of the cue to gain stimulus control.

- *Use time outs:* Dogs can become frustrated if the behavior they are offering (and believe is the one that the trainer wants) goes unrewarded. Simply breaking up the training session by taking a few minutes to play or train another behavior can reduce frustration and allow the trainer and dog to start fresh with the new behavior.

BEHAVIOR MODIFICATION TECHNIQUES

Behavior modification involves changing behaviors that are considered to be unacceptable and problematic to alternate behaviors that are acceptable. Although training and teaching new behaviors is almost always a component of behavior modification, the techniques that are used focus specifically on changing unwanted behaviors and often also altering the animal's emotional responses during problem situations. The three primary behavior modification techniques that are used with pet dogs and cats include extinction, systematic desensitization, and counter conditioning.

Extinction

The principle of extinction is that if no reinforcement is provided for an operantly learned response that is not desired, over time the response will decrease in frequency. Extinction can also be applied when attempting to change unwanted classically learned associations. This is accomplished by repeatedly presenting the conditioned stimulus by itself in the absence of the unconditioned stimulus. This results in a gradual decline (or extinction) of the conditioned response. For example, a dog who jumps up to go for a walk whenever his owner picks up his keys is demonstrating a classically conditioned association between picking up keys (conditioned stimulus) and going outside for a walk (unconditioned stimulus). Extinction of this behavior is accomplished by picking up the keys repeatedly without following this action with a walk. With classically conditioned responses, the use of extinction can only be successful if the pairing of the conditioned and unconditioned stimuli are permanently disassociated.

Extinction of operantly learned behaviors is accomplished by withdrawing reinforcement from the targeted response. In the absence of reinforcement, the frequency with which the dog or cat offers the behavior decreases, eventually leading to extinction. Typically, at some point after the reinforcement is first withdrawn, the pet shows a sudden increase in the behavior. This is called an **extinction burst**. If the behavior is subsequently never reinforced, extinction occurs. However, if reinforcement occurs during the extinction burst, or intermittently during an attempt at extinction, the behavior will again be reinforced and even strengthened. For example, barking for attention is a common problem in dogs that often has been unintentionally operantly conditioned. Reinforcement for this type of barking is the owner's attention. Ignoring the dog whenever he barks for attention will theoretically lead to extinction. However, although extinction techniques are often recommended for solving behavior problems, they are not effective in practical settings if used alone and without other forms of behavior modification.

Limitations of Extinction: When used alone, extinction has several serious limitations as a behavior modification technique. Removing an expected positive reinforcer inevitably causes frustration and agitation in the dog or cat, emotions that can lead to other problem behaviors in **an attempt to obtain the same reinforcer**. For example, a dog who barks for attention is simply asking for interaction with his social partners. If the barking is ignored (i.e. the owner turns and walks away) and no other behavior is reinforced with attention and affection, the dog is still lonely and still in need of social interaction. He may now instead begin to chew inappropriate items, steal objects and play "catch me if you can," or mouth and jump up. A second limitation is that extinction does not account for the self-reinforcing properties of many problem behaviors. A dog who barks when excited will not generally stop if ignored because barking is an expression of excitement and is often innately enjoyable to the dog. Finally, extinction is only useful as a training tool when the trainer has complete control over the reinforcers for the behavior. For example, if a dog is barking because he is chasing squirrels out of his yard, extinction is not an option since the reinforcer for this behavior is the squirrels running away, something that the owner cannot control. For these reasons, while extinction techniques can be a helpful addition to some behavior modification programs, they are not generally successful when used alone to attempt to reduce a problem behavior.

Adding Differential Reinforcement: The inclusion of differential reinforcement with extinction reduces the animal's frustration and enhances success at reducing the problem behavior. Differential reinforcement refers to selecting a competing (desirable) behavior that is incompatible with the problem behavior and reinforcing that behavior while at the same time extinguishing the problem behavior. Using the previous example, an appropriate and desirable behavior that could be differentially reinforced while attempting to reduce attention-seeking barking is "sit for petting." For maximum effectiveness, alternate behaviors should be reinforced *before* the dog begins to offer the behavior that the owner is attempting to extinguish, to prevent the establishment of a behavior chain. This means that the trainer should be aware of situations that typically lead to barking and then acts preemptively to engage the dog in an alternate, desirable behavior that is differentially reinforced. Trainers and behaviorists commonly refer to this as "reinforcing an alternate and incompatible behavior."

Systematic Desensitization

This is a technique that is often used to habituate established fearful or aggressive reactions. The fear-inducing (or aggression-inducing) stimulus is presented at a very low level of intensity until habituation is achieved. The level of intensity is then increased slightly and the animal is exposed to the stimulus until he is desensitized at that intensity. This is repeated in step-wise increments until the dog or cat is habituated to the full intensity of the stimulus. For example, a puppy who has been sensitized to the vacuum cleaner could undergo a program of systematic **desensitization** for his fear response. He would be gradually exposed to the cleaner at low levels of intensity, perhaps beginning with an inactive machine at a wide distance and gradually activating the vacuum cleaner and coming closer as the puppy habituates at each level. It is essential to this program that the stimulus is presented at a level that is always lower than the level that elicits a fearful response. A delicate balance of slowly increasing the intensity of the stimulus and avoiding a fearful response must be maintained for success. Systematic desensitization is the technique that is routinely used to treat pets who are shy or fearful with strangers. It is also helpful in the treatment of certain types of aggression (see Chapter 11 for a complete discussion). For maximum effectiveness and success, systematic desensitization is always paired with counter-conditioning (see below).

Counter-Conditioning (and Counter-Commanding)

Counter-conditioning refers to classically conditioning an animal to respond to a stimulus in a manner that is incompatible with the response that was previously evoked by the same stimulus. A new response that is behaviorally, emotionally, or physiologically incompatible with the previously undesired response is chosen. For example, eating is a pleasurable activity for dogs and is incompatible with feelings of fear or anxiety. Counter-conditioning during a systematic desensitization program for fear/flight responses includes feeding high-value food treats at each level of desensitization. As the dog or cat becomes habituated at each level of intensity, she begins to associate the presence of the stimulus with the pleasure of eating, rather than with fear (i.e. a classically conditioned relationship).

Counter-commanding is an operant form of the same method and involves teaching the animal to offer an alternate voluntary response that is incompatible with previously undesired behaviors. For example, if the sound of the vacuum cleaner causes a sensitized puppy to attempt to flee, a sit-stay with positive reinforcement could be counter-commanded during the desensitization program. Counter-commanding is used more often in dogs than in cats and works best with a dog who has had some obedience training and when the unwanted behavior has a relatively low level of motivation. Positive reinforcement should always be used to counter-condition a response, because the goal is to change a situation that was previously unpleasant into a pleasant experience for the dog. As with systematic desensitization, it is essential that the counter-conditioning and counter-commanding take place at levels of intensity at which the sensitized response is *not* elicited.

Flooding

Flooding is also called "response prevention" and can be used to extinguish certain types of avoidance responses. It is the direct opposite of systematic desensitization. The traditional procedure of removing the association between classically conditioned stimuli may be ineffective in solving some avoidance behaviors. This occurs because, once conditioned, the dog never allows himself to again encounter the aversive stimulus. As a result, it is not possible for the dog to learn that the aversive event is no longer present. For example, a dog who has become frightened by the sound of a loud truck begins to avoid the corner where the sound occurred (i.e. approaching the corner is now a conditioned stimulus that predicts the fear of the loud truck noise). By avoiding the curb, the dog never again is exposed to the truck, but also cannot learn that the truck is usually not present (i.e. the curb is actually not predictive of the truck's approach). It is also theorized that avoidance behavior may be reinforced by the fear reduction that occurs when the dog is able to avoid the fear-producing situation in the first place.

Flooding involves exposing the animal to the avoidance situation while preventing the avoidance behavior. The dog is "flooded" with the warning cues, but the aversive stimulus does not appear. In the case of the truck-fearful dog, this would involve repeated exposure to the curb, in the absence of loud trucks. Flooding can be a precarious technique because it usually produces an initial increase in fear. Therefore, it is extremely important to ensure that the aversive stimulus is not forthcoming while the flooding is taking place. In general, flooding can be successful in situations in which the animal's fear response is not severe, and in which the trainer is capable of having complete control over the fear-eliciting stimulus.

Risks Associated with Flooding: Flooding is not an effective tool, and in fact can be extremely damaging to the pet when it is used in an attempt to solve extreme fears or phobias. Moreover, some trainers and behaviorists have advocated the incorrect use of flooding by presenting the aversive stimulus itself to the dog or cat, rather than the cues that predict the aversive stimulus. In the case described above, this would entail restraining the dog on a lead while forcing her to endure the approach of many trucks (i.e. "flooding" her with the fear-producing stimulus). This "training technique" is not only a misapplication of flooding but is ineffective and cruel, and should never be supported. In the vast majority of situations, systematic desensitization is a more effective and humane method for dealing with fear or avoidance behaviors in dogs and cats.

REVIEW QUESTIONS

1. Identify the classical and operant components of teaching a dog to lie down and stay in position on command.
2. What is clicker training? Provide an example in dog or cat training.
3. Compare and contrast the concepts of prompting and free-shaping.
4. Discuss the benefits of successive approximation (shaping) in practical dog and cat training.
5. What is "back-chaining"? How is it used in dog training and what are its benefits?

REFERENCES AND FURTHER READING

Borchelt PL, Voith VL. **Punishment.** In: *Readings in Companion Animal Behavior*, VL Voith and PL Borchelt, editors, Veterinary Learning Systems, Trenton, NJ, pp. 72–80, 1996.

Burch MR, Bailey JS. **How Dogs Learn.** Howell Book House, New York, NY, 188 pp., 1999.

Donaldson J. **Culture Clash: A Revolutionary New Way of Understanding the Relationship between Humans and Domestic Dogs.** James and Kenneth Publishers, Oakland, CA, 221 pp., 1996.

Fukuzawa M, Mills DS, Cooper JJ. **More than just a word: non-semantic command variables affect obedience in the domestic dog (*Canis familiaris*).** *Appl Anim Behav Sci*, 91:129–141, 2005.

Johnson P. **Twisted Whiskers: Solving Your Cat's Behaviour Problems.** The Crossing Press, Freedom, CA, 154 pp., 1994.

Kubinyi E, Topal J, Miklósi I, Csányi V. **Dogs (*Canis familiaris*) learn from their owners via observation in a manipulation task.** *J Comp Psychol*, 117:156–165, 2003.

Ley J, Coleman GJ, Holmes R, Hemsworth PH. **Assessing fear of novel and startling stimuli in domestic dogs.** *Appl Anim Behav Sci*, 104:71–84, 2007.

Lindsay SR. **Handbook of Applied Dog Behavior and Training, Volume 1: Adaptation and Learning.** Iowa State University Press, Ames, IA, 410 pp., 2000.

Mills DS. **What's in a word? A review of the attributes of a command affecting the performance of pet dogs.** *Anthrozoos*, 18:208–221, 2005.

Miltenberger R. **Behavior Modification: Principles and Practices.** Brooks/Cole Publishing, Pacific Grove, CA, 1997.

O'Farrell V. **Dog's Best Friend: How Not to Be a Problem Owner.** Methuen Press, London, UK, 1994.

Pearce WD, Cheney CD. **Behavior Analysis and Learning,** 3rd edition. Lawrence Erlbaum Associates, Inc, Mahwah, NJ, 388 pp., 2004.

Pongrácz P, Miklósi A, Timar-Geng K, Csányi V. **Verbal attention getting as a key factor in social learning between dog (*Canis familiaris*) and human.** *J Comp Psychol*, 118:375–383, 2004.

Pryor K. **Don't Shoot the Dog.** Bantam Books, New York, NY, 187 pp., 1984.

Pryor K. **Karen Pryor on Behavior.** Sunshine Books, North Bend, WA, 405 pp., 1995.

Reid P. **Excel-Erated Learning: Explaining How Dogs Learn and How Best to Teach Them.** James and Kenneth Publishing, Oakland, CA, 172 pp., 1996.

Rogerson J. **Your Dog: Its Development, Behaviour, and Training.** Popular Dogs Publishing Company, London, UK, 174 pp., 1990.

Schilder MBH, van der Borg JAM. **Training dogs with help of the shock collar: short and long term behavioral effects.** *App Anim Behav Sci*, 85:319–334, 2004.

Spreat S, Spreat SR. **Learning principles.** *Vet Clin North Amer: Small Anim Pract*, 12:593–606, 1982.

Voith VL, Borchelt PL (Editors). **Readings in Companion Animal Behavior.** Veterinary Learning Systems, Trenton, NJ, 276 pp., 1996.

Yin S. **Classical conditioning: Learning by association.** *Compend Contin Vet, Small Anim Pract*, 472–476, June 2006.

CHAPTER 6

Training and Problem Prevention for Puppies and Kittens

Adopting a new puppy or kitten is an exciting time for most families. Whether the new pet is adopted from a local shelter or purchased from a breeder, the transition from mother and littermates to a new home and family can be stressful. A careful approach to introductions can facilitate a smooth transition to the new home and rapid integration of the puppy or kitten into the new family. Early training and socialization can prevent many common behavior problems from developing and contributes to a well mannered and well behaved adult pet.

THE NEWLY ADOPTED PUPPY AND KITTEN

Most breeders and foster homes wean puppies and kittens when they are between 7 and 9 weeks of age. For puppies, this age is well within the period of primary socialization period (5 to 12 weeks; see Chapter 2 for a complete discussion of sensitive periods of development). Conversely, kittens are entering their new homes near the end of their primary socialization period (2 to 9 weeks). During the time that puppies and kittens are with their mothers and littermates, they learn the species-specific behavior patterns that are needed for future communication with members of their own species. When puppies and kittens are adopted into new homes, they must also learn to communicate with human caretakers and must be taught behaviors that will facilitate integration into their human family.

The First Day

Owners should choose a relatively quiet period to bring a new puppy or kitten home. Although it may be tempting to adopt during holiday seasons, for many families the holidays are not a good time to obtain a new pet because their normal daily routine is disrupted, company is arriving, and family members may be traveling. During initial introductions, the puppy or kitten should not be allowed free access to the house, but should be confined to one room and allowed to explore that room. For puppies, it is often helpful to first introduce the puppy to the room in which his crate is located or where he will be spending a large amount of his time each day (see Sidebar 1 for a discussion of the appropriate use of crates). For kittens, it is important to introduce the

---

Sidebar 1 — CRATE TRAINING

Crating is an effective approach to safe confinement for newly adopted puppies (or adults) when they cannot be adequately supervised. A crate can also be used to teach a new kitten or adult cat to use a litter box. In addition, having a dog or cat who comfortably rides in a crate contributes to safe and enjoyable travel. Once an animal is reliably house-trained or litter box-trained and can be trusted not to damage household items when left alone, the crate should no longer be needed. Most owners use a crate as needed for confinement, with the ultimate goal being a pet who can be safely left alone loose in the home for several hours a day.

Several types of crates are available. A dog's crate should be large enough to allow the dog to stand up and turn around. For puppies, a large crate can be reduced to an appropriate size by blocking off part of the space. When needed with cats, the crate should be large enough to accommodate a litter box, a bed, and a water bowl, allowing for some space between these items.

A combination of classical and operant conditioning should be used to teach dogs and cats to accept being confined to a crate. The techniques used for "alone time" training can be used to pair the stimulus of the crate (neutral stimulus) with the stimulus of a highly desirable food treat (unconditioned stimulus). The initial introduction to the crate may involve simply tossing treats into the open door, allowing the pet to enter the crate (positively reinforcing entering behavior), and then come out. Both puppies and kittens can be classically conditioned to associate the crate with meal time by feeding them in the crate with the door open. When accustomed to being in the crate, the door can be closed. The time that the puppy or kitten is confined is gradually increased and is always paired with a food delivery toy or chew toy that is highly valued.

While crating is a highly effective way to keep a young dog safe when alone and to prevent house soiling and chewing damage, crates can be easily overused or abused. Studies show that isolation is stressful for puppies, especially those who are less than three months of age. Social isolation can be exacerbated by constraint to a small cage or crate. Although a crate environment can be enriched, it is still an impoverished environment, lacking in stimulation and restricting natural movement. A general rule of thumb is that a dog should never spend more than 4 to 6 hours of uninterrupted time in a crate. When extended isolation is unavoidable, owners must make other arrangements to ensure that the puppy's daily physical and emotional needs are met. Some solutions include having a noon-time dog walker or pet sitter come in, or using a doggy day-care service. Crating must also never be used as a form of punishment or as a place to confine the dog as an alternative to providing needed attention or exercise.

Frank D, Minero M, Cannas S, Palestrini C. Puppy behaviours when left home alone: A pilot study. *App Anim Behav Sci*, 104:61–70, 2007.

Hetts S, Clear JD, Calpin JP, Arnold CE, Mateo JM. Influence of housing conditions on beagle behavior. *App Anim Behav Sci*, 34:137–155, 1992.

---

litter box area and feeding area and to make certain that the kitten has easy access to these locations.

After the puppy or kitten has explored and settled in, family members can begin to interact with the newcomer by speaking to him quietly, sitting on the floor next to him and petting him, or enticing him into a game with a toy. It is often best to introduce each person singly to avoid overwhelming the new puppy or kitten with several excited family members at a time. Finally, owners should make sure that the new puppy or kitten is provided with a quiet place to rest often during the first day. Travel and introductions to a new family are exciting, but are also stressful for a new pet.

Establishing a Regular Daily Schedule

Because many owners have busy schedules and may be away from home during the work day, adjustments must be made to accommodate the new pet's needs. One of the best ways to do this is to set aside a portion of time every day that is spent exclusively with the puppy or kitten. This time can be spent a number of ways: exercising or playing, practicing obedience lessons (puppies), or just sitting quietly brushing and petting the puppy or kitten. This time may also be used to give the puppy and kitten exposure to new people and environments (see below). In addition, a regular daily routine should be established. Behavior problems that are associated with boredom, frustration, or separation stress are less likely to occur when a new companion animal learns to anticipate a regular schedule of attention, exercise, play, and feeding. Feeding should take place at approximately the same time and in the same area. Exercise, training and play periods should occur at around the same time, even on weekends when the family's schedule may change somewhat. For puppies, the same door should always be used for going outdoors to eliminate and kittens should be shown their litter box frequently during the first few weeks in the new home.

SOCIALIZATION

Both puppies and kittens enter the developmental period of primary socialization while they are still with their mothers and littermates. Therefore, it is important that breeders and foster homes handle puppies and kittens daily and introduce them to positive experiences and interactions early in life. Puppies who experience regular and varied opportunities for socialization while they are still young often maintain these positive relationships throughout life. Similarly, kittens who are handled for just several minutes a day during the first five weeks of life by a variety of people develop into more sociable adult cats. The opposite is also true – early studies of the canine primary socialization period found that puppies who are not properly socialized with people and other pets are more likely to be timid, inhibited and possibly defensively aggressive as adult dogs (see Chapter 2, pp. 33–34). Kittens who complete their primary socialization period (2 to 9 weeks of age) lacking human interaction have a greater risk of developing into timid adult cats.

Socialization procedures should be viewed as an enjoyable activity for both the pet and the owner, and include both informal training as well as community puppy and kitten classes (Sidebars 2 and 3). For puppies, regular walks and car trips to new places and to meet new and varied types of people allow habituation and also will establish a pleasurable routine for the dog's entire life. The proper use of classical conditioning techniques will support emotional responses to new people and places that are positive (friendly and outgoing)

PUPPY SOCIALIZATION CLASSES — Sidebar 2

Most communities offer several types of puppy kindergarten classes, either through private dog training schools, pet supply stores, or training clubs. A properly conducted socialization class provides puppies with non-threatening exposure to new dogs, people, and experiences. Because puppies are capable of learning at a rapid rate during the primary socialization period, this age represents a time during which obedience training should begin. Most puppy kindergarten classes teach household manners and include an introduction to basic obedience commands. For example, a study of the effects of a four-week puppy socialization class found that most puppies responded reliably to basic commands (sit, stay, come, and heel) at the end of the relatively short session of classes. In addition to introducing obedience commands, puppy classes teach puppies manners and encourage owners to thoroughly socialize their dogs through car rides, walks in the neighborhood and parks, and opportunities to greet friendly adults and children. All of these activities contribute to the development of a well-adjusted adult dog. Owners should be encouraged to continue with these habituation procedures after completing their class—during the juvenile period and, preferably, throughout the dog's life.

There is increasing evidence that puppy kindergarten classes are not only beneficial behaviorally, but that they enhance the probability that dogs will remain in their homes. A survey study of approximately 250 dog-owning homes reported that dogs who had attended a humane society puppy socialization class were more likely to still be in the adoptive home as adult dogs than were dogs who were adopted from the same agency but had not attended a puppy class. The study also found that dogs who were reported to have been handled frequently as puppies and were more responsive to their owner's commands had higher retention rates in homes. Although it is possible that more committed owners decide to attend puppy class in the first place, these results suggest that some of the skills and education that owners receive at puppy classes may have long-term benefits to the bond that owners develop with their dogs.

Seksel K, Mazurski EJ, Taylor A. Puppy socialization programs; short and long term behavioral effects. *App Anim Behav Sci*, 62:335–349, 1999.

Duxbury MM, Jackson JA, Line SW, Anderson RK. Evaluation of association between retention in the home and attendance at puppy socialization classes. *J Am Vet Med Assoc*, 223:61–66, 2003.

Sidebar 3 | NEW KITTEN CLASSES

Although not as common as puppy classes, kitten socialization classes can be both beneficial and enjoyable for cats and their owners. Kitten classes are typically offered by veterinary clinics and provide an opportunity for veterinarians and their staff to educate clients about their new kitten's basic healthcare, behavior, and training. Like puppy classes, kitten classes also provide a safe and structured setting for owners to socialize their kittens. The educational materials and information that are provided during the class aid in the prevention of future behavior problems such as furniture scratching, house soiling, and inappropriate play. Owners can also learn about methods for the proper and safe introduction of their kitten to other pets in the home.

Because the period of primary socialization occurs earlier in cats than in dogs, socialization classes usually target kittens who are between 7 and 14 weeks of age and include 3 to 4 weekly sessions. A properly conducted kitten class should provide kittens with non-threatening exposure to other kittens, to a variety of different types of toys and climbing structures, and with opportunities to meet and interact with a variety of people. The weekly car ride and visit to the veterinary clinic can be associated with enjoyable experiences, serving to classically condition the clinic setting and outings in the car as positive experiences rather than as threatening or unpleasant experiences to be avoided.

American Association of Feline Practitioners. *Feline Behavior Guidelines*, pp. 23–24, 2004.

---

as opposed to negative (timid, fearful, or defensively aggressive). Because classically acquired associations can be pleasant or unpleasant, owners must purposefully set up these learning situations so that they are enjoyable for the young puppy and kitten (Box 6.1 and Box 6.2).

Although the primary socialization period occurs early in life, continued socialization is also important throughout life to promote habituation and maintain social relationships. While 12 weeks in dogs is identified as the official "end" of primary socialization, the age of up to 4 to 6 months is generally accepted as a good period during which to socialize a new puppy. In fact, there is evidence suggesting that dogs benefit the most if socialization continues throughout the juvenile period. Similarly, primary socialization ends in kittens when they are between 7 and 9 weeks of age. Because kittens are close to the end of this stage when they enter new homes, kitten socialization classes that target kittens between the ages of 7 to 14 weeks are most beneficial.

Box 6.1   RILEY AND ALLIE OUT WALKING

Riley and Allie are littermate Border Collie mixes who were recently adopted by two families who live in the same neighborhood. Riley lives with the Smith family, while Allie lives several blocks away with the Jones family. Both puppies are very outgoing, energetic, and friendly. Their families take the puppies on daily walks and outings in the neighborhood for exercise and socialization. However, on one particular evening, although they were out walking just several streets apart, Riley and Allie had very different experiences:

Riley's Walk: Mr. Smith and Riley are walking along the sidewalk when a young boy approaches them from a nearby house. As Riley notices the boy, Mr. Smith immediately produces a yummy treat. He gives Riley the treat, and as the boy approaches, he asks the boy if he would also give Riley a treat. The boy is happy to help, and tells Mr. Smith that he loves dogs and would like to adopt a puppy someday. Mr. Smith hands a treat to the boy, who reaches down, speaks quietly and gently to Riley, and gives him the treat as he greets and pets Riley. Riley happily accepts the treat and the petting, and they continue their walk.

Allie's Walk: Mr. Jones and Allie are walking along another sidewalk when a young girl approaches them from a nearby house. As Allie notices the girl, Mr. Jones remembers that the last time Allie met someone new she jumped up to greet and put her paws on the person. Mr. Jones was embarrassed by her behavior and does not want to repeat that scenario. Anticipating Allie's jumping, Mr. Jones jerks on Allie's collar and very firmly tells her, "*No, Allie, No Jump!*" as the girl approaches. Allie cowers back and sits, staring nervously at the approaching girl. As the girl approaches without speaking, Mr. Jones repeats his reprimand to make certain that Allie remembers that she should not jump on people. Allie leans away from the girl as she passes, pinning her ears back and cowering slightly. After the girl has walked away, Mr. Jones and Allie continue on their walk.

What classical learning occurred in these two scenarios?
(See Box 6.2)

HOUSE-TRAINING PUPPIES

House-training takes advantage of a dog's innate tendency to urinate and defecate in an area that is located away from the daily living space. By five weeks of age, a litter of puppies usually has a preferred area away from their nesting site that is used for elimination. This natural tendency to urinate and defecate away from the living area can be enhanced by a breeder who takes the puppies outdoors to eliminate, starting at a young age. Once a puppy is weaned and enters his new home, the new owner's house-training strategy continues to capitalize on this innate predisposition and actively promotes the development of desirable location and substrate preferences.

A successful house-training strategy combines control of the puppy's environment to prevent elimination in the house and ample opportunity for urinating and defecating outdoors. Providing many and frequent trips outside early in the house-training process is essential for establishing a strong and positive association with elimination outdoors in a pre-selected location and on an appropriate

Box 6.2 CLASSICAL LEARNING: RILEY AND ALLIE OUT WALKING

Riley's Experience:

Appearance of Boy → Treats/Petting/Praise → Pleasant Emotions
(neutral stimulus) (unconditioned stimulus) (unconditioned response)

After Several Occurrences with New People

Appearance of Unfamiliar Person →→→ Pleasant Emotions
(conditioned stimulus) (conditioned response)

Allie's Experience:

Appearance of Girl → Collar Jerk/Reprimand → Unpleasant Emotions
(neutral stimulus) (unconditioned stimulus) (unconditioned response)

After Several Occurrences with New People

Appearance of Unfamiliar Person →→→ Unpleasant Emotions
(conditioned stimulus) (conditioned response)

---

surface (usually grass). Training should emphasize positive reinforcement of elimination outdoors and should not focus on punishment for mistakes made indoors. Sidebar 4 provides a set of guidelines for house-training young puppies and Sidebar 5 provides a useful daily schedule. The same approach can be used when house-training a newly adopted adult dog or when retraining a dog whose house-training has lapsed. (See Chapter 8 for a complete discussion of house-soiling problems.)

If a puppy is well-supervised and taken outside frequently to eliminate, accidents in the house should be rare. If elimination occurs in the house, the puppy should *not* be punished or reprimanded. The use of physical or harsh verbal punishment during house-training can cause fear and avoidance, and can result in a puppy who avoids being in the presence of the owner when eliminating. Punishing dogs for accidents causes them to associate the punishment with the owner and possibly with the location in the house, but does not necessarily affect future elimination habits. In fact, a history of punishment for in-home elimination is often the cause of house-soiling problems in which the dog eliminates out of sight of the owner, in another room of the home. (See Chapter 8 for a complete discussion.) If the owner observes the puppy eliminating in the house, the puppy should be calmly interrupted (*not* reprimanded) and carried outside to finish. The spot should be cleaned thoroughly with a biological cleaner designed for pet waste to reduce olfactory

STEPS FOR SUCCESSFUL HOUSE-TRAINING — Sidebar 4

A successfully house-trained dog eliminates reliably outdoors and does not eliminate indoors. This training can be accomplished through a set of simple management rules and a combination of both classical and operant conditioning:

- *Supervision, supervision, supervision:* Unsupervised puppies explore and, while they are exploring, they eliminate. Accidents can be prevented with this simple rule: The puppy is never alone and unsupervised (*never!*). A crate can be used to safely confine the puppy when he cannot be supervised. As the puppy matures and becomes more reliable, his freedom can be gradually increased.

- *Frequent outings:* One of the most common mistakes that new owners make with young dogs is to wait too long between trips outdoors. Puppy bladders are very small and young dogs have a limited ability to control their elimination habits. The Rule: When in doubt, Take him out! (see Sidebar 5)

- *Same door, same area:* The puppy should be taken to the same outdoor area where elimination is desirable and the owner should always accompany the puppy outdoors (*always!*). Many owners introduce a phrase such as "Do you want to go out?" immediately before taking their puppy outside to eliminate (classical learning).

- *Positively reinforce desired behavior:* Owners should accompany the puppy outside. Being present enables the owner to immediately reinforce elimination. This should be in the form of quiet praise or petting—exuberant praise should be avoided as it may interrupt or distract young dogs (operant learning).

- *Establish a regular feeding and exercise schedule:* Puppies should be fed 2 or 3 times a day, at approximately the same times, and walked and played with on a regular schedule. This promotes regular elimination habits that aid with house-training.

- *Avoid unrealistic expectations:* Puppies younger than 12 to 14 weeks have limited control over elimination and *must* be taken outside frequently. Unfortunately, some owners have been led to believe that an eight-week-old puppy can be fully house-trained. This is simply not true. More realistically, most puppies begin to become reliable when they are about 5 to 6 months of age. Once house-training is complete, young dogs and adults should still be provided with multiple opportunities per day to go outside for elimination.

---

attractiveness of the spot and the owner should intensify supervision of the puppy and house-training techniques.

Young dogs should not be allowed unsupervised freedom in the house until they are completely reliable in terms of house-training. Whenever the puppy or adolescent dog cannot be supervised, he should be confined to a crate or to a small "dog-proofed" area. As the puppy begins to demonstrate learning by eliminating promptly when taken outdoors, having no accidents

Sidebar 5: RECOMMENDED HOUSE-TRAINING SCHEDULE FOR NEW PUPPIES

Young puppies learn most rapidly if they have frequent opportunities to eliminate in the correct place (outdoors) and if they are prevented from making mistakes indoors. The best way to prevent accidents in the home is to take puppies outside regularly and often. For puppies between 7 and 12 weeks of age, this means every **20 to 30 minutes** during periods when the puppy is active and playing. This schedule can be gradually extended as puppy matures. Puppies should also go outside to eliminate at these times:

- In the morning immediately after waking
- After napping during the day—no matter how short the nap was!
- Following each meal and after long drinks of water
- After playing or training sessions (and periodically during extended play sessions)
- Immediately before going to bed in the evening
- Puppies less than 12 to 14 weeks of age: Once or twice during the night (small bladders are not yet able to last the entire night)

Final Note: Young puppies should not be *expected* to indicate their needs to go outside to owners. This is true for both puppies and for newly adopted adults who have never been house-trained. The time span between when the puppy feels the need to eliminate and when she must eliminate may be *very* short! Adults or adolescents who have not been adequately house-trained also need to *gradually* learn to wait to eliminate.

---

indoors, and occasionally approaching the door to indicate a need to go outside, the owner can gradually increase the amount of freedom that the puppy is allowed in the house. As puppies mature, owners can begin to rely more on the dog's signals and his or her elimination schedule to dictate the dog's elimination schedule. In general, all dogs, regardless of age, should be provided with an opportunity to go outdoors to eliminate every few hours during the day.

LITTER BOX TRAINING KITTENS

Similar to puppies, young kittens begin to voluntarily leave their sleeping and feeding area to eliminate at a very young age. Most cats develop strong surface preferences for elimination and prefer to eliminate in loose dirt. Kittens and adult cats of both sexes use a squatting posture for elimination. (The standing posture used for spraying is not typically used for elimination but is considered to be a marking behavior.) Prior to eliminating, a cat will dig and paw at loose dirt (or litter), sometimes digging a small hole. The cat then positions himself over the area in a squatting posture with the hind limbs directed outward. After eliminating, the cat covers the area with loose litter or dirt.

Because cats naturally eliminate in dirt and are attracted back to the same location and surface for elimination, teaching kittens and cats to use a litter box is accomplished by providing a suitable box and ensuring that the cat has access and repeated opportunities to use it. Because kittens are not yet capable of maneuvering throughout all of the rooms of a home (especially in multi-level houses), a new kitten should be confined to a small room or large crate that contains a resting place and a clean litter box whenever he is not supervised. Leaving a small amount of urine or feces in the box for several days when first introducing the box can help the kitten to identify the area for elimination. Most kittens begin to use the box reliably within several days. If initial training takes place in a crate, the box can be gradually moved to its permanent location once the kitten is regularly using it.

The location of the litter box in the home is an important consideration. Owners should select a location within the cat's core living area that is quiet and can be easily accessed by the cat at any time of the day. Although it may be convenient to the owner, the litter box should not be placed close to the cat's food and water bowls or near preferred resting areas. Some cats will not use the box if it is located in close proximity to the location where they eat or rest. Owners also should never force the cat to stay in the box; nor should they rub the kitten's paws in the litter as an attempt to encourage use, because this can cause fear and subsequent avoidance of the box.

In addition to location, important attributes of a litter box include its size, construction and the type of filler that is used (Sidebar 6). Keeping the box clean is of utmost importance because a dirty litter box is one of the most common causes of elimination problems in cats (see Chapter 8 for a complete discussion). The box should be scooped daily, and cleaned and disinfected on a regular basis to remove odors. If the household includes more than one cat, every box must be scooped and cleaned frequently because some cats refuse to use a box that contains the waste of another cat. Clay litter should be completely changed weekly while clumping litter should be replaced every few weeks. When a new kitten or cat is introduced to a home that already includes one or more cats, an additional box should be added. Boxes should be in multiple locations to accommodate the social dynamics of the cats. A basic rule of thumb is that the number of litter boxes in a multiple-cat home should be equal to the number of cats plus one.

Sidebar 6 | **LITTER BOXES AND LITTERS**

An appropriate cat litter box is sized so that an adult cat can comfortably fit inside, dig, and turn around. The box should be deep enough to contain at least 3 to 6 inches of litter and should be easy to clean, provide good ventilation, and constructed of a durable material, such as heavy plastic. The traditional box is an open plastic pan. These are easy to clean and most cats quickly learn to use this type of pan. Covered boxes are a good choice for homes that include a dog, because many dogs will consume cat feces if given the opportunity. These boxes include a pan and hood with an entry opening in one side, which also limits odor and prevents spillage. However, some cats may feel trapped or insecure in covered boxes and may be hesitant to use them. In recent years, automated self-cleaning boxes have been marketed. These include a timed sensor that activates a cleaning rake several minutes after the cat exits the box. Some cats seem to prefer this type of box, but others can be frightened by the mechanical noise of the rake. (Note: Although these boxes are helpful in keeping the box clean, they should not be expected to replace normal litter box hygiene.)

Suitable litter box fillers are easy to use and clean, have minimal dust or "tracking" properties, and are economical. The most common types of litter box fillers are clay litters, clumping litters, and silica litters. Many cats prefer unscented clay litters that have a relatively small granule size. These litters absorb urine well, have an agreeable texture, and control odor when good box hygiene is practiced. Although scented clay litters are available, these litters tend to mask but not eliminate litter box odor, and some cats will not use perfumed litters. Clumping litters are very popular today and are slightly more expensive than clay litter fillers. This type of filler reacts to liquid by forming "urine clumps" that are easily scooped and removed. Because clumping litters are fine in texture, they can be dusty and may adhere to cats' paws, resulting in tracking through the house. Textures vary however, and low-dust varieties are available. Silica pellet litters are a relatively new product, comprised of small and absorbent silica beads that change color when they are saturated. This provides a signal that it is time to replace the filler with fresh litter. Silica pellets control odor well but some cats do not tolerate the unusual texture of the silica beads. These products come in several different textures to try, so owners can experiment to find out what is most acceptable to their cat.

TEACHING PUPPIES TO ACCEPT ISOLATION

Isolation can be stressful for dogs, especially if time alone is not introduced gradually and with the proper use of classical conditioning. This is especially an important consideration with puppies, who have spent all of their life with their litter mates and mother (Sidebar 7). Once a puppy had arrived in her new home, she must gradually learn to be comfortable with being alone. Provided that regular and adequate time and attention is given to the puppy at other times of the day, puppies can learn that quiet time alone can be enjoyed chewing on a special "alone-time-toy" and resting. Conversely, if isolation is introduced without any attempt to classically condition a positive reaction, the puppy may learn the opposite – that isolation is associated with feelings of discomfort and loneliness. Following the simple rules of classical conditioning can teach most puppies to be comfortable and relaxed during times when their family is not with them.

DAP AND PRESENCE OF ANOTHER DOG HELP REDUCE NEW PUPPY ANXIETY — Sidebar 7

Dog Appeasing Hormone (DAP) is a synthetic form of maternal pheromones found in the mammary secretions of lactating dogs. These pheromones are believed to promote reassurance and attachment of puppies to their mothers early in life. As a commercially available product, DAP is delivered to a dog's environment via a plug-in air diffuser (similar to room fresheners). Studies of DAP as an intervention for the reduction of anxiety or stress have found DAP to effectively reduce noise-related anxieties and fears, separation-related stress, and stress associated with shelter environments.

One of the most recent studies hypothesized that the dispersion of DAP in a young puppy's environment may reduce the anxiety that newly adopted puppies experience as they adapt to their new homes. A group of 60 newly adopted puppies, ranging in age from 6 to 10 weeks, were housed in a room with their crate and a diffuser that emitted either DAP or a placebo. For a study period of 56 days, puppies were monitored for the number of times they cried and whined and how often they eliminated in their crates overnight. The researchers found that puppies who tended to be very stressed during the first few days of adoption were positively affected by exposure to DAP (i.e. were less stressed) when compared with similarly stressed puppies who were exposed to the placebo. Interestingly, sporting-breed puppies were most positively affected (and tended to be more likely than other breeds to cry and show anxiety during the night). Interestingly, the study also reported that an even more influential calming factor for new puppies was the presence of another dog. Puppies who slept with another dog rarely cried or disturbed their owners. The researchers hypothesized that this effect was due to both the physical presence of another dog and possibly to the production of appeasing hormone by the adult dog.

Therefore, when a puppy goes to a home with another dog, allowing the puppy to sleep with or near the dog is advisable, provided it is safe for the puppy and does not create anxiety in the adult dog. For puppies entering a home without other dogs, DAP may be an intervention that can help to reduce stress and anxiety as the puppy adapts to his new environment.

Taylor K, Mills DS. A placebo-controlled study to investigate the effect of Dog Appeasing Pheromone and other environmental and management factors on the reports of disturbance and house soiling during the night in recently adopted puppies (*Canis familiaris*). *Appl Anim Behav Sci*, 105:358–368, 2007.

It must be emphasized that puppies cannot be expected to quietly accept alone time if they are not receiving regular and daily periods of attention, exercise, socialization, and play. Once a regular schedule has been established, owners can carefully and deliberately begin to introduce scheduled periods of separation from the puppy. These training periods first occur when the owner is at home, and the puppy is crated in a separate room. The puppy should be first introduced to "alone time" when she has had a recent session of play or exercise and is tired. Alone-time sessions are paired with a specially selected toy or chew bone that is provided immediately before the period of isolation and at no other time. A durable rubber toy or bone stuffed with soft treats or a chewable food-delivery toy all work well for this training. Pet supply stores carry a variety of stuffable toys that are designed for this use.

As always with classical conditioning, the sequence of events is very important. The owner approaches the crate (or resting area) with the puppy and produces the highly desirable stuffed toy. In this sequence, the desired outcome is that after several repetitions, the crate is associated with the special treat and predicts an opportunity to chew on the high-value toy. Therefore, the puppy must always first be carried or walked to the crate and then immediately presented with the special toy. At this time, the owner can introduce a verbal cue such as "kennel" or "go to your bed." The puppy is then placed in her crate (or an exercise pen or puppy-safe room) with the toy. The owner leaves the room. If a highly desirable toy is provided and the puppy has received adequate attention and exercise, most puppies immediately become engaged with the toy. The owner returns within 5 to 10 minutes, removes the special toy, and

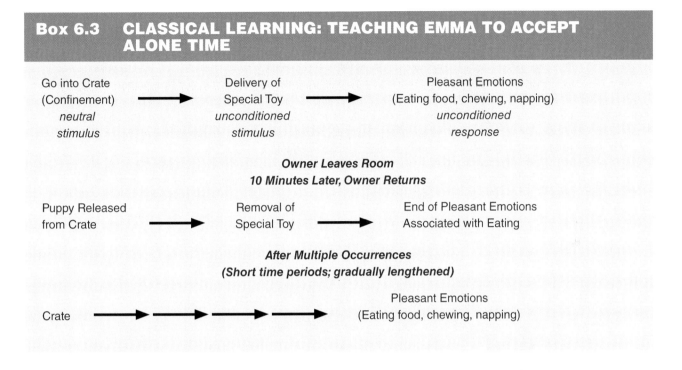

Box 6.3   CLASSICAL LEARNING: TEACHING EMMA TO ACCEPT ALONE TIME

releases the puppy from her crate or confinement area. The "alone training" is complete for that session (Box 6.3).

The time that the puppy is left alone is slowly increased, always pairing the crate or other area of confinement with the presentation of the special toy and pairing the owner returning with the removal of the special toy. While the puppy can and should have access to other toys, the "alone-time" toy must be provided only in association with periods of isolation. This serves to maintain the toy's high value to the puppy and its strength as an unconditioned stimulus. Although this training may not prevent the development of separation stress or boredom-related chewing in all dogs, it is an effective approach to reducing or completely eliminating isolated-related stress at an early age and teaching puppies that alone time can be pleasant rather than distressful. (See Chapter 10 for further discussions of separation stress in dogs.)

PREVENTING NIPPING IN PUPPIES

Gentle mouthing and nipping is normal play behavior for dogs. Puppies demonstrate these behaviors when they play with other puppies and when playing with human companions. Dogs also naturally investigate new articles with their mouths. This is normal exploratory behavior and is also a component of "teething" when the puppy's permanent teeth begin to erupt after four months of age.

Box 6.4 DODGER LEARNS TO PLAY GENTLY

Dodger is a five-week-old Brittany. He has two brothers (Max and Flyer) and one sister (Kelty). The puppies have started to play together more frequently during the past week. Dodger is a very active and bold puppy, and has been playing increasingly roughly with his brothers and sister. Luckily, puppies and mother dogs are well-equipped for teaching each other to inhibit the intensity of their bites and to play gently. (Interestingly, puppies use the same effective techniques as human trainers – negative punishment and redirection, with the occasional use of positive punishment.) Here is how Dodger learned to play well with other puppies:

Dodger is playing with Max. They wrestle and chase, yipping and play growling. After several minutes, Kelty and then Flyer join in. With all four playing, Dodger becomes increasingly excited and stimulated. He starts to play more roughly, and suddenly bites down (hard) on Flyer's ear. Flyer yips loudly in pain and runs away from Dodger. Play abruptly ends between Dodger and Flyer. Dodger continues to play with Max, gently at first, but again the play becomes increasingly rough as Dodger becomes more excited. While chasing Max, Dodger grabs Max's rear leg and bites too hard. Like Flyer, Max yips, but then turns back toward Dodger, snarling and bites Dodger's ear. Dodger yips loudly and stops playing.

All is quiet for several minutes. Then, tentatively, Dodger initiates play with Max and with Flyer, playing gently and inhibiting his bite. No biting incidents occur during the new round of play. Dodger has learned to inhibit his bite and to play gently with his littermates.

Puppies first learn to inhibit the intensity of their bite and to play gently from their littermates and mother (Box 6.4).

Once they are adopted into new homes, puppies may need to "relearn" these lessons as they begin to play with their new family and other pets in the home. Owners must teach puppies to inhibit the intensity of their nipping during play and to completely stop nipping and mouthing when asked to do so. Contrary to popular belief, puppies will not learn this on their own, nor will they "outgrow" nipping. A puppy who is not taught to inhibit his play biting while young can develop into an adult dog who uses his mouth much too roughly while he is playing. In some cases, this can even lead to using his mouth aggressively. Conversely, teaching puppies to play gently without biting leads to adults who play with humans and with other dogs gently and safely.

The two most effective training techniques for teaching gentle play are *redirection* and *negative punishment*. All play sessions with puppies should include a pre-selected chew toy or interactive toy that is durable and safe (see Chapter 9, pp. 216–217 for a complete discussion of toys). The toy is brought out at the start of every session. During play, all nipping and mouthing is redirected to the toy and away from hands or clothing. Gentle tug-of-war games can be introduced and used as an opportunity to teach the puppy "give" and "take it" commands (see Chapter 9, pp. 218–219). After several play experiences with the owner and the toy, the puppy will begin to associate the "special toy" with play and focus his nipping and tugging to the toy, not to hands or clothing. Negative punishment can complement redirection for puppies who persist in nipping at hands. In this case, nipping is negatively punished when the owner abruptly stops play whenever the puppy begins to play too roughly or use his mouth inappropriately. The owner reacts quickly, saying "Ouch – that hurts!!!," and immediately stops play. Play only begins again when the puppy is calm. Another command such as "Easy" or "Gentle" can be introduced in association with gentle and appropriate play.

As puppies mature, they can be taught alternate commands to use for redirection and to calm puppies who become too excited during play. The most appropriate are "sit," "down," and to come when called. Most puppy socialization classes include exercises that teach puppies to relax in a sit or down position. These exercises should be interjected sporadically into all play sessions, as they provide a positive way to moderate the puppy's energy and arousal levels, and provide an opportunity to positively reinforce relaxed behaviors. Coming when called is a very helpful command to use to call puppies away from play with other puppies, or to call the puppy to different family members. Although these exercises are helpful in redirecting nipping behavior, the trainer should be cautioned to avoid using a "sit" or "down-stay" only in response to nipping to avoid unwanted negative associations with these commands.

TEACHING KITTENS TO PLAY GENTLY

Similar to puppies, kittens first learn to play gently through interactions with their littermates and mother. Social play in cats includes behavior patterns that are inhibited or modified versions of feline predatory behavior, or behavior patterns typically observed during aggressive encounters. During play, kittens' bites are inhibited and they display facial expressions that appear to be important signals that the encounter is playful rather than agonistic. When litters are weaned and kittens are placed in new homes, social play will then be directed toward the new owner or other companion animals in the home. Studies of the development of play in cats have shown that social play peaks between 9 and 14 weeks of age, but can be extended by continued opportunities for play. Because social play in cats is more likely to escalate into aggression after 12 weeks of age, it is very important for owners to teach newly adopted kittens to accept handling (and grooming) and to play gently. Classical conditioning can be used to teach new kittens or recently adopted adult cats to accept gentle handling and grooming (Box 6.5).

The best approach to teaching a kitten to play gently is never to allow playful behaviors to be directed towards hands or feet. Because cats are highly

> ### Box 6.5 FLUFFY LEARNS TO ENJOY HANDLING AND PETTING
>
> Joe and Sally Smith recently adopted Fluffy, a nine-week-old kitten. Fluffy has a long, cream-colored coat that will matt and tangle easily if not brushed frequently. The Smiths know that they will need to groom Fluffy often and also would like Fluffy to enjoy brushing and handling. They use a classical-conditioning training program to ensure that Fluffy anticipates brushing positively rather than negatively. Here are the steps that they use:
>
> The Smiths select several primary positive reinforcers that Fluffy enjoys. These include Kitty Kat Liver Treats and small pieces of tuna.
>
> At a time when Fluffy is relaxing quietly in Sally's lap, Sally picks up the cat brush and immediately gives Fluffy a treat. She repeats this sequence 5 to 10 times on several occasions, petting Fluffy gently during each session, but not touching Fluffy with the brush.
>
> Once the association of "brush predicts treat and petting" has been established, Jane begins to brush Fluffy on areas of her body where she enjoys being touched and petted. Because Fluffy enjoys having her back stroked, Jane begins there. After each brush stroke, Fluffy is given a treat and quiet praise.
>
> After several strokes, Jane puts the brush and the treats to the side and gently removes Fluffy from her lap. She does not wait until Fluffy wants the handling session to end but always stops while the kitten is still enjoying the brushing and petting.
>
> As Fluffy began to anticipate and enjoy brushing, Jane gradually grooms other areas of her body. When Jane finds mats, she waits to remove these in a later session (and in a different part of the house). Jane does this to separate pleasant from unpleasant grooming procedures.
>
> *Results:* Fluffy enjoys brushing and has her "special" spot for petting and brushing that she shares with Jane several times each week. Because Jane discovered that Fluffy is very sensitive about having her ears handled, she has designated another area of the house and different time of day for ear care, thus preserving the positive associations that she has established in Fluffy with brushing.

stimulated by movement, kittens and adult cats readily will chase a moving hand or scurrying feet. Although some owners find it tempting to allow their kitten to chase and pounce at hands and feet, this teaches the kitten that hands are "toys to be pounced upon and bitten" and is counterproductive to teaching gentle play. From the cat's point of view, this is a form of object play, with the person's hands or feet being the object to hunt, stalk, and pounce (see Chapter 2, p. 33 for a complete discussion of object play). As with puppies, a kitten's play behavior can be directed towards an appropriate toy and away from hands or feet. Owners can select a variety of stimulating toys that the kitten can bat around, as well as interactive toys that are manipulated by the owner during play sessions. A common and highly popular interactive toy is a "fishing pole," in which a small toy is attached to the end of a pole with a long string or elastic line. Some cats also enjoy soft plush toys that are large enough to "wrestle" with and will flip over onto their backs to bite and rake the toy with their claws. Cats are stimulated by novelty and will lose interest in toys after repeated exposure. Therefore, an approach to maintaining interest in toys is to offer a variety and rotate them every few days. Owners can also store one or two "special" toys out of the cat's reach, bringing them out only during interactive play sessions.

Redirection can be used when a kitten attempts to ambush the owner's feet and legs as he or she walks past by tossing a desirable toy for the kitten to chase. Similar to puppy training, negative punishment can be used when a kitten persists in playing too roughly or attempts to stalk and pounce at people in the home. The owner should stop play immediately, remaining stationary to remove all reinforcement. Play only resumes once the kitten is calm. Although quickly squirting a cat with water from a squirt bottle is occasionally recommended for rough play, this approach is *not* advised. Cats quickly learn that the owner is the source of the spray and can become frightened or distrustful of the owner and other family members. Causing a cat to avoid interactions altogether out of fear or nervousness should not be confused with having taught the cat not to play roughly. (For a complete discussion of play-induced aggression in cats, see Chapter 11.)

TEACHING PUPPIES TO CHEW APPROPRIATE ITEMS

A primary underlying cause for chewing in young puppies is normal exploratory behavior. Puppies learn about their environment by smelling, touching, manipulating, and sometimes chewing on novel items. Chewing behavior in puppies is not classified as problem behavior, but rather a normal behavior that must be directed to appropriate items. (There are a variety of reasons that problem chewing behaviors develop in adolescent and adult dogs. These behaviors and methods for preventing and solutions are disused in detail in Chapter 9.) Management of the puppy's home environment is the most effective approach for preventing undesirable chewing and for teaching puppies which items are their own and are safe for chewing. An important rule is to keep all items that may attract the puppy secured and out of sight. The puppy must be supervised during all free time in the home, both to prevent inappropriate chewing and to monitor house-training.

Owners must provide a variety of appropriate chew toys for the puppy and have these available whenever the puppy is awake and active. Appropriate chew toys are hard enough to not be destroyed or torn apart quickly and are composed of a material that is known to be safe when swallowed. Because both dogs and cats enjoy novelty in their toys, owners should select a variety of types of chew toys and rotate them frequently. If the puppy does pick up something that is not his, the owner should simply remove it from his mouth and redirect him to one of his own toys, praising him quietly when he chews on appropriate items. Similar to house-training, using a harsh reprimand when the puppy is chewing on something undesirable can teach the puppy to chew out of sight of the owner or to run away whenever in possession of a novel item. Owners often mistakenly interpret this behavior as signifying that the puppy "knows he is wrong." However, avoidance behaviors such as these simply signify that the dog has learned though experience with punishment to avoid chewing in the presence of the owner or, in more extreme cases, to run away from the owner when in possession of a novel object.

Finally, in addition to managing the puppy's environment and providing plenty of varied and interesting chew toys, owners can teach their puppy the two invaluable commands of "leave it" and "give." The command "leave it" is used when the dog is intending to pick up something that is either dangerous or off-limits, and the command "give" can be used to retrieve an item from the dog once he has picked it up. These new exercise can be taught at a very young age. They are enjoyable for the puppy to learn and a great help to the owner when teaching their puppy about which toys are his and which items are not. (See Chapter 9, pp. 218–219 for a complete directions for teaching these commands.)

PREVENTING OBJECTIONABLE CLIMBING AND CLAWING IN KITTENS

Jumping and climbing onto elevated surfaces are normal and expected feline behaviors. Cats use elevated areas as vantage spots to view their surroundings and also climb to avoid unpleasant stimuli (such as a chasing dog). In kittens, normal exploratory behaviors include jumping up onto counters and climbing surfaces that provide a graspable surface such as draperies and furniture. Many cats also prefer to rest in an elevated space. Providing elevated resting and hiding spots for cats also helps to keep cats safe from harm in multiple-pet homes.

Although jumping and climbing are normal feline behaviors, most owners wish to prevent the kitten from jumping onto kitchen counters or climbing the living room curtains. First, adequate and desirable elevated resting and climbing areas in the home must be available to the cat. These should be introduced to the kitten on the first day and elevated at a level that the kitten can easily reach. Raised resting platforms in window sills and climbing structures that include carpet-covered climbing posts allow cats to choose the type of spot that they prefer. Owners can positively reinforce the use of these spaces by giving the kitten attention or a food tidbit whenever he uses a desired spot. In addition to providing several acceptable areas for climbing and jumping, owners can use environmental management techniques to prevent their kitten from undesirable jumping or climbing. Similar to puppies and chewing, a new kitten should not be allowed access to rooms that contain attractive furniture or draperies, except when supervised. Providing one acceptable elevated shelf or platform in each room that the cat frequents allows the owner to preemptively direct the kitten to that spot and away from counters and shelves that are off-limits. Finally, environmental punishers such as double-sided tape or motion-activated compressed air canisters can be placed on furniture or shelves that are off-limits to the cat. Cats do not like the feel of the sticky surface and most avoid the sound and feel of a blast of air. Because these aversive stimuli are mild, are delivered at the exact moment that the cat jumps onto the prohibited space, and are not associated with the owner, they can be effective at preventing cats from jumping or climbing onto objectionable surfaces.

Scratching behavior has several important functions for cats. These include providing an opportunity to stretch their limbs after resting, maintaining claw health, and as a form of communication (see Chapter 3 for a complete discussion of scratching behavior in cats). Because marking territory is an important social behavior and because all cats enjoy stretching, even those whose claws are clipped regularly or who are declawed demonstrate scratching behavior and will use a scratching post. If an attractive scratching post is not available, cats often choose to use the sides of couches and chairs as scratching surfaces. They use these areas because furniture is typically covered with fabric that is attractive for catching claws and depositing scent. Cats also tend to claw mark near favorite resting areas and around areas of entry and exit in the house. Regardless of these "natural" aspects of furniture clawing, most owners do not want their cats to use furniture or household items and prefer to train their cats to use scratching posts. Newly adopted kittens can be trained to direct all scratching towards a scratching post early in life, thus preventing the development of undesirable scratching behaviors.

There are many types, shapes, and sizes of scratching posts available. The most economical are vertical stand-alone posts, while other more elaborate posts are incorporated into climbing structures. Most cats prefer to use vertical posts, but a small number are attracted to scratching surfaces that lie flat on the ground and allow scratching on horizontal surfaces (Figures 6.1 and 6.2). Posts also come with a variety of coverings such as carpet, rope fabric (called "sisal"), burlap, or wood. Vertical posts should be sturdy and high enough for an adult cat to stretch the limbs when reaching upward (approximately 24 to 28 inches for most cats). Most cats are attracted to rough surfaces that they can shred and pull apart, and which catch their nails for claw care.

A favorite toy can be placed on the top of the post to attract the kitten. This often encourages young cats to play near the post and to reach up to bat at the toy with their front feet. Owners can also toss a small food treat to the kitten every time the kitten investigates the post. Once the kitten is regularly using the post, it can be moved to a spot that is agreeable to both the cat and the owner. However, owners must remember that the permanent spot for the scratching post should still be

figure 6.1 Cat using a vertical scratching post

figure 6.2 Cat showing horizontal scratching behavior

an area in which the cat spends time. Spots that are close to favorite sleeping and resting areas are attractive to most cats. Owners should avoid placing the post in spots that are out of the way, such as in the basement or laundry room; if the cat does not spend time in those areas, he will not use the post. (See Chapter 9, pp. 232–233 for a complete discussion of problem scratching.)

INTRODUCING A NEW PUPPY OR KITTEN TO OTHER PETS

Today, many homes include multiple animal companions, often of several species. In addition to meeting the new human family members, a new kitten or puppy may also be introduced to new animal companions. The approach that is used to introduce pets and the time that is allowed for acclimation depends a great deal upon the species of pet and upon each resident pet's temperament and tolerance for other animals. Because new puppies and kittens are not yet part of the resident pet's social group and are not (as some owners mistakenly believe) viewed as offspring, most adult dogs and cats do not immediately accept a new puppy or kitten. However, if proper introduction procedures are used, stress can be minimized and the new pet can be safely integrated into the existing social group.

Introducing a Puppy to a Resident Adult Dog

Before bringing the new puppy into the house, the resident dog's toys, food bowl, and bed should be picked up or secured in an area that is not accessible to the puppy. This will prevent any possession (resource-guarding) problems from occurring during initial introductions. The general rule with puppies is that the puppy is always controlled or confined when around the resident dog and is *never* allowed to jump on or harass the adult dog. If the adult dog's behavior is unpredictable, he should also be kept on a lead during initial introductions. Baby gates can be used to keep the puppy and the adult physically separated while still allowing safe and gradual introductions. All feeding should be conducted with separate bowls spaced far enough apart that the puppy cannot go to investigate the adult dog's food. Similarly, toys and sleeping areas of the adult dog should not be accessible to the puppy until he has been fully accepted by the adult dog and no possessive or resource-guarding behaviors are seen. The importance of gradual introductions and not allowing a young puppy to pester or harass the resident adult dog is of utmost importance and cannot be overemphasized.

Introducing a Puppy to a Resident Adult Cat

When introducing a puppy to a resident indoor cat, the cat must be allowed access to most of her territory and must be provided with escape routes that allow her to avoid the puppy. The puppy should be confined to one room while the cat is allowed to have the remainder of the house during the first few days or weeks. Cats often prowl around the doorway of the room in which the puppy is kept. Once the cat has shown signs of habituation to the puppy's presence in the house, she can be allowed to come into the room while the puppy is crated or held. Puppies should never be allowed to chase or torment a cat, even if the behavior appears to be in play. Most cats do not enjoy being chased and what appears to be play can rapidly turn into predatory behavior, putting the cat at risk for injury. Finally, the stress induced by a dog can lead to behavior problems in house cats such as defensive aggression or elimination problems. Owners should never insist that the cat remain in the same room as the puppy, and must provide safe areas in the home for the cat even after the puppy has been accepted by the cat.

Introducing a Kitten to a Resident Adult Cat

The best way to promote friendly relationships between cats is to introduce them gradually, while always ensuring that the resident cat feels comfortable and is in control of the level of interaction that occurs. Similar to puppy introductions, the new kitten should be confined to a room that contains his food and water bowls and a separate litter box. The resident cat is then allowed to have as much of his usual living space in the house as possible. The new kitten is kept confined to this room and out of sight of the resident cat until the resident cat becomes comfortable with the kitten's presence and smell (i.e. no hissing or posturing near the doorway). Visual exposure can be introduced once the resident cat is not stressed by the kitten's presence. The kitten should be held and the resident cat allowed to approach. The adult resident cat must be allowed as much time as needed to adjust and the cats should not be given physical access to one another until both are showing relaxed behaviors when in the same room together. Once the resident cat has completely acclimated to the sight of the new kitten, he can be allowed to enter the kitten's room. At first, the cats should be allowed physical access to one another for only very short periods of time. This time can be gradually increased as the cats habituate to one another. Although many cats may show an initial dislike for each other, with time some become playmates and bonded social partners. Others may not become companions, but learn to tolerate each other and live peacefully within the same household.

Introducing a Kitten to a Resident Adult Dog

Many kittens are naturally frightened by even the sight and smell of a dog, especially if they have not been previously socialized to dogs or if their mother showed a fear of dogs. Introductions are similar to those for a puppy, with the exception that some dogs view kittens and cats as prey and may show predatory reactions to a kitten. For initial introductions, it is helpful to confine the kitten to a crate or to place a screen or baby gate in the doorway so that the dog can see the new cat, but cannot yet come into physical contact. If the owner does not know how the dog will react to a cat, the dog should be kept on a lead and under control to prevent lunging or other unwanted behaviors. As they both become habituated, the kitten is released from confinement, while the dog is kept on a lead to prevent any chasing or unruly behavior. Each animal's behavior should be monitored carefully and increased time together can be allowed gradually. The dog should not be allowed to chase or play roughly with the cat and the kitten should never be forced to stay in the same room or area with the dog. (For dogs who insist on chasing cats, basic obedience training is often needed to prevent chase behaviors and to keep cats safe.)

REVIEW QUESTIONS

1. Describe a set of guidelines for successful house-training of young puppies.
2. What are important factors to consider when introducing a new kitten to an indoor litter box?
3. Discuss ways in which new owners can both prevent and stop playful nipping in puppies.
4. Describe a protocol for teaching a kitten to use a scratching post (and for preventing furniture clawing).
5. List guidelines for teaching puppies to chew on appropriate items (i.e. chew bones and toys).

REFERENCES AND FURTHER READING

Appleby DL, Bradshaw JWS, Casey RA. **Relationship between aggressive and avoidance behaviour by dogs and their experience in the first six months of life.** *Vet Record*, 150:434–438, 2002.

Beadle M. **The Cat: History, Biology, and Behavior.** Simon and Schuster, New York, NY, 1977.

Beaver BV. **Feline Behavior: A Guide for Veterinarians.** W. B. Saunders Company, Philadelphia, PA, 276 pp., 1992.

Bekoff M. **Play signals as punctuation. The structure of social play in canids.** *Behaviour*, 132:419–429, 1995.

Borchelt PL. **Behavioral development of the puppy.** In: *Nutrition and Behavior in Dogs and Cats*, RS Anderson, editor, Pergamon Press, Oxford, UK, pp. 165–174, 1984.

Bradshaw JWS. **The Behaviour of the Domestic Cat.** CAB International, Oxford, UK, 219 pp., 1992.

Caro TM. **Predatory behavior and social play in kittens.** *Behaviour*, 76:1–24, 1981.

Duxbury MM, Jackson JA, Line SW, Anderson RK. **Evaluation of association between retention in the home and attendance at puppy socialization classes.** *J Amer Vet Med Assoc*, 223:61–66, 2003.

Frank D, Minero M, Cannas S, Palestrini C. **Puppy behaviours when left home alone: A pilot study.** *Appl Anim Behav Sci*, 104:61–70, 2006.

Godbout M, Palestrini C, Beauchamp G, Frank D. **Puppy behavior at the veterinary clinic: A pilot study.** *J Vet Behav*, 2:126–135, 2007.

Hetts S, Clear JD, Calpin JP, Arnold CE, Mateo JM. **Influence of housing conditions on beagle behavior.** *App Anim Behav Sci*, 34:137–155, 1992.

Kanning M. **Socialization of puppies and kittens.** *Vet Tech*, 24:236–237, 2003.

McCune S. **The impact of paternity and early socialization on the development of cats' behaviour to people and novel objects.** *Appl Anim Behav Sci*, 45:109–124, 1995.

Mead-Fergus K. **Educating new puppy owners.** *Vet Tech*, 26:488–496, 2007.

Miller J. **The domestic cat: Perspective on the nature and diversity of cats.** *J Amer Vet Med Assoc*, 208:498–501, 1996.

Reisner IR, Houpt KA, Erb HN, Quimby FW. **Friendliness to humans and defensive aggression in cats: The influence of handling and paternity.** *Physiol Behav*, 55:1119–1124, 1994.

Rogerson J. **Your Dog: Its Development, Behaviour, and Training.** Popular Dogs Publishing Company, London, UK, 174 pp., 1990.

Seksel K, Maxurski EJ, Taylor A. **Puppy socialization programs: Short and long term behavioral effects.** *Appl Anim Behav Sci*, 62:335–349, 1999.

Serpell JA, Jagoe JA. **Early experience and the development of behaviour.** In: *The Domestic Dog: Its Evolution, Behavior, and Interactions with People*, JA Serpell, editor, Cambridge University Press, Cambridge, UK, pp. 80–102, 1995.

Slabbert JM, Rasa OAE. **Observational learning of an acquired maternal behaviour pattern by working dog pups: An alternative training method?** *Appl Anim Behav Sci*, 53:309–316, 1997.

Slabbert JM, Rasa OA. **The effect of early separation from the mother on pups in bonding to humans and pup health.** *J South African Vet Assoc*, 64:4–8, 1993.

Taylor K, Mills DS. **A placebo-controlled study to investigate the effect of Dog Appeasing Pheromone and other environmental and management factors on the reports of disturbance and house soiling during the night in recently adopted puppies (*Canis familiaris*).** *Appl Anim Behav Sci*, 105:358–368, 2007.

Thorne CJ, Mars LS, Markwell PJ. **A behavioral study of the queen and her kittens.** *Anim Tech*, 44:11–17, 1993.

Wilsson E, Sundgren P. **Behaviour test for eight-week-old puppies: heritabilities of tested behavior traits and its correspondence to later behavior.** *Appl Anim Behav Sci*, 58:151–162, 1998.

CHAPTER 7

Teaching Dogs and Cats Desirable Behaviors and Good Manners

PREVENTING PROBLEMS AND BUILDING BONDS

Dogs and cats are immensely popular as animal companions in the United States. Ask almost any pet owner about their dog or cat and they will gladly provide a list of the many attributes and marvelous personality quirks of their animal companion (as often as not, pictures will also be produced). In fact, the relationships that people have with their pets have become important enough to attract serious study. Over the last 30 years, this research has identified numerous benefits of pet ownership. Dog owners enjoy increased recreation and interactions with other people as well as with their dogs, while both cat and dog owners benefit from the companionship, opportunities for caregiving, and improved psychological health that a companion animal brings to their lives. The benefits of this relationship to the animal companion have also been studied. A recent study found that while most dog owners provide at least the basic level of care to their dogs, owners who are highly attached to their dogs are more likely to provide their dogs with an enriched environment in terms of opportunities to interact with humans and other dogs, provision of training, and offering their dogs a variety of toys and play opportunities.[1]

Unfortunately, there is also a negative side to keeping dogs and cats as pets. While strong attachments to pets often improve the welfare of the animal, the opposite is also true; people who are not strongly attached to their animals are more likely to neglect or abandon their pets. A large survey study examined owners who were relinquishing their pets to shelters in several different geographical areas of the United States.[2,3] Among other findings, the researchers reported that owning a pet for a short period of time and having unrealistic expectations about the pet's behavior put dogs and cats at significant risk of being abandoned. This information is important given the unfortunate paradox that, despite our professed love of dogs and cats, approximately two million dogs and cats are voluntarily relinquished by their owners to shelters each year. Behavior problems are the most common reason that dogs are relinquished; for cats, behavior problems are the second most common reason for relinquishment. Dogs who house soil, are destructive in the home, are "overly" active, or are reported to be fearful are at highest risk for relinquishment to a shelter. In cats, behavior reasons for abandonment include house soiling, damage to household items, and being "too active." Finally, these survey studies found that owners who relinquish their pets often have very poor general knowledge about dog and cat behavior and did not seek professional help for their pet's behavior problems when they occurred.

Many of these reported problems could be prevented through an enhanced understanding of normal dog and cat behavior and early training intervention. This chapter provides practical information for the application of the behavior information and training concepts that were discussed in the first sections of this book. These two complementary approaches – understanding normal dog and cat behavior, and teaching animal companions to show acceptable behaviors in homes and communities – can contribute to the development of positive and enduring bonds between owners and their pets and to a decrease in pet neglect and abandonment.

BASIC MANNERS TRAINING FOR DOGS

Teaching a dog to sit, lie down, come when called, stay until released, and walk on a loose lead are behaviors that are typically included in basic-level dog manners obedience classes. Training dogs to respond to these commands not only makes them more manageable and enjoyable pets to be with, but helps to prevent many behavior problems from developing in the first place. For example, teaching a young dog to sit quietly for greeting prevents jumping up, and training a reliable sit-stay can be used to interrupt and redirect dogs who become overly stimulated when playing. Teaching "come when called" reduces running away problems and also is used to teach dogs to return items that were picked up to be chewed and when teaching dogs to retrieve. Having a dog who walks on a loose lead without pulling makes walking with the dog a pleasant experience. This not only encourages more frequent outings and increased opportunities for exercise and socialization, but also reduces the risk of boredom-induced behavior problems.

Teaching Sit

The "sit" command is the first behavior that most dog owners teach to their puppy or dog. Both the "sit" and the "down" command can be introduced using a "lure" to induce part or all of the targeted behavior. These commands lend themselves well to lure training rather than to free-shaping because the movement of the hand that holds the food lure and induces the behavior can be faded to a hand-signal cue when the dog becomes proficient. To teach "sit," the trainer stands directly next to or in front of the dog. A food treat is held in the left hand and the hand is placed directly in front of the dog's nose (allowing the dog to smell the treat). The trainer slowly moves her hand up and forward slightly (in a diagonal direction) and commands "sit" in a pleasant voice. This movement of the lure encourages the dog to lean forward and to look up, shifting weight forward and off of the hind legs. Because this posture naturally makes sitting the most comfortable option, most dogs will immediately sit. The treat is provided immediately as the dog sits (for a refresher on the importance of timing with operant conditioning, see Chapter 4, p. 87). As the dog eats the treat, the trainer continues to praise quietly ("good sit . . . that's a beautiful sit").

When the dog offers a sit, he is immediately released with a pre-selected word. The verbal cue "okay" is commonly used. When released, all praise, petting, and treats stop. The release word is spoken in a calm and quiet tone. The trainer should avoid the common mistake of using an enthusiastic release word or using exuberant praise upon the release. Although this seems appropriate to novice trainers, using praise as the behavior terminates (i.e. as the dog stands up from the sit) unintentionally reinforces the behavior of "getting up" rather than the desired behavior of "sitting." During early training sessions, the use of the lure should be repeated several times. When the dog begins to spontaneously offer the behavior, the trainer removes the treat from the hand (the lure) and asks for a sit in response to the verbal and hand signal cue alone. The primary reinforcer (food treat) is now delivered immediately as the dog sits (and is not presented as a lure). Likewise, the primary reinforcer is *not* given if the dog does not offer the behavior. When lure training is used to induce a behavior, it is not uncommon for the dog to wait for the presence of the lure to offer the behavior. The risk of this happening increases if the trainer fails to remove the lure early in the training sequence. When this occurs, the trainer simply waits and if the dog offers even an approximation of sit, the dog is immediately reinforced (timing is crucial, once again). When the dog begins to offer "sit" in response to the trainer's verbal cue, the initial stages of learning this behavior have been accomplished. To attain reliability in this response, the trainer continues to train to achieve generalization, stimulus control, and fluency (Sidebar 1).

Sidebar 1: TRAINING FOR RELIABILITY (SIT EXAMPLE)

Step 1: Fading Extraneous Cues: Once the behavior of "sit" has been introduced and the cue has been added, the trainer must fade unwanted cues to which the dog may be responding. If the goal is a dog who responds only to the verbal command of "sit," then the hand signal that was used to initiate the behavior must be faded. Similarly, any body movements that the dog may be responding to must also be identified and removed. If the trainer wishes to emphasize the verbal command "sit" and fade the hand signal, this is accomplished with classical conditioning. The verbal cue is given immediately *prior* to the hand signal, each time the behavior is requested. After several repetitions, the verbal command becomes classically conditioned to elicit the behavior without being paired with the hand signal, which is then faded.

Step 2: Attaining Stimulus (Cue) Control: When the dog offers the behavior (sit) only in response to the conditioned stimulus (the verbal command "sit"), the behavior is said to be under stimulus control. In addition to reliably offering a sit behavior in response to the "sit" command, the dog must not offer "down" for the sit command nor offer "sit" for the down command. Stimulus control is part of the normal training process that must be consciously taught through repetition and the careful use of differential reinforcement (see Chapter 5, pp. 124–126 for a complete discussion of stimulus control).

Step 3: Generalizing the Behavior: Generalization refers to training the dog to perform the behavior (sit) in a variety of locations and in the presence of environmental distractions. Because dogs are highly sensitive to the context in which they learn, generalization involves retraining the "sit" in each different environment and in the presence of multiple forms of distraction. Sit is first taught in locations that are familiar to the dog and which have few distracting stimuli. As the dog becomes more proficient, the sit is taught in gradually more distracting and unfamiliar environments.

Step 4: Training for reliability (fluency): Reliability is simply the result of thorough training; providing adequate numbers of repetitions and opportunities for success and training in the presence of distractions and in different environments. If a dog has learned the behavior, has achieved stimulus control, and has generalized the behavior, reliability is simply a matter of adequate repetitions. The most important aspect of reliability is "real-life" training; the dog reliably offers "sit" in circumstances that are important to the particular family and lifestyle. For some owners this may mean sitting reliably when greeting visitors. Others may concentrate on sit stays during the dinner hour, while a competition obedience trainer will be most concerned with the precision and accuracy of the sit during a heeling routine.

Teaching Down

Dogs are initially taught to lie down on command from a sit position. Therefore, the trainer should teach a reliable "sit" command and have introduced "sit-stay" before teaching the down command. Once a dog is reliably offering the down behavior from sit, the trainer can begin to shape "down" from the stand position. Similar to teaching the sit command, the down command can be introduced using a lure. Although free-shaping can be used, waiting for an exuberant dog to spontaneously offer a down position (or even part of a down) so that the behavior can be positively reinforced is not always the best use of the trainer's time and can also be very frustrating for the dog if success if not achieved at a reasonable rate. Training begins with the dog sitting at the trainer's side. The trainer holds several small food treats in one hand. The trainer may be either standing or kneeling next to the dog, with the non-treat hand placed gently on top of the dog's withers or holding the back of the dog's collar. The treat-holding hand is placed in front of the dog's nose, allowing the dog to sniff at the treats in the hand (the lure). The hand is then moved slowly downward, from the dog's nose to a point on the floor just slightly in front of the dog's front toes. (The verbal command "down" can be introduced with the motion of the hand). If the dog readily follows the food and lies down, the trainer immediately opens the hand and reinforces with one treat. The behavior can also be positively reinforced with calm verbal praise ("good girl, that's a good down") and with gentle petting along the dog's back (Figure 7.1).

figure 7.1 Teaching down

When teaching the down command, positive reinforcement is withheld if the dog moves his head toward the ground but does not lie down (i.e. the "vulture pose"), if the dog stands up, or if the dog moves away from the trainer. In other words, all forms of positive reinforcement (treats, petting, and praise) are withheld if the dog offers a behavior other than lying down in response to the "nose to toes" movement of the trainer's hand. Unwanted responses are extinguished when they do not earn any reinforcement. Because the down position is similar to a submissive body posture and perhaps because this position places the dog in a vulnerable body position, some dogs learn to sit very rapidly but resist learning the down position. In these cases, the trainer can use successive approximation (shaping), first reinforcing a simple foot coming forward or head lowered toward the ground. Holding several food treats in the hand at once allows rapid and continuous shaping as the trainer gradually shifts the criteria for reinforcement in the direction of a complete down position.

After several training sessions, most dogs begin to respond consistently to the movement of the hand from "nose to toes" and the voice command of "down." The primary reinforcer (food treats) are then moved from the trainer's hand and placed in either the petting hand or in a treat pouch. The trainer continues to use the treat hand signal for down, reinforcing immediately when the dog lies down with a food treat. (Note: Either the left or the right hand can be used as the hand signal for "down," but the same hand should be used consistently). It is important to remove the lure as early in training as possible, to avoid an association between "food in the hand" and the down position (i.e. the food becomes the cue for the dog to lie down). This unintentional learning can occur more rapidly with the down than with the sit because the hand remains close to the dog's nose as he executes the behavior. An effective approach to teaching dogs that "it is the behavior that leads to reinforcement, not the presence of the treat in the hand" involves switching the contingencies of the behavior (Sidebar 2). This technique can also be used to solve the "treat in the hand" problem when training other behaviors. If the trainer wishes to train the dog to respond primarily to a voice command, the visual cue (hand movement) must be gradually faded and the voice command alone is emphasized. Alternatively, the hand movement used for the lure can be faded to a shortened version of a brief movement of the hand downward if the trainer desires to use both a verbal and a hand signal command for the down. The same sequence for gaining proficiency that is used with the sit command is used for the down command (see Sidebar 1).

Teaching Stay

The "stay" command has many practical uses. In the home, a dog who stays quietly in a down position next to the dinner table has the benefits of being with the family during dinnertime without begging or being disruptive. The stay command is also used when teaching a dog to "go to her spot" when visitors arrive,

SOLVING "LURE" PROBLEMS – WHEN THE LURE BECOMES THE CUE | Sidebar 2

Although using a lure has many benefits in dog training, a common error is to fail to remove the lure early in the learning process. If the lure (for example, food in the hand that is serving as the hand signal for "down"), is not removed shortly after the behavior has been initiated, the dog learns to recognize the lure as one of the cues that elicit the behavior. Because food is a potent positive stimulus and because it is very easy to detect (smell), food in the hand rapidly becomes the dog's cue for offering the down behavior. When this problem occurs, the trainer can use a technique called "reversing the contingency" to teach the dog that food in the hand is not a relevant (discriminating) stimulus:

Goal: To teach the dog that the presence of food in the signaling hand is not part of the cue and is not necessary for the opportunity to earn reinforcement. Rather, it is moving into the down position (the targeted behavior) that earns the reinforcer, not perceiving the presence of a treat in the trainer's hand. The trainer will teach the dog that it is the empty hand, signaling the down, that is a reliable predictor for reinforcement and conversely, a food lure in the signal hand represents a poor predictor for food.

Step 1: Warm-up Repetitions: The trainer trains 4 to 6 repetitions of "down," holding food treats in the signal hand. If he has been sufficiently taught "down," the dog should reliably follow the hand signal with the lure into the down position. The dog is reinforced each time. There repetitions serve to build a reinforcement history for the training session.

Step 2: One more repetition (no food in hand): Food is removed from the signaling hand and placed either in the opposite hand or in a treat pouch. The dog is again signaled to lie down, but this time with an empty signal hand. (Because there was a rapid rate of positive reinforcement during Step 1, the majority of dogs will lie down for one repetition without food in the hand). The dog is *immediately* reinforced with several food treats from the opposite hand when she lies down.

Step 3: Changing the contingency: The trainer once again places several food treats in signal hand and requests the dog to lie down using the lure in the hand for three repetitions. However, during this set of repetitions the dog is *not* reinforced for lying down with the lure (verbal praise is given, but no food treat). After three repetitions, the treats are again removed from the signal hand, the dog is asked again to down, and is *immediately* reinforced with a food treat from the treat pouch or opposite hand upon lying down.

The new rule is this: Food in the signal hand *never* predicts an opportunity for positive reinforcement; no food in the signal hand *always* predicts an opportunity for positive reinforcement.

Step 4: Proficiency: After several repetitions, the trainer begins to phase out having treats in the signal hand altogether for the down signal, and just reinforces responses to the signal hand when it is not holding treats. The dog has now learned that "it is the behavior (lying down) that predicts positive consequences, not the presence of treats in the trainer's hand."

or to stay in position for brushing and petting. Moreover, the stay is an exercise that allows owners to repeatedly reinforce calm and relaxed behaviors in a young or exuberant dog. Because the command "stay" may mean different things to different people, the trainer must first be certain what it is that she wishes to teach. Some owners consider the "stay" exercise to mean "go to a designated spot, such as a bed, and stay there." For others, stay means "do not move out of the position that you are in until you are released." For the purposes of this section, teaching stay refers to the latter definition (stay in position until released). Teaching a dog to "go to his place" is actually a form of target training (Sidebar 3).

Sidebar 3 — INTRODUCTION TO TARGET TRAINING

Target training is helpful for a variety of behaviors that owners are interested in teaching to dogs and cats. The pet is taught to identify a specific target and to touch the target with either her nose or her paws. Two commonly used targets are the touch stick (the animal is trained to touch the end of the stick with the nose), and a carpet square or mouse pad (the subject is trained to touch the pad with her paw).

Nose Touch: Most dogs and cats are curious and will intuitively examine a novel object with their noses. A touch stick can be easily introduced by presenting the end of the stick to the dog or cat and immediately reinforcing when the subject sniffs the end of the stick. The exact moment that the dog or cat touches the stick is marked with either a conditioned word ("Yes!") or a click, followed in both cases by a treat. (For more about clicker training see Chapter 5, p. 115). Once the behavior is initiated, a verbal cue is added ("Touch!") and the trainer teaches stimulus control (touch the stick only when requested, not just when the stick is presented). When stimulus control is attained, the touch stick can be used as a lure to guide the dog to different locations and to teach different body positions. For example, the nose touch can be used to teach dogs a variety of tricks, such as spin in place, put your head down, and crawl. Obedience competition trainers use the nose touch to teach the go-out exercise, while agility trainers use it to teach direction when targeting obstacles. For cats, the nose touch can be useful for teaching a cat to "go to her bed."

Paw Touch: The paw touch can be more challenging to teach because not all dogs are naturally aware of how they use their feet. Dogs who use their feet to hold and manipulate toys or who paw when they want attention are usually more "foot aware" and more likely to quickly learn "hit it" using their feet. Similarly, cats who bat at their owners with their front paws for attention or affection can be taught paw touch. Flat targets such as a small carpet square or a computer mouse pad work well as paw touch pads. (Note: The target must differ from the target that is used for nose touch, as it is a discriminative stimulus signaling a different behavior). The target is placed on the floor near the dog and any interaction with the target, with the exception of touching with the nose, is reinforced (yes-treat or click-treat). After several repetitions, the target is moved several inches to help to focus the dog's attention on the object. Shaping is used to train "front paw on the pad" and the verbal cue ("Hit it!") is introduced once the dog is placing one or both front feet on the target. When the pad is reliably targeted, the trainer can shape duration, reinforcing an increasing number of seconds that the dog's foot remains on the target. The "hit it" targeting behavior is useful when teaching dogs to "go to your bed" (place the target on the bed), directed retrieving, and agility contacts (Figure 7.2). For cats, "hit it" can be used to teach shake hands and wave.

figure 7.2 Target training – Go to your bed

When teaching a dog to stay in position, two separate criteria must be shaped – the duration (length of time) that the dog stays in position and the space (distance) that separates the dog from the trainer. These two criteria must be clearly separated in the training protocol. Duration is trained first, and distance is introduced once the dog has reliably offered "stay in position" for more than 1 to 2 minutes. The combination of time and distance are introduced once each criterion has been thoroughly shaped. While some dogs are more comfortable in a sit position, others prefer a down. To facilitate early success (and allow for the continued use of positive reinforcement during shaping), the trainer should select the position that is most acceptable to the dog and train that stay position first.

Shaping duration is accomplished by positively reinforcing very short (1 to 2 seconds) periods of time during which the dog maintains the prescribed position (sit or down). The trainer reinforces with a continuous schedule using food treats (a primary positive reinforcer for the majority of dogs), praise, and petting. Because stay exercises are intended to encourage calm and relaxed behaviors, petting should be gentle and soothing (Figure 7.3). Just as dogs have their own set of tastes in terms of food treats, so too do they vary in the types of petting and praise that they enjoy. The trainer should select an area of the dog's body and a style of touch that the dog enjoys (i.e. finds to be positively reinforcing). Similarly, the tone of voice used to reinforce "stay" should be calming and quiet in tone.

When the release word is given ("okay"), all treats, petting, and praise stop. De-emphasizing the release draws the necessary distinction between "stay" (targeted behavior) and movement. As explained previously, although it is

figure 7.3 Gentle and calm praise for sitting and staying

quite common for dog owners to believe that they are "rewarding a job well done" when they praise their dogs at the release of the exercise, this in actuality reinforces the *release* from the stay, not the behavior of staying in position. If this mistake is made repeatedly, the stay becomes simply something to get through in order to earn a reward for the release. Concentrating on providing pleasing stimuli during the stay and removing pleasing stimuli upon the release targets the desired behavior and prevents reinforcement of the release. This is especially important because for most dogs, being released from a stay has several self-reinforcing properties (e.g., moving around is innately more enjoyable than sitting or lying in one place). The time that the dog remains in the down position (and is consistently reinforced) is increased gradually. The criterion (time) is only increased when the dog shows proficiency at each lower level. Movements out of position are ignored (extinguished). The dog is simply re-cued to sit or lie down if he moves out of position. The trainer stays close for duration training and the dog is kept on lead. The trainer always shifts the criterion (i.e. increases time that the dog remains in position) at a rate that allows the dog to be successful.

The second criteria, increasing distance, is introduced when the dog is consistently staying in position for 1 to 2 minutes. The trainer begins by simply taking one step away from the dog and immediately returning to reinforce the stay. The trainer stands close to the dog, reinforces for several seconds and then takes a step away again, in another direction, again immediately returning to reinforce the dog. If the dog stands or follows the trainer each time that the trainer steps away, this is information to the trainer that distance is being introduced prematurely. Retraining on "stay in position" is needed, working on duration alone. This "yo-yo" training, with the trainer moving a few steps away and immediately returning, is gradually shaped by increasing the distance (not time) that the trainer moves away from the dog. As the dog becomes reliable with increasing distances, the trainer begins to leave the room or moves to an area that is outside of the dog's line of vision. Similar to training duration, any movement out of position is ignored (reinforcements are immediately withheld) and the dog is repositioned. Repeated errors signify the need to drop down to a previous criterion and to retrain at that level.

Duration and distance are combined when the dog reliably stays in position for a predetermined duration of time (with the trainer close) and for a predetermined distance (with the trainer leaving and immediately returning). As with all training, combined criteria are shaped slowly, starting with incremental increases in both time and distance. When the dog is reliable for time and distance together, training then focuses on generalizing the stay to other settings. Similar to the initial training, changes in context (commonly referred to as "distractions") are introduced individually and shaped by starting at the lowest level of intensity possible. This minimizes the risk that the dog will make repeated errors and sets up a training session that allows for numerous opportunities for positive reinforcement. Examples of contextual criteria that can be introduced for generalization of the stay command include training in new locations, in the presence of familiar and unfamiliar people, near other dogs, and in indoor and outdoor environments. The owner should identify the various contexts in which it will be important for his dog to respond reliably to a stay command and can train specifically for those criteria (Box 7.1).

Box 7.1 TEACHING JAKE TO STAY WHEN GREETING VISITORS

Jake is an 11-month-old Australian Shepherd. As a puppy, Jake completed a puppy kindergarten class, where he was socialized with other puppies and people and was introduced to basic obedience exercises. Jake is very friendly and loves to meet visitors at home and while out walking. Jake's owners, Steve and Janet Black, have tried to continue with Jake's obedience training, but they have also allowed Jake to jump up when greeting visitors. The Blacks realize now that Jake needs to be controlled when greeting visitors because, although he is very friendly, he is a big dog. They decide to teach Jake to sit-stay for petting during greeting visitors to their home. Here are the steps that the Blacks use in their training program to teach Jake to maintain a sit-stay when greeting visitors in their home:

- **Select a strong primary reinforcer:** The Blacks choose small bits of cheese. During this training, cheese bits will only be available to Jake as a reinforcer for sit-stay and at no other time.
- **Choose the lowest possible intensity of the competing stimulus:** Janet has noticed that Jake becomes most excited with friends whom he knows well and when the visitor directly greets him. Therefore, she and Steve begin Jake's training with Sam, a friend who has not met Jake previously. Sam is instructed to ignore Jake during the initial training sessions.
- **Train "sit-stay" without the visitor, in the area where greeting will take place:** Several minutes before Sam arrives, Janet attaches a lead to Jake and trains several repetitions of sit-stay in the foyer. Even though Jake is well-trained to stay when no distractions are present, the purpose of this training is to build a recent reinforcement history for Jake. This is repeated every session.
- **Introduce the low-intensity stimulus:** When Sam arrives, Jake is commanded to sit-stay and is continually reinforced while he stays. Sam walks into the area and then immediately leaves. When he leaves, treats stop and Jake is released. This sequence is repeated 5 to 10 times, with Sam appearing (Jake sit-stay; high value treats) and then leaving (Jake released, high value treats disappear).
- **Increase intensity of a single attribute of the stimulus and train at each level:** The Blacks decide to repeat this training with different individuals, gradually bringing in people whom Jake knows well. Each person arrives and then leaves immediately.
- **Introduce another attribute:** When Jake is reliably staying when visitors come into the house but do not speak to him, Janet asks Sam to come over again. This time Sam briefly says hello to Jake. Jake is reinforced for maintaining a sit. When Jake is calm and sitting, Sam is given a cheese bit and reinforces Jake for sitting as he says hello. Sitting for staying during greeting is repeated with other visitors. If Jake breaks his stay, the visitor immediately walks away, leaving the area (negative punishment).
- **Practice, Practice, Practice!** The Blacks realize that because Jake is a friendly and outgoing dog, the competing reinforcers of social interaction will always be challenging for Jake to ignore. Therefore, sit-stay for greeting will be a behavior that the Blacks will practice with Jake frequently, and they will expect mistakes to be made now and then!

Teaching Wait

"Wait" is a context-specific behavior that is used to stop a dog's forward movement. The most common uses of this command include keeping dogs from rushing out of doors or gates and achieving control when getting a dog into or out of a car. The "wait" command is also useful with dogs who try to push their way ahead of their owners when walking up or down stairs, or even when moving from one room to another in the home. Some trainers also use the wait command for walking with a dog who is on a retractable lead or is off-lead in a park. Upon hearing "wait," the dog is trained to stop and wait for the owner to approach, not proceeding until released. The wait exercise differs from a stay exercise in that the dog's position is not important; the dog may remain in a stand, sit, or down position. The final teaching criterion is a dog who stops moving forward when asked to "wait."

Because dogs are sensitive to many environmental cues that are associated with a command, most do not confuse "wait" with "stay" if the wait command is taught and used in the specific situations (contexts) that are important for a particular dog and family. For this reason, the wait command is initially taught in the specific places that it will be used, but without the exciting stimuli that typically cause problems. For example, a dog who has learned to bolt out of the front door whenever guests arrive is trained to "wait" at the front door, but initially in the absence of any competing stimuli such as guests or family members. Similarly, a dog who has been rushing down the stairs and endangering family members is initially taught to wait at the top and bottom of the stairs without distractions present. The instructions that follow describe teaching the wait exercise at an exterior door. These can be modified to fit other contexts and situations.

The dog is on-lead and the trainer approaches the door with the dog at his left side. The lead is kept snug to prevent the dog from rushing ahead and the dog is reinforced for maintaining a position next to the trainer's left hip. The trainer stops near the door and then reaches forward to open the door, without moving to go through the door. The dog is commanded to "wait" and the lead is kept snug to prevent forward movement. The trainer immediately reinforces "no movement forward" with treats and calm verbal praise. The trainer reinforces for several seconds and then reaches forward and closes the door. This sequence is repeated this several times within each training session. The next criterion is introduced when the dog begins to stop when the door opens (classical conditioning again!) in anticipation of treats and praise from the trainer for remaining stationary at the trainer's side. The next step (criterion) is to ask the dog to wait for several seconds, followed by "let's go" or "okay," and the trainer and the dog proceed out of the door together. (This type of "wait" training is important for getting in and out of the car, because the owner wishes for the dog to wait until his lead is attached before jumping from the car.)

The final criterion that is introduced involves teaching the dog to wait at the door while the owner (and other family members and visitors) pass through the door without the dog. This is used either when exiting the home or yard without the dog, or when visitors arrive and the dog is expected not to attempt to bolt out of the door. This is the most difficult and challenging level of training, given that most dogs love to accompany their owners and because having visitors is a highly distracting and exciting time for many dogs. As previously stated, the dog is on lead and the lead is kept snug as the trainer approaches the door. After opening the door, the dog is commanded to "wait" as the trainer steps through the door to the outside. The lead is kept on the dog and the trainer immediately walks back to the dog and reinforces the wait. The trainer then leaves and shuts the door briefly, returning after a few seconds to reinforce. When this level of the wait command has become reliable, owners can begin to train "wait" with the distracting stimuli of visitors or family members coming and going.

Teaching Walk on a Loose Lead

One of the most enjoyable activities that dog owners share with their dogs is going for walks together. Walking and hiking with dogs provides exercise; visual, olfactory, and mental stimulation; socialization to new places and people; and the opportunity to habituate to new stimuli. An added benefit that has been reported in early studies is that people who are walking with their dogs experience more opportunities for positive interactions with other people than do people who are out walking alone.[4] Because walking a dog who is pulling or dragging his owner along is not enjoyable, teaching loose-lead walking is an important goal for most dog owners. The primary objectives of this training are to teach the dog to walk on a loose lead without pulling and to sit at the owner's side when the owner stops.

Training Challenges: Paradoxically, while loose-lead walking is of utmost importance to the bond between owners and their dogs, it can be one of the most challenging behaviors to train. There are several reasons for this difficulty. The first is that pulling on-lead often has a strong reinforcement history. Dogs pull because they have learned that pulling forward while going on a walk brings new smells, sights, and opportunities for interactions with other people. These environmental reinforcers are highly rewarding for most dogs. A dog who is walked regularly to a community park forges ahead (pulls) more persistently as she approaches the park entrance. This occurs because arriving at the park is a strong reinforcer. Because arriving immediately follows pulling, pulling behavior is positively reinforced every time that it is offered. Many dogs have experienced a long and highly satisfying reinforcement history for pulling by the time that their owners decide that they no longer enjoy walking with a dog who almost dislocates their shoulder. The second reason that loose-lead walking is a difficult exercise to teach is related to the type of behavior that it is. Essentially, walking on a loose lead is a *place* rather than a discrete behavior. Teaching a dog to walk at the owner's side within a "no-pull" zone requires that the dog learns to maintain a position that keeps the lead loose relative to wherever the trainer is located and whether the trainer is moving or not. Because this is a relatively complex concept, it is more difficult to train and to achieve proficiency. (To illustrate this point, owners can be asked to attempt "heeling" next to another person, constantly changing their rate of walking to match their trainer's pace and direction!)

Changing Reinforcement Contingencies: Because many dogs have been allowed to pull their owners towards other dogs (greeting a dog = positive reinforcement), towards visitors (greeting a person = positive reinforcement), and to investigate smells in the grass (more positive reinforcement), pulling behavior is well-established in many dogs and is associated with a long reinforcement history. It is impossible to eliminate these varied and highly potent environmental reinforcers that dogs receive during walks. However, careful training can change the contingencies for reinforcement and use these stimuli to positively reinforce *loose lead* rather than *tight lead*. Remember that the primary reinforcement that a dog receives for pulling on lead (i.e. maintaining a tight lead) is "getting places" and attaining all of the desirable stimuli that are found in those places. The trainer can modify this response-reinforcer relationship by setting up a rule that the behavior that will lead to reinforcement is "loose lead" while the behavior that results in the *loss* of an opportunity for reinforcement is "tight lead." Some trainers refer to this training technique as "Pushing the Go Button" because the dog is taught that the way to get the owner to walk forward is to offer loose lead (go button), while pulling and maintaining a tight lead causes the trainer to stop and refuse to move forward (stop button).

Training Criteria: Like any other new behavior, loose-lead walking is trained in small increments, using shaping. For many dogs, switching to a new type of training collar is helpful in facilitating loose-lead walking (Sidebar 4). Training begins in a quiet area that offers minimal opportunities for environmental reinforcement. The owner's living room works well. The first criterion is that the dog is positively reinforced for keeping a loose lead while the trainer is not moving. The trainer puts the lead on the dog (which, because it is often a conditioned stimulus, usually evokes great excitement in the dog), plants her feet, and waits. Any movement that the dog makes toward the trainer's left hip and to keep a loose lead is positively reinforced. The goal of the trainer is to build a strong reinforcement history for staying within a "no-pull" zone located next to the trainer's left hip and with no tension on the lead. Remaining stationary while teaching this minimizes external stimuli that would be attractive to the dog, allowing for a high rate of reinforcement and success. The trainer can take one or two steps in various directions during this part of the training, reinforcing "loose lead" when the

Sidebar 4 | **TYPES OF TRAINING COLLARS**

A training goal for most owners is to teach their dog to walk on lead without pulling. Because many dogs have already experienced a strong reinforcement history for pulling when this training begins, a training collar or harness is often used as an aid to facilitate learning. Commonly used training collars include

- *Buckle Collar:* Fixed-circumference buckle collars are best suited for dogs who do not pull vigorously or who have had some previous training. This type of collar is usually all that is needed for training small and toy breed dogs and is the collar of choice for dogs once they are trained to walk on lead without pulling.

- *Limited Slip Collars:* These collars lie flat on the dog's neck, like a buckle collar, but include a loop of added material that keeps the collar loose when not pulling and tightens if the dog pulls. When tight, the collar will not come back off over the dog's head, which is a benefit for dogs with narrow heads. A limited slip collar is most appropriate for dogs who do not pull strongly or who have had basic training.

- *Head Halter:* This collar is similar in appearance to a horse's halter and functions by physically controlling the dog's head and subsequent ability to pull forward. Head halters restrict head movement and exert pressure around the dog's muzzle when the dog pulls ahead. Most dogs perceive this as an aversive stimulus and stop pulling into the collar to relieve this discomfort (i.e. not pulling is negatively reinforced). Head halters are also highly effective in controlling dogs who are reactive to other dogs, as they allow the trainer to redirect the dog's head and attention. Although head halters are effective training collars for many dogs, some do not tolerate the feel of the nose piece and respond better to another type of training collar.

- *Pinch Collar:* The pinch collar is a metal collar comprised of small connected joints. When the dog pulls into the collar, the joints contract and pinch skin around the dog's neck. In the same manner in which not pulling is negatively reinforced by a head collar, this aversive stimulus (pinching) is relieved by moving back into a loose-lead position. This type of collar is recommended for dogs who do not tolerate halter collars or whose head type will not allow fitting a halter collar. It is *not* recommended for dogs who are reactive to other dogs, as the negative stimulus on the neck can inadvertently become associated with the presence of another dog.

- *No-pull Harnesses:* These harnesses are designed to discourage pulling by physically restricting the dog's ability to fully extend the front legs. Although forging ahead is reduced, the harness does not provide control of the dog's head. Some dogs who do not tolerate a head halter response well to no-pull harnesses. However, because this type of aid is relatively new, the long-term effects of restricting front leg movement are not known. For this reason, these harnesses should be used only during early stages of training and not as a long-term solution to pulling.

- *Metal Slip Collar:* These collars (also called "choke collars") exert an aversive stimulus on the dog's neck and trachea when the trainer "jerks" the collar in response to pulling. Because it is difficult to time the aversive correction accurately and because slip collars exert sudden and concentrated pressure on the dog's neck, trachea, and esophagus (not just on skin), they are less effective than other options and present a risk of injury. For these reasons, slip collars are not recommended as a training aid.

dog moves along to keep the leash slack. Because there are few distractions at this stage, most dogs quickly learn this and enjoy the game of keeping the lead loose to earn treats. Pulling behavior can be rapidly extinguished in this setting by ignoring pulling and providing high value food treats, verbal praise, friendly eye contact and petting when the lead is loose. The dog learns that maintaining a position next to the trainer's left hip is the behavior that "works" (i.e. has positive consequences). Finally, it is important for the trainer to maintain a very high rate of positive reinforcement when teaching loose lead to avoid opportunities for the dog to pull into the lead and be reinforced by environmental stimuli. Multiple small treats should be provided and every attempt that the dog makes to stay close to the left hip is reinforced.

Once stationary loose lead has been taught, the trainer shifts the criterion and begins to reinforce loose lead while walking. A commonly used verbal cue for loose-lead walking is "Let's go" or "Heel." The trainer provides the verbal cue in a pleasant voice and steps forward. Each step that the dog takes while keeping the lead loose is reinforced. Similar to stationary training, a high rate of reinforcement should be provided to prevent the dog from pulling forward and attaining reinforcement in the form of getting closer to desired environmental stimuli. Whenever the dog moves out of the no-pull zone and causes the lead to become tight, the trainer immediately stops moving forward and remains motionless. This reaction removes all opportunity for environmental reinforcement to the dog. When the dog looks back or moves back toward the trainer, the trainer immediately reinforces the behavior and if the lead is loose again begins to walk. The trainer only continues to move forward if the dog stays close and loose lead is maintained. Accurate timing is essential because the goal is to communicate that the behavior that earns reinforcement is maintaining a loose lead. The dog learns that the behavior that works to move the trainer forward and toward the reinforcing properties of the walk is to keep her lead slack. Stopping when the dog is pulling uses negative punishment because the opportunity for something positive to occur is removed when the unwanted behavior of pulling is offered. Likewise, only moving forward when the lead is slack delivers immediate positive reinforcement (getting to go, treats, and praise) for proper heel position.

A final way in which environmental rewards can be used to the trainer's (and dog's) benefit is to include them in daily walks as periodic reinforcers for loose lead. This can be used when a dog generally walks well but will occasionally pull toward something specific such as the park entrance or the driveway of a doggy friend who will be joining the walk. The trainer reinforces loose lead with food treats, praising, and walking toward the desired stimulus, for example, the park entrance. If the dog pulls, the trainer not only stops, but walks backwards several steps and only starts forward again when the dog is heeling nicely. When the dog offers a few steps of loose lead, the owner releases the dog to enter the park or to meet the dog friend, using these opportunities as positive reinforcers.

The environmental reward that is earned is an opportunity to sniff, play, or romp with a doggy friend. This approach is attractive to owners who would like their dog to walk nicely at heel for part of the walk but also wish to provide their dogs time to just "be a dog"; allowing sniffing, exploring and playing. When environmental rewards are used, it is important to introduce another cue, such as "go play" when the targeted spot is reached and the dog is allowed a period of "just be a dog" time.

Teaching Come When Called

Teaching a dog to respond to his name and to come when called is not only good manners training, it has the potential to save a dog's life. In addition, having a dog who will reliably stay close during walks in parks and who comes when called while out playing in the yard allows off-lead exercise and a more stimulating and enjoyable life for the dog (Figure 7.4). However, similar to walking on a loose lead, the unwanted behavior of not coming when called or even worse, running away, often has a strong reinforcement history. It is not uncommon for owners to make the mistake of calling their dog to them only in circumstances that the dog is doing something undesirable, or when they are going to administer something unpleasant. Something as simple as putting the lead back on the dog and ending the romping time is unpleasant. For example, a dog who is only called to the owner at the end of play sessions with other dogs quickly learns that being called signals the end of social time with other dogs. In this case, coming when called is being negatively punished, and will decrease in frequency! And, naturally, a dog who comes when called and is then punished for digging or barking in the yard will rapidly learn to not respond to a come command (Box 7.2).

figure 7.4 Coming when called in the park

Box 7.2 TEACHING MAX TO RUN AWAY....

Without realizing it, Paul Crane taught his young dog Max to not respond to his commands of "come" and, even worse, to run away when he was called! Here is what happened:

Puppy: As a young puppy, Max was a bit timid and was very attached to Paul. When Max was between 8 and 12 weeks of age, he followed Paul everywhere and always responded to his name and to Paul's command "come" by running happily to Paul. Max also stayed close to Paul in the park, almost tripping him to maintain physical contact. As a result, Paul assumed that Max understood the command to come when called, and so he began to allow Max to be off lead in many situations. This worked fine – for about two months.

Five months old: When Max was five months old, he became increasingly confident and curious. While out in the park, he started to investigate more, leaving Paul's side to sniff, chase, and play. Max even met a few friendly dogs on several of their walks. The day came when Max was off-lead playing and Paul called him: "Max, come!" However, Max was much more interested in sniffing the grass. Paul called again . . . and again . . . and again. Max learned that responding to "come" was optional, and that the environment has the potential to be much more rewarding than his human pal. Each time that this scenario was repeated, "not coming when called" built a stronger reinforcement history.

Learning to run away: In exasperation, Paul finally chased and caught Max and reprimanded him physically and verbally. Paul tells a trainer later that he is certain that Max understood this because he was "remorseful." The trainer explains that Max's behavior was not remorse; it was submissive and possibly fearful. Unfortunately, Paul assumed (wrongly) that Max had "learned his lesson." The next time that this scenario played itself out, Paul called, Max ignored, Paul approached, Max remembered, "unpleasant things happen when I am caught," and Max chose an *alternate behavior* – running away. Paul was unable to catch Max. Max's running away behavior was negatively reinforced (i.e. it was a behavior that allowed Max to avoid an impending unpleasant stimulus).

More learning: Max steals Paul's slipper. Paul chases Max, yelling "Come here now!" Max learns that the fun of having the slipper ends if he allows himself to be caught (and he may be punished as well).

Lessons learned: Paul could have avoided these inadvertent learning situations by teaching Max to "turn to his name" and by positively reinforcing (and shaping) coming when called. The result would have been a young dog who enjoyed coming when called and did not learn to run away.

Therefore, the first step of training is to ensure that the word "come" becomes associated (classically conditioned) only with positive stimuli. The trainer selects a set of highly valued reinforcers. These are reserved for coming when called training alone (i.e. withheld in all other circumstances). Examples include highly valued food treats such as cheese or hotdog bits, a favorite squeaky toy, or a highly desired ball. These are to be paired only with the behavior of coming when called and are not used in any other circumstances. The behavior of coming when called is comprised of two component parts: turning in response to the dog's name, and returning to the owner close enough to be touched (i.e. touching the dog's collar). Shaping the behavior of coming when called is achieved by shifting criteria of distance and environmental distractions and reinforcers.

Turn to Your Name: Although this is a step that is often overlooked, teaching a dog to turn to his or her name is an important first step in teaching come when called. This is especially important in homes in which the dog's name is over-used or has been used as an aversive stimulus, such as yelling the dog's name in a harsh tone to interrupt an unwanted behavior or as a form of verbal reprimand. The goal of teaching "turn to your name" is a dog who turns and gives the owner her attention and anticipates another command to follow. The trainer selects a quiet area in the home. Standing close to the dog, the trainer quietly speaks the dog's name. If the dog turns his head to look, the trainer touches the dog's collar and immediately reinforces with a high-value treat. This sequence is repeated 10 to 20 times, with the trainer standing close to the dog and maintaining a high rate of reinforcement. Once the behavior of "turning to your name" is reliable, the trainer shifts the criteria (distance) and steps behind or to the side of the dog, repeating the training sequence. Touching the collar when the dog comes close has two important functions. It ensures that the dog comes close enough to the trainer each time (i.e. a target) and also teaches the dog that someone reaching for the collar always predicts a treat rather than an aversive outcome (classical conditioning). The "turn to your name" behavior is then gradually generalized to other parts of the home, when out on walks, and in the yard.

Teaching "Come": Teaching "come" begins once "turn to your name" is reliable. When teaching a dog to come when called, it is important that the trainer never calls the dog in circumstances in which the dog can easily choose not to respond (see Box 7.2). The dog's environment is manipulated to ensure that the dog is highly motivated to come to the trainer each time that the command "come" is given. Examples include calling the dog to come at mealtime as the dog's food is being prepared, when the owner comes home after being away for work or on errands, or when the dog is approaching the trainer for attention and affection. This early training allows the trainer to classically condition the "come" command with positive stimuli (food, affection, petting) and ensures that all uses of the command "come" are successful (i.e. the dog comes when called). Other approaches that owners can use include having one family member hold the dog while another calls the dog from another room of the house. This "back-and-forth" training often becomes an enjoyable game for the dog and provides opportunities for practice and repetition.

Once "come" has been introduced in these settings, the dog must also be prevented from *not* responding (or, worse, running away) when she is called to the owner. To prevent this, all situations in which the dog is free in a yard (or even in the home with a dog who has learned to run away) and in which the dog can choose not to respond, the dog should be either dragging a long line or is attached to a retractable lead. When the come command is given, the lead can be picked up and the dog guided into the owner if she does not immediately respond. Off-lead training can be conducted in a safe area such as a fenced yard or in the home. Just as with stays, distance, time away, and distractions are introduced separately and shaped slowly.

When teaching a dog to come when called, it is important to realize that no matter how well-trained a dog is, she will always potentially respond to competing environmental stimuli that are highly attractive. A running squirrel, another dog, or frightening sounds all have the potential to cause even the most well-trained dog to not respond, even if she has been trained with many distractions and has generalized her response to many settings. Although trainers attempt to minimize the chance of this happening, just like people, dogs do make mistakes. It is the owner's responsibility to ensure the safety of his or her dog and not to put the dog in situations in which a failure to respond will put the dog in physical danger. Moreover, responsible pet ownership means not allowing a dog to roam freely in a neighborhood, nor to randomly run up to neighbors or other dogs.

TRAINING A WELL-MANNERED CAT

Although pet owners generally do not train their cats to the same extent to which they train their dogs, teaching a cat to rest in a designated area, to come when called, and to walk on a harness are behaviors that can contribute to a happy, well-exercised, and well-behaved feline companion. Just like dogs, cats respond best to operant conditioning techniques that use positive reinforcement. A variety of reinforcing stimuli are desirable to cats. These include food treats, petting, praise, and an opportunity to play a favorite game. Each cat has individual preferences – owners should experiment and find out what things are most enjoyable to their cat. These desirable items can then be selectively managed for use during training sessions.

The ease or difficulty of teaching a new behavior to a cat depends somewhat upon the extent to which the behavior is in harmony with the cat's natural instincts. For example, it is very easy to teach cats to use scratching posts because scratching rough surfaces is a normal marking behavior and also serves as a way for cats to keep their claws healthy (see Chapter 6, pp. 153–154). Similarly, it is easy to train a cat to jump up onto a stool or platform in response to a command because cats naturally enjoy sitting on high places within their territories. Conversely, while most cats very naturally chase moving objects such as toys, they do not have a naturally strong motivation to return to their owner with a "captured" toy (i.e. to retrieve). As a result, teaching a complete retrieve to a cat usually requires more concentrated training for most cats.

Training Cats to Use Specific Sleeping Areas

Most cats naturally enjoy investigating and resting on elevated surfaces. Unfortunately, this normal behavior often results in unwanted behaviors such as jumping onto counters or dining room tables. An approach to preventing the use of areas that are unacceptable to the cat's owner is to train the cat to jump up onto a bed or platform in a selected resting area. By establishing a strong positive reinforcement history of one or more acceptable spots, the cat's use of other areas can be reduced. This training begins first by observing the cat to see what types of areas she naturally selects for resting. For example, some cats enjoy sitting in window ledges with an outdoor view, while others prefer elevated platforms on the back of couches or chairs. A cat bed or a platform that is located near a preferred spot and composed of a desirable material will have the highest likelihood of success as a resting spot (Figure 7.5). When the new bed or platform has been placed in the desired location, the cat is carried to the platform and fed a high value treat while on the bed. The cat is reinforced with the select treats each time she approaches the new bed or is close to the area. A large "jackpot" reinforcement can be given when the cat jumps up onto the bed. If the site and the bed surface are attractive to the cat, the owner should notice that the cat is using the bed after several repetitions of reinforcement.

Alternatively, if the cat does not begin to use the new resting platform, the owner can select another site that is attractive and try again. Some cats are

figure 7.5 Cat resting on a raised platform located near a window

very particular about their preferred resting spots, but once an acceptable place is found they rapidly begin to use it. If possible, several acceptable resting spots should be provided in different rooms of the house. At the same time environmental corrections (i.e. aversive stimuli) can be placed on or around the locations that the owner wishes to prevent the cat from using. Effective aversive stimuli that cause avoidance but do not usually produce fear include double-sided sticky tape or motion-activated compressed air canisters. Making all of the previously used areas unattractive while providing equally appealing new resting places usually shifts the cat's use patterns very quickly. Finally, if the owner wishes to take this training one step further and teach the cat to "go to her spot," this can be accomplished using target training (see Sidebar 3).

Training a Cat to Come When Called

The first step of teaching a cat to come is to ensure that the cat knows and responds to his name. This is achieved using the same techniques that were outlined for dogs (see pp. 178–179). The trainer should select a highly valued treat; a bit of canned food or minced liver work well with cats. The trainer stands immediately in front of the cat, and speaks the cat's name in a soft and friendly tone. The cat is immediately given a bite of treat. This is repeated 10 to 15 times, with the trainer not moving and always immediately following the cat's name with a treat (classical conditioning). When the cat begins to respond to her name by looking for the treat (and the owner), the trainer can take one step away, while still facing the cat. The trainer speaks the cat's name and immediately reinforces the cat when she takes one or two steps to the trainer. Many owners do not include the word "come" with their cats, but simply train their cat to come to their name. If the trainer prefers to teach the "come" command, it should be inserted as soon as the cat begins to move toward the trainer, always using "[Cat's name], come." Once this training has been initiated, it is important that owners never use their cat's name as an aversive to stop the cat from doing something unwanted. When aversive sounds are used, a loud clap or a mechanical sound is preferable to the cat's name and prevents the development of negative associations with the owner's voice.

Training a Cat to Walk on a Harness

Training a cat to walk on a harness is an excellent way to provide a cat with enjoyable and safe excursions outside. Walking or allowing a cat to explore outdoors also allows owners and their cats to spend time together. A harness that is specifically designed for cats should be selected and fitted snuggly to prevent escape. Training the cat to tolerate and walk while wearing a harness should take place indoors in an area in which the cat feels comfortable and secure. The first step is to habituate the cat to wearing the harness. An effective approach is to put the harness on the cat immediately before feeding. This will classically condition the presence of the harness with the pleasant associations of eating. As soon as the cat finishes his meal, the harness is removed and the feeding bowl is picked up and put away. Owners should be patient, as some cats require several weeks to become habituated to wearing a harness. When the cat is comfortable with the harness, a lead can be attached and the classical conditioning training sequence is repeated. The cat is then allowed to drag the lead around the house, without any resistance. Once the cat is accustomed to the lead, the trainer can pick up the lead and will follow the cat around, positively reinforcing a loose-lead. For some owners, this level of training is acceptable if they simply desire that their cat can go outside and explore with the lead and harness being held by the owner to keep the cat safe. If walking on lead with the owner is desired, the trainer begins to lure the cat to follow using a highly valued food treat. (It is sometimes helpful to use a long wooden spoon, using a small bit of canned food in the end, to reinforce walking.) This training is gradually shaped and walking criterion is shifted to require the cat to follow the owner. Although many cats can be lead trained, most will not trot at your side the way a dog will. Rather, an enjoyable walk for a cat involves stopping frequently to sniff and examine everything around him.

REVIEW QUESTIONS

1. Describe how positive reinforcement and successive approximation can be used to train a dog to stay in place (sitting).
2. Discuss the training challenges that are associated with teaching dogs to walk comfortably on a loose lead.
3. Why is teaching a dog to "turn to your name" an important component of teaching "come when called"?
4. Describe a training program for teaching an adult cat to sleep on a raised cat bed placed near a favorite window.
5. Identify the key components that are needed to attain reliability (fluency) in a newly trained behavior.

REFERENCES AND FURTHER READING

Alexander MC. **Click for Joy.** Sunshine Books, Inc, Waltham, MA, 208 pp., 2003.

Beadle M. **The Cat: History, Biology, and Behavior.** Simon and Schuster, New York, NY, 1977.

Beaver BV. **Feline Behavior: A Guide for Veterinarians.** W. B. Saunders Company, Philadelphia, PA, 276 pp., 1992.

Burch MR, Bailey JS. **How Dogs Learn.** Howell Book House, New York, NY, 188 pp., 1999.

Bradshaw JWS. **The Behaviour of the Domestic Cat.** CAB International, Oxford, UK, 219 pp., 1992.

Case LP. **The Dog: Its Behavior, Nutrition and Health,** 2nd edition. Blackwell Publishing, Ames, IA, 479 pp., 2005.

Case LP. **The Cat: Its Behavior, Nutrition and Health.** Blackwell Publishing, Ames, IA, 391 pp., 2003.

Csanyi V. **If Dogs Could Talk: Exploring the Canine Mind.** North Point Press, New York, NY, 334 pp., 2005.

Donaldson J. **Culture Clash: A Revolutionary New Way of Understanding the Relationship between Humans and Domestic Dogs.** James and Kenneth Publishers, Oakland, CA, 221 pp., 1996.

Messent PR. **Social facilitation of contact with other people by pet dogs.** In: *New Perspectives on Our Lives with Companion Animals*, pp. 37–46, 1983.

Miller J. **The domestic cat: Perspective on the nature and diversity of cats.** *J Amer Vet Med Assoc*, 208:498–501, 1996.

Miklosi A. **Dog: Behavior, Evolution and Cognition.** Oxford University Press, Oxford, UK, 274 pp., 2007.

New J, Salman M, King M, Scarlett J, Kass P, Hutchingson J. **Characteristics of shelter-relinquished animals and their owners compared with animals and their owners in U.S. pet-owning households.** *J Appl Anim Welfare Sci*, 3:179–210, 2000.

Pryor K. **Clicker Training for Cats.** Sunshine Books, Waltham, MA, 84 pp., 2001.

Pryor K. **Don't Shoot the Dog.** Bantam Books, New York, NY, 187 pp., 1984.

Pryor K. **Karen Pryor on Behavior.** Sunshine Books, North Bend, WA, 1995.

Rogerson J. **Your Dog: Its Development, Behaviour, and Training.** Popular Dogs Publishing Company, London, UK, 174 pp., 1990.

Sanders CR. **Understanding Dogs: Living and Working with Canine Companions.** Temple University Press, Philadelphia, PA, 201 pp., 1999.

Shore ER, Douglas DK, Riley ML. **What's in it for the companion animals: Pet attachment and college students' behaviors towards pets.** *J Appl Anim Welfare Sci*, 8:1–11, 2005.

Stafford K. **The Welfare of Dogs.** Springer Publishing, Dordrecht, Netherlands, 280 pp., 2007.

Salman ME, New JC, Scarlett JM, Kass PH, Ruch-Gallie PH, Hetts S. **Human and animal factors related to the relinquishment of dogs and cats in twelve selected animal shelters in the United Stats.** *J Appl Anim Welfare Sci*, 1:207–226, 1998.

Footnotes

[1] Shore ER, Douglas DK, Riley ML. What's in it for the companion animals: Pet attachment and college students' behaviors towards pets. *J Appl Anim Welfare Sci*, 8:1–11, 2005.

[2] Salman ME, New JC, Scarlett JM, Kass PH, Ruch-Gallie PH, Hetts S Human and animal factors related to the relinquishment of dogs and cats in twelve selected animal shelters in the United States. *J Appl Anim Welfare Sci*, 1:207–226, 1998.

[3] New J, Salman M, King M, Scarlett J, Kass P, Hutchingson J. Characteristics of shelter-relinquished animals and their owners compared with animals and their owners in U.S. pet-owning households. *J Appl Anim Welfare Sci*, 3:179–210, 2000.

[4] Messent PR. Social facilitation of contact with other people by pet dogs. In: *New Perspectives on Our Lives with Companion Animals*, pp. 37–46, 1983.

SECTION 3

Understanding and Solving Common Behavior Problems

CHAPTER 8

Elimination Problems in Dogs and Cats

Most pet owners have the expectation that their dogs can be trained to reliably eliminate outdoors and their cats can learn to dependably use a litter box. Therefore, when house-soiling problems occur, they are often very disturbing to owners because of these expectations and because of the damage that is done to home furnishings. Over time, the inability to resolve a house-soiling problem may cause the owner to isolate the pet, house the pet outdoors, or even abandon the pet at a shelter. Recent shelter statistics indicate that house-soiling problems are a common reason that owners relinquish pets.[1] As with all behaviors that people consider to be problematic, effective treatment of house-soiling problems relies upon correctly identifying its cause. Common causes of house soiling in dogs include incomplete house-training, marking behavior, submissive or excitable urination, and anxiety due to separation or fear. Cats who house soil either stop using their litter box for elimination altogether or are exhibiting marking behavior. Each of these behaviors may be triggered by several causes. In both dogs and cats, the presence of an underlying medical problem must always be ruled out prior to addressing the behavioral aspects of a house soiling problem (Sidebar 1).

MEDICAL CONDITIONS THAT MAY CONTRIBUTE TO HOUSE-SOILING PROBLEMS

Sidebar 1

In both dogs and cats, medical causes must be considered when previously trained pets begin to house soil. Medical conditions that contribute to urination in the house or outside of the litter box include any disorder that causes increased urination (polyuria), a frequent need to urinate (pollakiuria) or pain during urination. Although less common, health problems can also contribute to inappropriate defecation:

Urination House Soiling

- **Increased urine production:** Diabetes mellitus, chronic renal disease, hyperadrenocorticism (Cushing's disease), and psychogenic polydipsia (excessive water consumption)
- **Increased frequency of urination:** Urolithiasis, cystitis, enlarged prostate, urinary tract tumors
- **Painful urination:** Urolithiasis, cystitis, urinary tract infection
- **Incontinence (involuntary urine leakage):** Bladder or urinary tract infection, neurologic problems (weak bladder sphincter or spinal cord disease), psychogenic polydipsia
- **Mobility problems:** Arthritis, spinal cord diseases, hip dysplasia

Defecation House Soiling

- **Increased frequency of defecation:** Colitis (inflammation of the large intestine), intestinal parasites (whipworm)
- **Painful defecation:** Anal gland infection, constipation, obstipation
- **Incontinence (involuntary defecation):** Spinal cord injury or disease
- **Mobility problems:** Arthritis, spinal cord diseases, hip dysplasia
- **Cognitive dysfunction:** Injury or age-related changes

ELIMINATION PROBLEMS IN DOGS

Puppies younger than four months old have limited bladder control and must be provided with frequent opportunities to urinate and defecate in an outdoor location. They are also unable to sleep through an entire night without a need to urinate, so must be given the opportunity to relieve themselves at least once during the night. For both puppies and adult dogs who have not been house-trained, successful house-training relies upon providing the dog with frequent and regular opportunities to eliminate in an appropriate outdoor area and preventing accidents in the house. This is accomplished by closely supervising the puppy or dog at all times and using confinement to a crate or "safe area" when the dog cannot be closely watched. Basic classical and operant conditioning techniques are used to put elimination on cue and to reinforce correct behaviors, respectively (Chapter 6, p. 141). Although most companion dogs become successfully house-trained, a substantial proportion do not. For example, almost 80 percent of house soiling problems presented to a veterinary behavior clinic were attributed to inadequate house-training.[2] Other causes of house soiling in dogs include marking behavior, submissive or excitable urination, and anxiety-induced elimination.

Incomplete House-Training

The use of ineffective house-training methods coupled with unrealistic expectations can result in inadequate house-training. For example, owners should expect to maintain vigilant supervision and a frequent outdoor schedule with puppies up to five to six months of age. Young dogs must also be confined to a crate or another suitable area during times when they cannot be closely supervised. Unrealistically expecting a puppy to be completely trained within the first few weeks in the new home leads to repeated incidents of urination or defecation in the home. This can cause a surface or location preference for the indoor area, which further impedes house-training efforts.

Surface and Location Preferences: Substrate preferences develop very early in puppies; by five weeks of age puppies in a litter are leaving their sleeping area to urinate and defecate, and begin to choose specific locations for elimination. If no guidance is provided and repeated use of a particular indoor area occurs, the puppy can develop a substrate preference for carpet or linoleum surfaces. A related problem is seen in dogs who are confined for long periods of time and forced to eliminate in their crate or kennel. In this case, elimination behaviors may not develop normally and instead the puppy acquires a substrate preference for the smooth surface of a crate or kennel. In extreme cases, the dog learns to tolerate lying or standing in his own waste. These behaviors are most frequently seen in puppies who were purchased from pet stores where they lived confined to a cage and in shelter dogs who were housed in kennels with a concrete surface. Retraining a dog who has a preference for an indoor area or surface is much more challenging than training a young puppy who was encouraged to develop outdoor preferences early in life. Constant supervision prevents accidents and opportunities for the development of unwanted preferences and enhances the owner's ability to learn to recognize their dog's signs of needing to eliminate.

Learned Avoidance: Verbal or physical punishment for house soiling can result in a dog who is not only poorly house-trained but who also has learned to avoid eliminating in the presence of the owner. This behavior is a direct result of the misconception that it is acceptable to punish a dog for eliminating in the house whenever the dog is "caught in the act." This oft-repeated myth is not just an ineffective house-training technique, it is counterproductive and potentially inhumane. When a dog is punished for eliminating, the aversive stimulus (yelling, hitting) causes anxiety and fear. Some dogs learn to avoid this consequence by eliminating out of sight of the owner. The operant behavior the dog learns is that eliminating near the owner has a negative consequence (yelling, hitting, and not being allowed to finish eliminating), while eliminating in private has a positive consequence (relief associated with emptying the bladder, no exposure to yelling). Although the owner may believe that his dog "knows he is wrong" because the dog eliminates in another room and shows submissive or fearful body postures when "caught," the underlying reason for these behaviors are the dog's expected reaction to negative reinforcement and punishment. Essentially, in this case, the owner has very efficiently trained his dog to eliminate in the house, out of sight, and in another room (Sidebar 2).

EFFECTS OF PRE-ADOPTION EDUCATION ON OWNER PERCEPTIONS OF HOUSE-TRAINING SUCCESS

Sidebar 2

Although owners rarely euthanize their dogs for house-soiling problems, it is an unfortunate fact that they do frequently relinquish them to animal shelters for this behavior problem. The frequency with which house soiling is listed as a reason for shelter relinquishment suggests that owners perceive elimination problems to be very serious. This is especially tragic in light of the fact that most house soiling problems are due to incomplete house-training. Therefore, shelter staff and veterinarians who work with new dog owners are in a position to encourage complete house-training using proper methods through education and to help reduce relinquishment of dogs for house soiling reasons.

With this objective in mind, a group of researchers examined the effects of using pre-adoption counseling to encourage proper house-training techniques to new adopters at a Midwest animal shelter. A group of 113 newly adopted adult dogs and their owners were randomly assigned to two groups. The treatment group received a short counseling session that provided guidelines for house-training. The control group received all other normal adoption procedures with the exception of the house-training program. Instructions to the house-training group included providing consistent indoor supervision to prevent accidents, offering frequent trips outdoors, using positive reinforcement for outdoor elimination (treats, praise), feeding meals at scheduled intervals to promote regular elimination habits, and crate training for confinement when no one was at home to supervise the dog. In addition, the use of punishment for accidents, regardless of when they were discovered, was strongly discouraged. Owners were also advised about using appropriate enzymatic cleaners to clean soiled areas.

One month following adoption, owners were interviewed regarding their perceptions of house-training success. Overall, a high proportion were satisfied with the degree to which their dog was house-trained, but significantly more owners in the house-training group reported success than owners in the control group (98.1 percent vs. 86.4 percent, respectively). More notably, owners who received pre-adoption counseling were less likely to use verbal punishment with their dogs during house-training and were more likely to use enzymatic cleaners on soiled areas than owners who had not received the counseling. An unexpected finding was that although more owners in the counseled group reported satisfaction with their dog's level of training after one month, the actual frequency with which dogs continued to have accidents in the house did not differ appreciably between the two groups. This disparity suggests that owners who received counseling not only were more likely to use appropriate techniques for training and cleanup, but were better prepared to understand the normal progression of house-training and were possibly more tolerant when mishaps occurred. Because it is the owner's perception of the seriousness of a behavior problem that affects the bond that they have with their dog (and risk of relinquishment), these results suggest that providing information about house-training to new owners both improves house-training success and encourages realistic perceptions about the house-training process.

Herron ME, Lord LK, Hill LN, Reisner IR. Effect of preadoption counseling for owners on house-training success among dogs acquired from shelters. *J Amer Vet Med Assoc*, 231:558–562, 2007.

Treatment: Treatment for dogs who have not been completely house-trained follows the same protocol for initially house-training puppies and adults. The dog is confined to a small area of the house and taken outside at very frequent intervals. Areas of the home that have been soiled are thoroughly cleaned. The dog is prevented from revisiting the areas during the retraining program using baby gates or closed doors. The most important rule that the owner must follow is that the dog is only allowed unsupervised access to the house if he has been taken outside and has eliminated within the last 30 minutes. If the dog is taken out and does not urinate, he must be confined or closely supervised in the house, and taken outside again 10 to 20 minutes later. An alternative approach is called the "umbilical cord" technique. The dog is attached to the owner by a leash or is tethered to a piece of furniture in the same room as the owner. This approach prevents the dog from wandering into another room to eliminate and has the added benefit of encouraging owners to learn their dog's signals for needing to eliminate. Typical body signals that dogs show include panting and pacing, staring at the owner, sniffing, pawing, and circling. Because dogs who are perceived as being incompletely house-trained often give signals that their owners do not recognize, this form of retraining can help the dog to become more reliable and the owner to learn to respond to the dog's signals.

During retraining, a designated area of the yard should be used consistently, and a specific cue word ("Go potty" or "Hurry up") is used to classically condition elimination behavior. If the dog indicates a surface preference in the home, such as carpet, placing a piece of the preferred material in the designated outdoor area can be helpful in encouraging elimination in that area. Owners should also avoid ending daily walks as soon as the dog eliminates. This habit can inadvertently cause the dog to delay eliminating when he learns that urination and defecation always predict an end to an enjoyable outdoor excursion. For this reason, trips outside for elimination only should be separated from trips outdoors for walks and exercise. Using different leashes or exiting different doors for each type of event can be helpful. In the majority of cases, simply establishing a regular routine, providing an adequate number of trips outside, reinforcing correct elimination habits, and preventing indoor accidents will complete the dog's house-training (Box 8.1).

Box 8.1 CASE STUDY – CHAMP'S HOUSE-TRAINING PROBLEM

Description of problem: Champ is a neutered, 1-year-old, male, Boxer mix. His owners, Sue and Brad Brown, adopted Champ from a local shelter three weeks ago. Champ had been living at the shelter for 4½ months prior to adoption. Champ is a very quiet dog in the home and is affectionate toward Brad and Sue and to visitors. He is not destructive at all, and has a number of toys that he chews and plays with. He is crated when the Browns are not at home, but they would like to begin leaving him out free with their other dog, Mandy, as Champ becomes more reliably house-trained.

The Browns did not know if Champ was house-trained and instituted a complete training program upon bringing him home. Their program included confining Champ to the kitchen or closely supervising him when he was with Brad or Sue in another room of the house, and frequently taking Champ into a designated spot in the yard to eliminate. At first, this seemed to work well and Champ did not have any accidents in the kitchen or when he was supervised. However, as the Browns allowed Champ to have more freedom in the house, he began to go into the basement once or twice a day to defecate. The owners then closed off his access to the basement, and there were no accidents for several days, with Champ using the backyard when he was taken out. However, Champ recently began to defecate on the hardwood floor in the dining room. With the exception of the dining room and kitchen, the rest of the house is carpeted. Champ has never eliminated in the carpeted areas or in the kitchen. The Browns have also never observed Champ defecating in the house; he has only done this when they are not directly supervising him. On one occasion, he left the living room where the family was sitting, went into the dining room, defecated, and returned.

Diagnosis: Incomplete house-training; learned surface preference for hard, non-absorbent surfaces; possibly learned avoidance of owner when eliminating.

Treatment: Increase supervision and block access to all areas that are potential elimination sites in the home. Clean all soiled areas with a biological (enzymatic) cleaner. Introduce cue word ("Hurry Up") for elimination and positively reinforce all eliminations in the owner's presence. Use "umbilical cord" approach for supervision in home.

Outcome: When Champ was prevented access to soiled areas and supervised closely, no further accidents occurred. After two weeks of strict supervision and no accidents, Sue and Brad were able to gradually increase Champ's freedom in the home. At the one-month follow-up, the Browns report that Champ is reliably house-trained.

Marking Behavior

Marking is characterized by the frequent voiding of small amounts of urine, usually targeting socially significant locations (see Chapter 3, pp. 46–48 for a complete discussion). Less frequently, feces may be deposited in conspicuous areas. All dogs are capable of marking behavior, but it is most commonly reported as a behavior problem in intact males. With the exception of sleeping areas, parts of the home or yard in which the dog spends most of his time are frequently marked by a dog who exhibits urine marking (Figure 8.1). When marking occurs in the home and the dog has been determined to be fully house-trained, the marked areas often signify a site that is associated with anxiety or stress. For example, a dog who suddenly begins to mark the resting spot of a newly introduced cat is presumed to be expressing anxiety over the change in social structure in the home. Similarly, a dog may begin marking when a new dog or human family member has been introduced to the home.

figure 8.1 Marking behavior in male dog

Because marking behavior has both a hormonal and a social basis, neutering eliminates or significantly decreases marking behavior in more than 90 percent of intact male dogs.[3] Contrary to popular belief, this effect is observed even in mature dogs who have a history of marking. If the dog is marking because of anxiety or a change in social status, the underlying cause of anxiety must be identified and treated. Counter-conditioning and systematic desensitization are used to condition the dog to accept a new member of the household (human or non-human) and to change classically conditioned associations. In some cases, the use of anti-anxiety medications may be helpful as adjunctive therapy. Because dogs who mark in the house usually revisit the same areas, blocking access to previously marked areas or using the umbilical cord approach discussed previously helps to prevent a recurrence of marking behavior during retraining.

Submissive/Excitable Urination

Submissive and excitable urination are actually two separate disorders but are often discussed together because it can be difficult to distinguish between the two behaviors. Some dogs, especially if young, may show both types of urination but in different contexts.

Submissive Urination: Submissive urination is seen in dogs who show excessively submissive body postures during greeting. Although this is most commonly reported as a problem during interactions with their owners or other people, dogs can also submissively urinate when greeting another dog. The dog greets with lowered head, averted eyes, and retracted lips, and dribbles urine as the person or dog moves closer (Figure 8.2). Submissive greeting is normal for

figure 8.2 Submissive urination during greeting

young puppies who are interacting with adult dogs and for many puppies when they greet humans. Most dogs gain confidence as they mature and no longer demonstrate this level of deference. However, some dogs either do not develop adequate confidence as adults or have learned to offer extreme appeasement in response to harsh reprimands for jumping up or for misbehaving during the owner's absence. A common but very unfortunate scenario is the owner who has verbally or physically reprimanded the dog each time that the dog has been destructive when left alone. This reaction not only causes the dog to show extreme appeasement and possibly to urinate when greeting, but also increases the dog's level of anxiety, which further exacerbates anxiety-induced destructive behaviors.

Treatment: Submissive urination can be prevented by decreasing the intensity of greetings and avoiding standing or leaning over the dog while interacting. An effective approach to diffuse intense interactions is to redirect the dog by tossing a biscuit or toy to the side as the dog approaches to greet. The dog's movement to the side positions her out of direct eye contact and prevents the owner or visitor from standing directly over the dog. After the dog has consumed several treats in this way, she is greeted with the owner or visitor crouching down and positioning himself laterally (from the side) to avoid leaning over the dog (Figure 8.3). It is also helpful to train the dog to offer a reliable sit-stay for petting in non-greeting contexts (i.e. when excitement level is low). The sit-stay for greeting is then incorporated into greeting situations, with the treat offered to the side as a reinforcer when the dog sits (see Chapter 9, pp. 212–216 for a complete discussion).

Excitable Urination: Excitable urination differs from submissive urination in that, while it can occur during greeting, it may also happen when a young dog becomes excited or during play. Dogs who show excitable urination often do not squat, but rather dribble small amounts of urine as they walk or jump around. Body postures and communication signals that accompany excitable urination do not reflect excessive submission but rather are those of an excited and possibly

figure 8.3 Properly greeting a submissive dog

over-stimulated dog. Because excitable urination is often caused by a lack of complete neuromuscular control of the urinary sphincter, most dogs stop showing this form of urination as they mature.

Treatment: As with submissive urination, treatment involves decreasing the intensity of greeting and play to prevent over-arousal. Redirecting the dog's attention to toys or training the dog to retrieve a ball can provide exercise and play while reducing excitement. Greeting intensity can also be decreased by providing the dog with several minutes to romp outside before interacting. This also allows the dog to empty her bladder prior to greeting. Similar to the treatment for submissive urination, teaching a sit-stay for greeting can help to prevent excitable urination, provided the sit-stay is first taught in a quiet environment. Distractions that typically excite the dog are then gradually introduced. When behavior modification alone is not effective, the medication phenylpropanolamine can be used as an adjunctive therapy. This must be prescribed by a veterinarian and functions to increase sphincter tone in the urethra. Although phenylpropanolamine is more typically prescribed to curtail urinary incontinence, it can also help to reduce excitable urination. Behavioral modification should continue and the drug is gradually reduced and eliminated as the dog improves.

Other Causes of House Soiling in Dogs

Other potential causes for elimination problems in dogs include medical conditions, separation stress, and environmental fear or anxiety (for example, fear of fireworks). The most distinguishing factor of separation-induced elimination is that the problem occurs only when the dog is isolated and never when the owner is present. Treatment of separation-related house soiling is directed toward solving the underlying cause of separation stress (see Chapter 10 for a complete discussion of separation anxiety in dogs). Similarly, house soiling that is related to environmental or situational fears or anxiety is treated by either eliminating or reducing the intensity of the causative stimuli and attempting to desensitize the dog (see Chapter 10, pp. 250–255).

ELIMINATION PROBLEMS IN CATS

House soiling problems in cats can be divided into two major categories: cats who stop using the litter box for elimination purposes (hereafter referred to as inappropriate elimination) and cats who are exhibiting marking behavior. Each of these classifications may have one or more causes.

Inappropriate Elimination

Cats who have stopped using their litter box for elimination squat to urinate on a horizontal surface, voiding a large volume of urine and leaving a large puddle. The location may be close to but not actually in the box (suggesting a box aversion) or in a particular location (suggesting a substrate or location preference). Some cats who stop using the box select sites that are quiet and remote, such as behind a couch or in the basement. Others demonstrate a preference for novel or seemingly unusual surfaces such as in bathtubs, sinks, or over a heating vent. Regardless of the location or surface that is chosen, failure to use the litter box is differentiated from marking in that the cat stops using the litter box for most or all urinations (and, less commonly, for defecation as well).

Health Causes: Cats develop learned aversions to their litter box for environmental, social, or health reasons. Medical problems that cause pain during urination or defecation are common initiating causes of litter-box aversions. Cats who experience discomfort while eliminating associate the pain with the litter box or the location of the box, and subsequently avoid the box. Common medical causes of litter box problems include lower urinary tract infections or disease, declawing surgeries (because of the discomfort caused by digging when paws are tender), gastrointestinal disease, and in elderly cats, cognitive dysfunction (see Sidebar 1). Although a substantial number of cases of inappropriate elimination develop originally in response to a medical disorder, curing the medical problem does not always solve the litter-box problem because the cat has an established association between using the box and pain. Therefore, while medial disorders must always be ruled out with litter-box problems, behavior modification and environmental changes are often still needed to retrain the cat to use the box even when these problems have been resolved.

Environmental and Social Causes: The most common environmental cause of litter-box aversion in cats is the owner's failure to keep the litter box clean. Cats vary in their willingness to use a soiled litter box. While some cats are relatively tolerant of a dirty box, others will avoid a box that has even small amounts of waste. In multiple-cat homes, the need for a clean box may be amplified if one or more of the cats dislike using a box that contains another cat's waste. Aversions in multiple-cat households can also be caused by social conflicts between cats that prevent a cat from using or going near the box. (It is also important in multiple-cat homes to correctly identify the cat who has the elimination problem.) Even when a cat's negative experience does not directly involve using the litter box, the cat may begin to avoid the location of the box to reduce the risk of repeating an unpleasant experience. Experiences that can precipitate an aversion include being frightened by a loud noise while using the box or having the box located to a high-traffic area of the home.

Litter Box Location and Type of Filler Material: The location of the box is also an important consideration. Cats will not use a box that is difficult to get to or that is located in an area of the home that they rarely use. Placing the box too close to food bowls or resting areas may also cause aversions. The type of litter substrate that is used can be important. A sudden change in the texture or scent of the litter that is used can trigger an aversion. Some cats dislike scented litters or filler that has an unusual texture. Cats who have developed a litter substrate aversion may use the box occasionally but will fail to dig and cover wastes, dig for a shorter period of time, or exit the box abruptly (Sidebar 3). Other signs of an aversion to litter substrate include straddling the edge of the box to avoid touching the litter, digging on the floor outside of the box, or eliminating immediately next to, but not in, the litter box.

LITTER BOX BEHAVIORS IN CATS WITH AND WITHOUT INAPPROPRIATE ELIMINATION

Sidebar 3

Litter-box aversions are a significant cause of inappropriate elimination in housecats. These aversions may develop indirectly as a result of an unpleasant or painful experience while using the box, or directly in response to attributes of the box or its location. A recent study was conducted to examine the types of behaviors that cats with and without elimination problems show around and inside of their litter box. It was theorized that cats with inappropriate elimination due to an aversion to the box or litter material might show different behaviors when using the box than cats without inappropriate elimination.

A group of 40 healthy cats was studied: 20 who reliably used their litter boxes and 20 who were diagnosed with inappropriate elimination. Camcorders were used to record behavior prior to and during elimination for a continuous 72-hour period. The duration of each elimination-associated behavior (pawing, sniffing, digging, and covering waste) was recorded. In addition, information about the type of litter box and filler material and its location were collected. Each cat's core area (i.e. the smallest area of the home that accounted for 75 percent of the cat's activity) was also identified.

The researchers found that the type of box (hooded vs. open), number of boxes available to the cat, type of litter used, frequency with which owners cleaned the box, and the cat's claw status (declawed or not) did not differ between cats who had problem elimination and those who did not. In 37 of the 40 homes, the litter box was located in an area that was outside the periphery of the cat's core living area. The measured distance of the box from the center of the cat's core living area was inversely related to the duration of time that cats spent sniffing prior to eliminating. In other words, cats spent more time sniffing when the box was located close to their primary living area than when it was located further away.

Perhaps the most noteworthy finding of this study was that cats with problem elimination spent significantly less time digging in their box prior to elimination and tended to spend less time covering their waste than did cats without inappropriate elimination. Together, these differences suggest that cats with inappropriate elimination spend less time in and around their litter box. It is possible that failure to use the litter box is related to box attributes that affect digging, such as the type of filler material or the size of the box. This information may be helpful when attempting to determine if a box or litter-material aversion is the underlying cause of inappropriate elimination. Future studies will examine whether a cat's actual digging times increase or decrease when offered different substrates and if differences in digging behavior signify aversions or preferences of individual cats.

Sung W, Crowell-Davis SL. Elimination behavior patterns of domestic cats (*Felis catus*) with and without elimination behavior problems. *Amer J Vet Res,* 67:1500–1504, 2006.

Type of Box: Finally, the size and type of box can cause problems for some cats. A general rule of thumb is that a cat's litter box should be large enough for the cat to scratch and easily turn around. This translates to a length that is approximately 1½ times the length of the cat, measured from the nose to the base of the tail. Given the large size of some housecats, many commercially available litter boxes are too small. A second issue for some cats is the use of a covered litter box. Covered boxes include a pan and hood with an entry opening in one side. This type of litter box offers privacy and controls odor and spillage and are often used in homes with dogs to prevent litter box scavenging. However, some cats may feel trapped or insecure in covered boxes and are reluctant to use them. In recent years, automated self-cleaning boxes have become increasingly popular. These boxes include a timed sensor that is activated several minutes after the cat exits the box. A cleaning rake slides along the litter and scoops urine and feces into a container. Although these boxes are convenient for owners, some cats are frightened by the mechanical noise of the rake and begin to avoid the box. The self-cleaning properties of these boxes may also cause owners to be less fastidious about box hygiene. Because the rake cannot thoroughly remove waste, some cats stop using these boxes as they become increasingly soiled or odorous.

Learned Surface Preferences: Cats who stop using the litter box because of an associative or direct aversion need an alternate and acceptable place to eliminate. A cat will select a location and surface that is acceptable and, if the problem persists, will develop a preference for this new location or surface. Cats typically select areas that are quiet and somewhat secluded. As the spot becomes saturated with urine, it attracts the cat back and becomes a preferred elimination site. Absorbent surfaces such as carpet or a particular piece of furniture are frequently used, although some cats seem to prefer smooth and cool surfaces such as bathroom sinks or bathtubs. When surface or location preferences develop, they are usually a result of a litter-box aversion, as opposed to being the initial problem. Therefore, treatment is directed toward increasing the attractiveness of the litter box and preventing use of the newly selected surface.

Treatment for Inappropriate Elimination

Because medical disorders are a common initiating cause of house-soiling problems in cats, a complete veterinary examination should be conducted. When a medical condition is found, it must be treated or managed before starting behavioral treatment. It is not uncommon for house soiling to continue after the health problem has been resolved because of an associative aversion to the box or a developed surface preference (Box 8.2). Therefore, even after medical treatment, litter-box aversions and location preferences often persist and must be treated behaviorally.

Box 8.2 CASE STUDY – LUCY'S LITTER BOX PROBLEM

Description of problem: Lucy is a seven-year-old domestic longhaired cat owned by the Parker family. She has a history of struvite urolithiasis (lower urinary tract disease caused by crystals composed of struvite), but all of her symptoms have been controlled for more than nine months with a therapeutic diet. Recently, the Parkers decided to transition Lucy back to a normal over-the-counter maintenance cat food. Lucy has been eating and behaving normally, and has not shown any signs of pain while urinating in the litter box. However, she has stopped using her litter box for some urinations, and has used the kitchen and bathroom sinks several times. At first, the Parkers prevented Lucy from using these areas by placing metal pans in the sinks, and this seemed to be effective for several days. However, Lucy has started to urinate in the bathtub of an upstairs bathroom. Lucy continues to defecate in her box, and uses the box for at least some urinations, but also uses the sinks or bathtub once or twice every few days.

Additional background: Joe Parker heard from his neighbor (who knows a lot about cats) that he should "catch Lucy in the act" and harshly reprimand her. When Joe saw Lucy in the sink, he yelled at Lucy sharply and swatted her. Lucy stopped urinating and ran away from him. Joe chased her and caught her, and then carried her to her box and firmly placed her in her litter box. He also gave her another "swat" at the same time (so that she would know that what she did was wrong). Joe relates that Lucy was "very upset" and struggled all the way to the box. He believes that Lucy must have known that she had done something wrong because she immediately ran away from him again. Anna Parker agrees because since that happened, they have not seen Lucy urinate in the sinks or the tub, but are still finding urine in those places.

Diagnosis: Litter-box aversion initiated by pain associated with lower urinary tract disease and exacerbated by owner's use of punishment. Concomitant development of a surface and location preference (smooth, cool surfaces; out of owner's presence).

Treatment: Complete veterinary checkup for presence of struvite in urine; recommence diet therapy for struvite urolithiasis; block all access to soiled areas and retrain to use litter box using confinement; discontinue all use of punishment. Attempt to repair relationship with Joe by having Joe provide all of Lucy's meals and engage in regular daily petting and play sessions. Conduct regular urinalysis for recurrence of lower urinary tract disease.

Outcome: One month following treatment, Lucy is again reliably using her litter box (and Lucy's trust in Joe has been restored).

Litter Box Hygiene and Management: In all cases, the litter box must be kept scrupulously clean. Dirty litter should be removed from the box once or twice daily and fresh litter provided every 2 to 3 days. If clumpable litter is used, soiled clumps are removed daily and the litter is completely replaced every two weeks. The entire box should be washed with unscented cleaners and thoroughly rinsed once a month. For some cats, simply improving litter box hygiene solves inappropriate elimination problems. Additional management approaches include moving the box to a new location or increasing the number of litter boxes that are available to the cat. Adding additional boxes is especially important in multiple-cat homes. A general rule of thumb is that the number of boxes should be equal to the number of cats plus one. If the form or scent of the litter material is suspected to be a cause, a variety of litters can be offered in different boxes. The cat is then carefully observed to determine which is preferred. Many cats avoid heavily scented litter, even though the owner may find these appealing (Sidebar 4). Similarly, some cats dislike the tactile feel of certain types of litter, and so a variety of textures can be offered and the cat's use monitored.

Sidebar 4 — ODOR IN THE LITTER BOX – TO WHOM IS IT IMPORTANT?

Improving litter-box hygiene practices is a regular recommendation when treating inappropriate elimination in cats. Increasing the frequency of scooping and replacing used litter reduces the amount of waste in the box and decreases box odor. Although owners appreciate a reduction in fecal and urine odor generated from a dirty box, the effect that these odors have on a cat's use of the litter box had not been studied. It was postulated that both cleaning the box regularly and applying a commercial litter-box spray designed to eliminate odors would enhance the box's attractiveness to cats. A series of studies was conducted to test this hypothesis.

When the odor-eliminating spray was tested with healthy cats who did not have a house-soiling problem, cats' box preferences were not affected by the application of the spray. However, when tested with a second group of cats who showed behaviors suggestive of litter-box aversion (scratching or pawing near the box, avoiding the box, absence of digging or covering behaviors), the cats had a significant reduction in these behaviors when their litter boxes were treated with the odor-reducing product. Cats who regularly showed inappropriate elimination by urinating or defecating outside of their boxes significantly reduced the number of eliminations outside of the box when the litter box was treated with the odor-reducing product. On average, a 40-percent reduction in inappropriate elimination was reported during the two-week period that the product was used.

These results provide strong support for the belief that odor is an important factor involved in litter-box aversions in cats with inappropriate elimination. They also suggest that cats who show behaviors that are associated with litter-box aversion but who do not eliminate outside of the box may benefit from a reduction in litter-box odor. Cleaning the box more often and applying an odor-reducing product are two approaches that owners can use to control litter-box odor. These changes may increase the attractiveness of the litter box to cats and potentially prevent the development of inappropriate elimination problems.

Cottam N, Dodman NH. Effect of an odor eliminator on feline litter box behavior. *J Feline Med Surg*, 9:44–50, 2007.

Treating Surface Preferences: Surface preferences are treated at the same time by thoroughly cleaning all areas that the cat has used for elimination. For absorbent surfaces, a biological cleaner formulated to degrade the odors of pet waste should be applied liberally. In some cases of prolonged use, it is necessary to completely remove and replace the surface. Baby gates and closed doors can be used to temporarily block the cat's access to these areas during litter-box retraining. Keeping a cat from returning to an area of preference that was previously used can be a long-term challenge for some cats. Some must be permanently prevented from returning to those areas of the home.

Litter-Box Retraining and Removing Social Causes: Once a suitable box and litter material have been identified, it is often necessary to confine the cat to a small room or crate with the new box for retraining. The cat should have ample room for sleeping and eating, but freedom in the home is limited. Confinement allows the owner to carefully monitor litter box use and also prevents the cat from revisiting soiled areas. In multiple-cat homes, confinement also can provide a temporary reprieve from underlying social tension that may have contributed to the house-soiling problem. As the cat is observed using the box and begins to be reliable again, freedom to other parts of the house can gradually be allowed. If social tension is suspected as an underlying cause, the cat should be provided with a separate living space that eliminates or reduces the need for contact between cats who do not get along. Examples include allowing cats to live on separate levels of the home or having their core living areas centered in different rooms of the home.

Ineffective Treatments: Approaches to treating inappropriate elimination that are occasionally used but are *not* effective include punishing the cat for inappropriate elimination or confining the cat without making any box or environmental changes. Verbal and physical punishment administered after the cat has eliminated does not teach the cat to eliminate in the litter box and can cause fear or avoidance. Cats who are punished when they are "caught in the act" can become frightened and learn to avoid the aversive stimulus by eliminating in privacy, out of sight of the owner. Similarly, simply confining the cat to a small space with the same litter material and box may force the cat to use the box while confined, but will not address the underlying cause of the problem. Although confinement is helpful for preventing accidents when the cat is left unsupervised, if it is the only approach that is used, most cats will revert to the original problem once allowed access to the home again.

Marking Behavior in Cats

Cats who are exhibiting marking behavior continue to use the litter box for most eliminations while also voiding urine in small quantities in specific locations. Most cats spray when urine marking, using the characteristic standing body posture to spray urine backwards toward a vertical surface (see Chapter 3, p. 46). In addition, some cats adopt a normal squatting posture and void small amounts of urine on a horizontal surface. These cats are still considered to be marking if small amounts of urine are found, if the location is significant, and if the cat continues to use the litter box for most urinations. (Note: Urine marking of this type can be difficult to distinguish from inappropriate urination caused by a mild box aversion or surface preference.)

It is important for owners to understand that urine marking does not function for elimination but rather represents a form of social communication (see Chapter 3, pp. 46–48). Therefore, most cats who mark continue to use the litter box and do not demonstrate behaviors that suggest an aversion to the box or its location. Marking is most frequently exhibited by intact males or by cats living in multiple-cat homes. Typically marked spots are vertical surfaces such as walls, drapes, or the legs of furniture. Some cats target very specific items such as the owner's clothing or locations near a window where the cat can see an unfamiliar cat prowling outside. When horizontal surfaces are urine marked, the target is often clothing, the owner's bed, or a piece of furniture. Finally, although feral cats reportedly deposit feces in conspicuous areas within their hunting ranges, housecats do not appear to use fecal marking for social communication.[4] Therefore, the presence of feces outside of the box usually signifies a litter-box problem rather than marking behavior.

Identifying the Cause of Marking: Urine marking by housecats is classified as being either a sexual or reactional (stress-induced) behavior. Sexual marking is most commonly seen in intact males who urine mark to designate territory and attract mates. However, cats of both sexes are capable of sexual urine marking and reproductively intact females often mark during the estrus stage of their estrous cycle. In both sexes, sexual marking behavior increases during the cat's breeding season and is effectively eliminated or reduced by neutering. Contrary to popular belief, neutering is an effective treatment for sexual-marking behavior regardless of the age of the cat or the duration of spraying.[5]

When urine marking occurs in neutered cats, the underlying cause is almost always reactional, resulting from social conflict among cats in a multiple-cat home or from anxiety due to a sudden environmental change. Typical causes of stress-induced marking include the introduction of a new cat or cats, a move to a new home, a change in the cat's or the owner's daily routine, decreased attention from the owner, or overcrowding with other cats. The areas that are marked can provide information regarding the underlying cause. Spraying near windows and doorways indicates territorial marking in response to the presence of cats outside of the home, while marking walls and furniture within the home is suggestive of conflict with other cats or anxiety related to the social climate of the home. It is theorized that in both types of spraying, depositing the cat's own scent around the home has a calming effect and increases the cat's feelings of security and confidence.[6]

Treatment for Marking

Treatment for urine marking includes determining the cause and either eliminating it or managing the cat's surroundings to minimize stress and anxiety. If another cat has been introduced to the home and is suspected as the cause, the new cat should be confined and gradually reintroduced to the resident cat or cats. Behaviors indicating tolerance and acceptance by the resident cat are used to determine the rate of reintroduction. In cases in which the cause is the cat seeing outdoor cats through a window or glass door and spraying in response, the cat must be prevented from seeing the outdoor cat by blocking access to windows or rooms where the marking occurs. When the cause of spraying is social conflict or overcrowding within a group of cats, the cat's anxiety level can often be reduced by providing additional vertical space for resting and cubby holes for hiding, and, in some cases, separating cats in different parts of the home. There is also some evidence that the delivery of aerosolized synthetic feline facial pheromones (FFP) into cats' living environments helps to reduce anxiety and spraying behavior (Sidebar 5). Finally, although having several cats is increasingly common today, not all cats are comfortable living in larger cat groups. In these cases, the owner must either find a way to provide the affected cat with his own secure area within the home or, if that is not possible, consider finding a new home that does not have multiple cats.

Sidebar 5 | FELINE FACIAL PHEROMONES AND URINE-SPRAYING BEHAVIOR IN CATS

Feline facial pheromones are a group of compounds secreted by the cat's facial glands that are deposited on furniture and other areas of the cat's living area when cats bunt and allorub. In recent years, synthetic forms of these compounds have been manufactured and tested as an adjunctive therapy for reducing anxiety and stress-related behavior problems in cats. It is theorized that depositing synthetic FFP in a cat's living environment causes the cat to increase his or her own facial marking, leading to increased security and deceased anxiety. Initial reports of the use of FFP with cats who were urine marking in homes reported a high success rate. However, controlled studies were needed to corroborate those claims.

Daniel Mills conducted such a study with 22 cats diagnosed with problem urine spraying. Cats were randomly assigned to either a treatment group or the control group. FFP was administered to the treated group using a heated diffuser that continuously delivered FFP vapor into the cat's primary living quarters. A diffuser that delivered a placebo was used with the control group. At the start of the study, frequency of spraying in the two groups of cats was similar. Over a four-week period, frequency of spraying decreased in both groups, but the reduction was only statistically significant in the FFP-treated group. In addition, the effectiveness of the FFP as a treatment in reducing urine spraying tended to increase with time. This rate change was not observed in the control group. No behavior therapy was provided to the cats in this study and owners were specifically instructed to not alter their routines or behavior with their cats during the study period. Therefore, the decrease in spraying behaviors that was observed in the FFP-treated cats was presumed to be in response only to the use of the synthetic pheromone.

In another study, telephone interviews were used to determine long-term effects of FFP therapy in a group of 43 cats who had been treated successfully by the author's clinic approximately 10 months previously. Fourteen percent of owners reported that their cats had completely stopped spraying and an additional 63 percent reported that their cat's spraying behavior was still under adequate control. Most of the owners used the occasionally or as needed, but none felt the need to use FFP on a daily basis. Finally, other studies have reported that FFP is most effective at reducing non-sexual (i.e. reactive) marking in cats, but is less effective when overt inter-cat aggression and frequent fighting between cats occurs in the home. In these cases, FFP may act synergistically with behavioral therapy and management techniques to reduce anxiety and tension between cats.

These results support the hypothesis that synthetic FFP delivered continuously into a cat's environment helps to reduce anxiety-related marking behavior. Because the use of FFP is non-invasive, is easy for owners to administer, and does not have the risks of negative side effects that are associated with conventional pharmacotherapy, FFP provides an adjunctive therapy that may be very beneficial for cats who show anxiety-related urine spraying.

Mills DS. Pheromonatherapy: Theory and applications. *In Practice*, 27:368–373, 2005.

Mills DS, Mills CB. Evaluation of a novel method for delivering a synthetic analogue of feline facial pheromone to control urine spraying by cats. *Vet Rec*, 149:197–199, 2001.

Mills DS. Long-term follow up of the effect of a pheromone therapy on feline spraying behaviour. *Vet Rec*, 147:746–747, 2000.

Just as with failure to use the litter box, all marked areas must be thoroughly cleaned with a biological cleaner to eliminate urine odor. Common household cleaners should not be used because the perfumes of these cleaners simply mask urine odor and the cat will continue to detect marked areas.[7] Initially, the cat is prevented from revisiting areas that were previously marked. The owner can also place mildly aversive stimuli over or near marked spots, such as citrus scents or aluminum foil. In cases in which only one or two specific areas of the home were marked, relocating litter boxes close to the spot can change the cat's motivation from marking to urinating and may encourage the cat to use the box. Regardless of these retraining approaches, however, it is still essential that the underlying cause of the initial spraying behavior is identified and eliminated before allowing the cat to have free access to the home and near previously marked areas.

Pharmacotherapy for House Soiling in Cats

When house-soiling problems are directly related to territorial behaviors or social conflict, cats who do not respond to behavior therapy or household management solutions may respond to drug therapy. Because all pharmacological agents are associated with side effects, the use of drug therapy for urine marking or for inappropriate elimination should only be considered after all other approaches have failed. The three types of medications that are used for marking and litter-box aversions in cats include selective serotonin reuptake inhibitors, tricyclic antidepressants, and azaperones (Table 8.1). While many cats respond to drug therapy and behavior modification and can subsequently be weaned off medical treatment, other cats relapse upon cessation of treatment. In these cases, long-term therapy with the lowest effective dosage of an effective drug may be necessary.

table 8.1 Medications Used to Treat Refractory House Soiling in Cats

CLASS OF DRUG	EXAMPLES	MODE OF ACTION	SIDE EFFECTS	COMMENTS
Benzodiazepines	Diazepam, Alprazolam	Anti-anxiety; reduced state of arousal	Lethargy, sedation, ataxia, increased appetite, liver toxicity (rare)	Controls spraying in ~75% of cats but problem often resumes when discontinued
Azaperones	Buspirone	Neurotransmitter effects; reduces social anxiety	Behavioral changes; gastrointestinal upset	May take up to 3 weeks to affect behavior; non-sedating
Tricyclic Antidepressants	Amitriptyline, Clomipramine	Neurotransmitter effects; reduces arousal, anxiety	Increased drinking and urination, GI upset, urinary or fecal retention, glaucoma	Should not be prescribed in cats being treated with monoamine oxidase inhibitors
SSRIs	Paroxetine, Fluoxetine	Enhance availability and effects of serotonin	Loss of appetite, diarrhea, vomiting	May take several weeks to affect behavior; used effectively for urine spraying

REVIEW QUESTIONS

1. Explain the relevance of "learned avoidance" to house-soiling problems in dogs.
2. Provide a set of tips for dog owners to aid in the completion of house-training in a dog who has not been reliably house trained.
3. What health problems should be considered in a cat who has stopped using the litter box?
4. What are the most common causes of litter-box aversion in cats?
5. What characteristics of the litter box are important to consider when treating a cat with a house-soiling problem?

REFERENCES AND FURTHER READING

Blackshaw JK. **Feline elimination problems**. *Anthrozoos*, 605:52–56, 1992.

Borchelt PL. **Cat elimination behavior problems.** *Vet Clin North Amer: Small Anim Pract,* 21:257–264, 1991.

Borchelt PL, Voith VL. **Diagnosis and treatment of elimination behavior problems in cats.** *Vet Clin North Amer: Small Anim Pract,* 12:673–681, 1982.

Cooper LL. **Feline inappropriate elimination.** *Vet Clin North Amer: Small Anim Pract,* 27:569–600, 1999.

Cottam N, Dodman NH. **Effect of an odor eliminator on feline litter box behavior.** *J Feline Med Surg*, 9:44–50, 2007.

Crowell-Davis SL. **The litterbox blues.** *Compend Cont Ed Pract Vet*, Jan:34–37, 2007.

Feldman H. **Methods of scent marking in the domestic cat.** *Can J Zoology*, 72:1093–1099, 1994.

Halip JW, Vaillancourt JP, Luescher UA. **A descriptive study of 189 cats engaging in inappropriate elimination behaviors**. *Feline Pract*, 26(4)18–2, 1998.

Hart B. **Objectionable urine spraying. Urine marking in cats: Evaluation of progestin treatment in gonadectomized males and females.** *J Amer Vet Med Assoc*, 177:529–565, 1980.

Hart BL, Barrett RE. **Effects of castration on fighting, roaming, and urine spraying in adult male cats.** *J Amer Vet Med Assoc*, 163:290–296, 1973.

Hart BL, Leedy M. **Identification of source of urine stains in multi-cat households.** *J Amer Vet Med Assoc*, 180:77–78, 1982.

Hart BL, Eckstein RA, Powell KL, Dodman NH. **Effectiveness of buspirone on urine spraying and inappropriate urination in cats.** *J Amer Vet Med Assoc*, 203:254–258, 1993.

Herron ME, Lord LK, Hill, LN, Reisner IR. **Effects of pre-adoption counseling for owners on house-training success among dogs acquired from shelters.** *J Amer Vet Med Assoc*, 231:558–562, 2007.

Kendall K, Ley J. **Cat ownership in Australia: Barriers to ownership and behavior.** *J Vet Behav*, 1:5–16, 2006.

Koroffsky P. **Identifying source of urine on rugs.** *J Amer Vet Med Assoc*, 191:917, 1987.

Landsberg GM, Wilson AL. **Effects of clomipramine on cats presented for urine marking.** *J Amer Anim Hosp Assoc*, 41:3–11, 2005.

MacDonald DW. **Patterns of scent marking with urine and faeces amongst carnivore communities.** *Symp Zoolog Soc London*, 45:107–114, 1980.

Melese P. **Detecting and neutralizing odor sources in dog and cat urination problems.** *Applied Animal Behavior Science*, 39:188–189, 1994.

Mills DS. **Long-term follow up of the effect of a pheromone therapy on feline spraying behaviour.** *Vet Rec*, 147:746–747, 2000.

Mills DS. **Pheromonatherapy: Theory and applications.** *In Practice*, 27:368–373, 2005.

Mills DS, Mills CB. **Evaluation of a novel method for delivering a synthetic analogue of feline facial pheromone to control urine spraying by cats.** *Vet Rec*, 149:197–199, 2001.

Olm D, Houpt KA. **Feline house-soiling problems.** *Appl Anim Behav Sci*, 20:335–345, 1988.

Overall KL. **Diagnosing and treating undesirable feline elimination behavior.** *Feline Pract*, 21:11–15, 1993.

Pal SK. **Urine marking by free-ranging dogs (*Canis familiaris*) in relation to sex, season, place and posture.** *Appl Anim Behav Sci*, 80:45–59, 2003.

Salman ME, New JC, Scarlett JM, Kass PH, Ruch-Gallie PH, Hetts S. **Human and animal factors related to the relinquishment of dogs and cats in twelve selected animal shelters in the United States.** *J Appl Anim Welfare Sci*, 1:207–226, 1998.

Simpson BS. **Feline house soiling. Part I. Inappropriate elimination.** *Compend Continuing Ed Pract Vet*, 20:54–66, 1998.

Simpson BS. **Feline house soiling. Part II. Urine and fecal marking.** *Compend Continuing Ed Pract Vet*, 20:301–309, 1998.

Sung W, Crowell-Davis SL. **Elimination behavior patterns of domestic cats (*Felis catus*) with and without elimination behavior problems.** *Amer J Vet Res*, 67:1500–1504, 2006.

Taylor K, Mills DS. **A placebo-controlled study to investigate the effect of Dog Appeasing Pheromone and other environmental and management factors on the reports of disturbance and house soiling during the night in recently adopted puppies (*Canis familiaris*).** *Appl Anim Behav Sci,* 105:358–368, 2007.

Tynes VV. **Canine house-training challenges: From dogs soiling in their crates to dogs afraid of eliminating in front of people.** *Vet Med*, 102:254–261, 2007.

Wirant SC, McGuire B. **Urinary behavior of female domestic dogs (*Canis familiaris*): influence of reproductive status, location and age.** *Appl Anim Behav Sci,* in press, 2003.

Wright JC, Amoss RT. **Prevalence of house soiling and aggression in kittens during the first year after adoption from a humane society.** *J Amer Vet Med Assoc,* 224:1790–1795, 2004.

Wright JC, Nesselrote MS. **Classification of behavior problems in dogs: distributions of age, breed, sex and reproductive status.** *Appl Anim Behav Sci,* 19:169–178, 1987.

Yeon SC, Erb HN, Houpt KA. **A retrospective study of canine house soiling: Diagnosis and treatment.** *J Amer Anim Hosp Assoc,* 35:101–106, 1999.

Footnotes

[1] Salman ME, New JC, Scarlett JM, Kass PH, Ruch-Gallie PH, Hetts S. Human and animal factors related to the relinquishment of dogs and cats in twelve selected animal shelters in the United States. *J Appl Anim Welfare Sci*, 1:207–226, 1998.

[2] Yeon SC, Erb HN, Houpt KA. A retrospective study of canine house soiling: Diagnosis and treatment. *J Amer Anim Hosp Assoc*, 35:101–106, 1999.

[3] Hart BL. Castration and urine marking in dogs. *J Amer Vet Med Assoc*, 164:140–144, 1974.

[4] MacDonald DW. Patterns of scent marking with urine and faeces amongst carnivore communities. *Symp Zoolog Soc London*, 45:107–114, 1980.

[5] Hart BL, Barrett RE. Effects of castration on fighting, roaming, and urine spraying in adult male cats. *J Amer Vet Med Assoc*, 163:290–296, 1973.

[6] Simpson BS. Feline house soiling. Part II. Urine and fecal marking. *Compend Continuing Ed Pract Vet*, 20:301–309, 1998.

[7] Melese, P. Detecting and neutralizing odor sources in dog and cat urination problems. *Appl Anim Behav Sci*, 39:188–189, 1994.

CHAPTER 9

Unruly and Disruptive Behaviors in Dogs and Cats

Most problem behaviors that pet owners report in their dogs and cats are normal canine and feline behaviors that are being expressed in circumstances that are not appropriate or in situations that distress the owner or other animals in the home. Behaviors that are classified as unruly or disruptive often reflect the exuberance of a young dog or cat who has not been thoroughly trained or who is experiencing a lack of social interaction, exercise, or mental stimulation. Meeting a pet's physical and emotional needs is an important component of animal care and a required part of responsible pet ownership. These may not be satisfied because the owner lacks understanding of the pet's behavioral needs, has insufficient time or resources available, or is not sufficiently committed to the pet. The paradox is that spending time interacting with and training a pet not only helps to prevent unruly and disruptive behaviors, but enhances the owner's affection for and commitment to the pet.[1,2] Training dogs and cats to be well-behaved family members should be viewed as one of the ways in which owners provide for their companion animal's behavioral needs, rather than in the more traditional view of training as discipline that is used to "correct" a poorly behaving animal. Basic training can prevent problems such as jumping up, excitable barking, boredom-induced destructiveness, and nocturnal hyperactivity. It has the added benefits of improving communication between owners and their pets, enhancing the development of the human-animal bond, and providing an enjoyable activity or pastime.

DOGS: JUMPING UP

When dogs greet their owners and other people, they often put their paws up on the person in an attempt to engage in physical contact. Some attempt to lick at the person's face and mouth. Licking near the face is part of normal submissive greeting behavior but may also be a learned response that has been reinforced in other situations (i.e. teaching "kiss"). In addition to greeting, some dogs also jump up and paw at their owners to solicit attention or as an invitation to play. Again, these behaviors are normal communication signals that are used during play between dogs and that dogs learn to direct toward people. When jumping up is used to solicit play, it may be accompanied by non-aggressive mouthing and barking.

Although all of this physical contact reflects normal communication patterns of dogs, many owners find jumping and mouthing to be annoying and wish to prevent these behaviors. However, once it is established, jumping up to greet can be difficult to extinguish. Owners have often inadvertently reinforced jumping up by being inconsistent with when they allow or disallow the behavior. For example, a dog who is allowed to jump on the owner during play, but not when the owner arrives home from work, is receiving inconsistent information. Similarly, some owners allow their dog to jump up on family members but are then not pleased when the dog jumps on visitors. Because jumping up is inherently reinforcing to the dog, every episode in which the behavior occurs provides positive consequences (physical contact, licking, interaction) for the dog and strengthens the behavior. Inconsistent responses from the owner put successful jumping attempts on an intermittent reinforcement schedule, serving to further strengthen the behavior. However, the self-reinforcing properties of jumping up can be used in the trainer's favor, as the dog can be taught that interaction and petting during greeting or play can only be attained by offering an alternate behavior: sitting for greeting.

Factors to Consider

Prior to teaching the alternate behavior, several other factors should be considered. First, the degree of excitement that the dog is showing should be evaluated. Dogs who are uncontrollable and appear to be hyper-stimulated during greeting are often not receiving adequate attention or exercise. For example, it is unreasonable to expect a dog who spends 16 to 20 hours a day in an outdoor kennel to sit quietly for greeting during the few minutes that he spends with his family each day. Although living in a kennel may provide adequate shelter and food, it is a socially impoverished environment that does not meet a dog's physical or emotional needs. In this case, changes must first be made to provide for the dog's social and exercise needs before attempting to gain control and reduce problem jumping. Another factor that must be evaluated is consistency within the family. If one family member believes that jumping and mouthing is acceptable and encourages the behavior, while another dislikes jumping and attempts to discourage or punish it, teaching the dog to consistently "sit for greeting" will be unsuccessful. In these cases, a possible solution (provided all family members agree) is to put the behavior of jumping on cue ("paws up") and to teach "sit for greeting" as a second behavior (Box 9.1).

Box 9.1 TEACHING SOPHIE TO "PAWS UP" (AND TO SIT FOR GREETING)

Sophie is a one-year-old Golden Doodle (Golden Retriever/Poodle cross) owned by Jan Wilson. Sophie is a very exuberant and outgoing dog and is well-socialized to people and other dogs. Jan walks with Sophie every day and meets several friends with their dogs to go walking several days a week. Jan and Sophie completed a basic obedience class at a local dog training school and have just recently started training in agility using a clicker as a conditioned reinforcer. Although Jan lives alone, she has many visitors and entertains frequently. Jan's friends all love Sophie and enjoy interacting and playing with her when they visit.

The only persistent behavior problem that Jan reports with Sophie is that she jumps up on Jan's friends when they visit. Sophie does this during initial greeting and also when visitors engage her in play. Jan has trained Sophie to "sit for greeting" and Sophie is relatively reliable with this behavior when she greets Jan and Jan's parents. However, Sophie continues to jump on Jan's friends. Jan is aware that several of her friends enjoy having Sophie jump on them and even encourage her to do this when they play. Because they all love Sophie and because Jan does not want to discourage her friends from playing with her dog, she decides to use clicker training to teach Sophie to put her paws up on a person on command. She then will ask her friends and visitors to always use this command when they wish to allow Sophie to jump up and to positively reinforce Sophie for "sitting for greeting" at other times. Here are the steps that Jan uses to train Sophie's new behavior:

- **Luring the behavior:** Jan realizes that Sophie may be hesitant to jump on her because she has a strong positive reinforcement history for sitting for greeting with Jan. Therefore, to induce the behavior, Jan first asks Sophie to sit. She then gently pats her chest and commands "Paws up!" in a friendly tone. When Sophie's feet come up (even a bit), Jan clicks and treats (CT).
- **Shaping:** As Sophie becomes more confident about putting her paws up, Jan gradually shapes the behavior by shifting criteria toward paws touching Jan's knees and then finally her chest. She also shapes putting paws on her extended forearm, CT for each successful attempt.
- **Gaining stimulus control:** Jan fades the visual cue of patting her chest and begins to reinforce only responses to the verbal command of "Paws up." Because she recognizes that Sophie is naturally more motivated to jump up than to sit, she reintroduces "sit for greeting" and trains the two behaviors together in each session. Atleast two or three repetitions of "sit" are trained for each single opportunity to practice "paws up."

- **Using differential reinforcement:** Jan also uses differential reinforcement to ensure that Sophie continues to offer "sit for greeting" and to promote stimulus control of each exercise. She selects a very high-value food treat (small pieces of hot dog) to use only when positively reinforcing "sit for greeting" (the less desirable behavior for Sophie). Once she has "paws up" on verbal cue, she decides to use just verbal reinforcement for this exercise, recognizing that "paws up" itself is inherently reinforcing for Sophie.
- **Time to generalize:** When Sophie reliable offers both "sit for greeting" and "paws up" with Jan, she begins to enlist her friends to help with Sophie's training. She maintains the differential reinforcement schedule with her friends, giving them the yummy treats to reinforce Sophie when she sits and asking them to always use the command "Paws up" when they wish to allow Sophie to jump up.

Training an Alternate and Incompatible Behavior

Teaching a dog to sit quietly for petting while he is greeting is incompatible with jumping up. In other words, it is impossible for a dog to be sitting and jumping up at the same time. The commands for "sit" and "stay" are first taught in a non-distracting environment and in situations in which the dog does not normally attempt to jump up (see Chapter 7, pp. 161–170 for instructions on training these commands). Once the dog has achieved a reliable sit and stay, the commands are introduced when the dog is greeting the owner or a family member. A lead is attached to the dog and very exuberant greeters also benefit from the use of a head halter to increase control. One person acts as the greeter and a second person handles the dog. The greeter is provided with several high-value food treats. As the dog approaches, the greeter holds his hand with the treats at the dog's eye level and asks the dog to sit *as the dog approaches*. (Note: Many novice trainers wait too long to cue the sit. If the dog is already "air-bound" when the command is given, the training is not effectively teaching an alternate behavior.) As the dog sits, the greeter provides one treat, keeping several treats still in his hand. The handler maintains a snug lead to prevent jumping up and the dog is reinforced verbally (quiet, calm praise) and with food treats by both the handler and the greeter (Figure 9.1).

After several repetitions have been reinforced and the dog has started to offer "sit for greeting" to family members, the greeter can begin to concurrently use negative punishment. If the dog tries to jump up, the greeter immediately backs away several steps, saying nothing and abruptly withdrawing the treat-holding hand. Jumping up is negatively punished in this situation because the desired stimuli (the greeter and the treats) are withdrawn in response to the undesired behavior (jumping up). The greeter steps forward to greet only when the dog is controlled and responds to the sit command. It is very important the dog is first introduced to the alternate behavior (sit) before attempting to negatively punish

figure 9.1 Teaching "sit for greeting"

jumping up, because the dog must have the option of offering a behavior that will allow him to earn positive reinforcement. If negative punishment is used alone or *before* the dog understands that sitting quietly has positive consequences, jumping up may be punished, but the dog does not know that "sitting" will avoid negative punishment and lead to interaction with the greeter. The repeated use of negative punishment (i.e. backing away) should always be avoided because frequent episodes of losing an opportunity to greet and interact are extremely frustrating to the dog and can result in redirected behaviors such as mouthing, barking, or nipping in frustration.

Approaches to Avoid

Unfortunately, various forms of positive punishment continue to be recommended for jumping up, despite their ineffectiveness and the risks associated with their use. Some traditional advice includes "kneeing" the dog in the chest, stepping on the dog's back feet, jerking the dog off of the greeter with a training collar, or yelling "no" in a harsh voice as the dog jumps. None of these approaches are as effective as the method of teaching dogs to greet in a controlled and friendly manner, and they all have the potential to be abusive. Even when used carefully, subjecting a dog who is greeting in a friendly manner to such aversive stimuli causes discomfort and anxiety at the very least and in many cases fear. It cannot be overemphasized that avoidance behaviors and extreme submission should not be confused with learning to greet without jumping up. The repeated use of positive punishment for jumping results in an insecure dog who greets nervously and submissively or may become fearful and avoid greeting altogether. For these reasons, the use of positive punishment for jumping up is never recommended.

Shaping and Practicing

The described approaches: first teaching an alternate and incompatible behavior (sit for greeting) and prudently adding negative punishment (removing social interactions) effectively communicates to the dog that the way to gain all of the positive stimuli (treats, petting, praise, and interaction) is to offer a sit as a greeting and to remain sitting for petting. This training is practiced repeatedly with different visitors and in all of the settings in which the dog typically jumps up. Owners must also be informed that because jumping up is a normal communication signal for greeting and because physical contact and interaction are powerful reinforcers for dogs, sitting for greeting is a training exercise that owners must practice consistently. Although not all dogs learn to sit for greeting in every distracting situation that they encounter, this training improves overall control and effectively reduces jumping up problems in the majority of dogs.

DOGS: DESTRUCTIVE BEHAVIORS

Chewing, scratching, and digging are the most common forms of destructive behaviors reported by dog owners. Once again, these are normal behaviors that become problem behaviors when they are expressed in an undesirable place or toward forbidden objects. Destructive behaviors can have several underlying causes. The most common cause of destructiveness in young dogs is normal exploratory chewing that the dog has not been taught to direct toward appropriate items. Other causes of destructiveness in the home (chewing, scratching) and in the yard (digging) include a lack of exercise or mental stimulation, insufficient or inappropriate training, boredom, and as an expression of learned attention-seeking behavior. Destructiveness that occurs as a result of separation stress or anxiety is a final classification and is discussed in detail in Chapter 10 (pp. 246–258).

Exploratory Chewing

Regular chewing is a normal part of the dog's feeding behavior and is also one way in which dogs explore and learn about their environment. Although puppies tend to chew most vigorously and frequently, all dogs enjoy chewing and should be provided with a variety of appealing chew toys throughout life. Puppies and many newly adopted dogs come into homes without an understanding of which items are acceptable chew toys and which are not. Training includes preventing access or redirecting the dog away from unacceptable items, and positively reinforcing the use of appropriate toys. As they explore their new home, puppies will be especially attracted to items that can be shredded (pillows, children's stuffed toys, clothing) and objects that can be gnawed (furniture, shoes). These objects are enticing because this type of chewing mimics the predatory feeding behaviors of evisceration, dissection, and gnawing on bones. Happily for dog owners today, an understanding of the types of materials that are attractive to dogs and recognition of the dog's natural need to chew has led to the development and marketing of a variety of suitable and safe chew toys. These include hard natural bones; synthetic bones; chew devices constructed of hard nylon, plastic, or rubber; and a variety of interactive "food delivery toys" (Sidebar 1). Owners can present several different types of chew toys to determine their dog's particular preferences, and can also rotate toys occasionally to enhance the dog's interest. Chew toys should be offered frequently to the dog, and the dog should be quietly praised whenever she voluntarily selects a toy or bone to chew.

Sidebar 1: INTERACTIVE, FOOD DELIVERY, AND CHEW TOYS FOR DOGS

Interactive Toys: This category includes toys that owners and dogs play with together. These are available in many forms and have a variety of functions. Some act as a bridge between the owner and dog during play, allowing the dog to focus on the toy while engaging in positive social interactions with the owner. This is beneficial for young and exuberant dogs as they learn that it is acceptable to use their teeth on a toy during play, but not acceptable to mouth or nip the person who is playing with them. Interactive toys can also provide mental stimulation and physical exercise. Teaching dogs to retrieve is an excellent way to provide both exercise and, when sit-stay is incorporated, to reinforce control exercises. Providing several different types of retrieving toys that move and bounce in different ways also help to maintain a dog's interest. Interactive toys can also be used to teach "find it" games in which dogs search using scent and vision. Again, basic training exercises can be incorporated into the game, such as a sit-stay as the toy is hidden and "come when called" when it is found. Finally, a pre-selected rope or pull toy can be used to play tug-of-war games. Tug of war provides an opportunity to teach dogs to play gently (no growling or grabbing allowed), to "give" the toy when commanded, and to remain under control when trained to "get it," "tug," and "give."

Food Delivery and Puzzle Toys: Items such as stuffed hard bones and stuffed Kong toys are effective tools to use during classical conditioning exercises. For example, a bone stuffed with soft-moist treats is presented during grooming sessions to condition the dog to enjoy being brushed and handled. A stuffed toy can also be used as a positive reinforcer for calm behaviors or when reinforcing control exercises such as a down-stay. Food delivery products for dogs are also available in a variety of mentally stimulating designs. Examples include treat balls and hard rubber or plastic products that contain slits and holes that hold treats. Some of these deliver individual kibble pieces as the dog manipulates them or pushes the toy along the floor, while others anchor a hard treat to part of the toy to encourage manipulation and chewing. Puzzle toys include plush toys with smaller pieces that can be pulled out and played with, some of which mimic small prey animals. While all of these toys can provide entertainment and stimulation for dogs, caution is advised. Just as a crate can be overused to confine dogs, food delivery toys can be overused and should never replace positive interactions with people, exercise, or training.

Chew Bones and Toys: Hard bones for dogs are constructed with a variety of materials and provide opportunities for extended periods of chewing and gnawing. The most popular chew bones include rawhide chews, hard sterilized bones, and hard plastic bones. The type of chew bone that is provided to a dog should be selected to match the dog's interest and intensity of chewing. In general, a safe and appropriate chew bone is one that takes a dog several weeks to destroy or consume, does not damage teeth or gums, and the dog does not break apart into large chunks. Similar to other types of toys, rotating chew bones can help to maintain interest and novelty.

In addition to providing attractive chew toys and bones, and encouraging their use, owners must also proactively prevent their dog's undesirable chewing. Management of the environment is the best approach to preventing undesirable chewing and for teaching dogs which items are their own and are safe for chewing. First, any object that might be a potential chew target must be secured out of sight and out of reach. Second, young dogs must be supervised whenever they are loose in the home, both to prevent inappropriate chewing and to monitor house-training. If the dog picks up something that is not hers, the owner should simply remove it from her mouth and redirect her to one of her own toys, praising her quietly when she chews on appropriate items. Although a loud noise or "no" can be used to interrupt chewing, the repeated use of positive punishment should be avoided. Similar to house-training, the continual use of harsh reprimands when the dog is chewing on something undesirable can teach the dog to chew out of sight of the owner or to run away whenever in possession of a novel item. Also similar to house-training beliefs, owners often mistakenly interpret this behavior as signifying that the puppy "knows he is wrong." However, avoidance simply indicates that the dog learned through repeated experience with punishment to stop chewing in the presence of the owner or, in more extreme cases, to avoid the owner when investigating new objects.

Finally, in addition to managing the puppy's environment and providing varied and interesting chew toys, owners can teach their dog the two invaluable commands of "leave it" and "give." The command "leave it" is used when the dog is intending to pick up something that is either dangerous or off-limits. The command "give" is used to retrieve an item from the dog once he has picked it up (Sidebar 2). These invaluable commands can be taught to puppies and to adult dogs who have a chewing or stealing problem. They are enjoyable to learn and a help to the owner when teaching a new puppy or dog about which toys are his and which items are not.

Sidebar 2 — TEACHING "LEAVE IT" AND "GIVE"

Picking up, chewing, and playing with novel items is an activity that almost every young dog enjoys. Part of learning the rules of a home include identifying which items are acceptable to chew (i.e. dog toys) and which items are not acceptable to chew (everything else). Teaching the commands of "leave it" and "give" allows owners to effectively redirect a dog from taking something that is not his and to retrieve stolen objects. Here are the basic steps for training these essential commands:

LEAVE IT: Desired response: The dog turns away from an item of interest and toward the owner, and does not attempt to pick up the item. **Steps:**

- A lead is attached to the dog's collar and the dog is commanded to sit/stay. The trainer then places an item that is mildly attractive to the dog on the floor, approximately 2 feet in front of the dog. The trainer stands on the dog's right side, slightly behind the dog.

- Holding the lead snugly, the trainer shows the dog a high-value food treat in his right hand, held at a position that requires the dog to look away from the object and toward the trainer (i.e. over his right shoulder).

- The trainer then pleasantly commands "leave it," takes one step backward and lures the dog toward him (and away from the object). As soon as the dog rises and moves toward the treat, the trainer positively reinforces this movement.

- If the dog attempts to reach for the forbidden object, the trainer simply stops the dog with the lead and redirects him with the food lure.

- Repeat. The lure is removed when the dog offers "turn" spontaneously. The trainer continues teaching "leave it" from the sit position until the dog readily turns towards the trainer and away from the object. The response is shaped from the stand position and then when the dog is walking toward the object. All turns to the trainer and away from the object are reinforced.

- The behavior is practiced with other objects, gradually increasing in their desirability to the dog (start with low-value items and work toward high-value items).

GIVE: Desired response: The dog releases the item in his mouth into the owner's hand. This can be used both with items that the dog is not supposed to have as well as with his toys when playing games of fetch. **Steps:**

- Teach with a highly valued toy that is used for fetch games. The toy is tossed and when the dog returns to the owner (a retractable lead can be used, if needed), the dog is asked to sit.

- As the trainer reaches toward the toy with the left hand, a food treat is presented in the right hand. (Initially, the food treat is shown to the dog, as a lure.) When the dog focuses on the food treat, the trainer commands "give" in a pleasant voice. As the dog opens his mouth for the treat, the owner takes the toy in the left hand and reinforces with the food from the right hand (i.e. "make a trade").

- The dog is not allowed to jump for the toy or to try to take the toy back after eating the treat. If necessary, the dog is commanded to sit again before resuming the game.

- Note: The purpose of using "make a trade" is to teach the dog to open his mouth and release an object into the trainer's hand, as opposed to the trainer *pulling* the object from the dog's mouth.

- A food treat is used to reinforce "give," but is removed as a lure when the behavior has been initiated.

Chewing as a Result of Boredom

Dogs whose social and behavioral needs are not being met often develop a barrage of boredom-related behaviors. Chewing is innately enjoyable to dogs and appears to aid in the reduction of stress and anxiety. As a result, destructive chewing is a common outlet for boredom or a lack of exercise. Destructive chewing, digging, and scratching are commonly observed in dogs who are left isolated in yards or kennels for extended periods of time, are crated extensively without sufficient opportunity for exercise, or who are frequently isolated from their family in a garage or separate room. Although owners often present these dogs to behaviorists and trainers with complaints about problem behavior, in essence these cases involve either owner ignorance or negligence regarding provision of a dog's basic needs (Sidebar 3). Treatment involves eliminating the cause of boredom and providing increased opportunities for exercise and for positive interactions with people. If the dog is spending a large portion of her day outdoors, bringing her inside and instituting a basic manners-training program to teach acceptable indoor behavior eliminates the cause (boredom) and the environment (the yard) in which the destructive behaviors are occurring. When a dog must spend many hours alone because of the owner's work schedule, hiring a professional dog walker to come in around noon or using community doggy day care can provide adjunct opportunities for exercise and socialization. Providing a variety of interactive food delivery toys when the dog must be confined is also helpful but should not be expected to replace human interaction and exercise.

Sidebar 3: OWNERS' RESPONSIBILITIES – PROVIDING FOR DOGS' AND CATS' BASIC NEEDS

In 1965, the Brambell Committee in the United Kingdom proposed a set of minimal standards for domestic animal care called the "Five Freedoms." Although these standards were originally written to address the needs of farm animals, Irene Rochlitz modified them to address the needs of companion animals. The basic five freedoms mandate that owners of companion animals are responsible for providing the following to their dogs and cats:

- *Food and water:* A balanced diet that meets the pet's nutritional needs at every life stage and fresh, clean water.

- *Suitable living environment:* Pets need adequate space, shelter from inclement weather and protection from temperature extremes, adequate light, low noise levels, and a clean living area.

- *Healthcare:* For dogs and cats, optimal healthcare includes vaccinations, neutering, parasite control, identification, and prompt veterinary care when injured or ill.

- *Opportunities to express normal, species-specific behaviors:* This includes, but is not limited to, social interactions with humans and other pets, opportunities for regular exercise and for mental stimulation, and opportunities to experience new environments and settings.

- *Protection from conditions that may cause fear or distress:* For dogs and cats this includes preventing excessive isolation or multiple re-homings and not being exposed to abusive or irresponsible training practices.

In recent years, studies have been conducted to examine the needs of dogs and cats in homes and the quality of life that they experience. Factors that have been found to positively influence the quality of care that a pet receives include the owner's past experience with pets, his or her education level, and the number of emotional bonds the owner has with other people. For dogs, level of care also tended to be higher when the owner was single, but decreased as the dog aged. In cats, quality of life was higher for cats living in multiple-cat homes compared with single-cat homes, suggesting that social interaction with other cats is more important for domestic cats than has been commonly believed. In addition, while the majority of cat owners were found to provide "good" to "excellent" care for their cat's physical needs, few considered the behavioral needs of their cats. One-third of the cats were reported by their owners to be poorly socialized and to be fearful or timid toward visitors. Similarly, a study of dogs reported that while owners typically provided toys to dogs in fenced yards, the dogs rarely played with the toys but rather spent most of their active time traveling between doors or windows of the home. This finding suggests that social interaction with people is more important to dogs than the opportunity to play with toys. Together, these studies provide important insight to factors that affect the level of care that people provide to their pets and suggest areas to target when educating potential adopters and current owners about the behavioral and social needs of their dogs and cats.

Adamelli S, Marinelli L, Normando S, Bono G. Owner and cat features influence the quality of life of the cat. *Appl Anim Behav Sci*, 94:89–98, 2005.

Kobelt AJ, Hensworth PH, Barnett JL, Coleman GJ, Butler KL. The behaviour of Labrador retrievers in suburban backyards: The relationship between the backyard environment and dog behavior. *Appl Anim Behav Sci*, 106:70–84, 2007.

Marinelli L, Adamelli S, Normando S, Bono G. Quality of life of the pet dog: Influence of owner and dog's characteristics. *Appl Anim Behav Sci*, 108:143–156, 2007.

Rochlitz I. A review of the housing requirements of domestic cats (*Felis silvestris catus*) kept in the home. *Appl Anim Behav Sci*, 93:97–109, 2005.

Chewing (or Stealing) as Attention-Seeking Behaviors

Occasionally, dogs learn that a successful method for gaining attention is to pick up something that is forbidden and entice the owner into an invigorating game of "catch me if you can." (Another common attention-seeking behavior is barking; see p. 222) Following the basic rules of operant conditioning, dogs who steal or chew items for attention have learned that these behaviors "work" for them. Stealing results in desired interaction with the owner, exercise, and the opportunity to play a game. Attention-seeking chewing can be distinguished from exploratory chewing or chewing due to boredom in that it is always conducted in the owner's presence. The dog may even approach the owner while holding the forbidden object, play bow, and then dance teasingly away. There may also be a history of the owner chasing the dog when he has taken a forbidden item.

When dogs chew or steal for attention, the first step of treatment is to ensure that sufficient attention and exercise are being provided to the dog on a regular basis. If walks, opportunities for play, and mental stimulation such as games and training are not part of the dog's daily routine, these must be established. A dog who is receiving strenuous exercise and who has learned to expect regular interactions each day with his owner is less likely to develop or maintain problem attention-seeking behaviors. For many dogs, simply increasing exercise and introducing some form of daily mental stimulation such as manners training, teaching retrieving, or find-it games, or walking in a new area several times a week, reduces or eliminates problem attention-seeking.

Preemptive redirection is an important component of treatment for attention-seeking chewing. Dogs who chew, steal, or bark for attention usually have very specific circumstances that trigger the problem behaviors. For example, a young dog may show attention-seeking behaviors every evening when the family sits down to dinner or later at night when the owner is watching television. Preemptive redirection involves providing the dog with an interactive toy *before* the problem attention-seeking occurs. For example, before each meal, a hard bone is stuffed with cheese or soft treats and given to the dog to chew. The dog is attached to a lead and is trained to maintain a down-stay during dinner as he chews on the special bone. For dogs who steal items when the owner is watching television or reading, the owner can hide several food-delivery toys around the living room just before sitting down (i.e. *before* the problem behavior is triggered). These toys provide an appropriate chew toy to the dog and effectively prevent attention-seeking chewing when they are provided before the problem behavior is triggered. In addition, the cues that previously signaled "opportunity to steal/chew for attention" is gradually reconditioned to predict an opportunity to chew on a favorite bone.

Preemptive redirection differs significantly from waiting until the dog offers the problem behavior and then replacing the toy with an appropriate toy, a technique that is commonly recommended. Waiting for the behavior to occur before redirecting the dog is not generally effective because this approach does not prevent the problem behavior and if used repeatedly may actually reinforce it by giving the dog attention and something that is desired as a "make a trade." Finally, basic manners training is very important for dogs with any type of problem attention-seeking behavior. A reliable "come when called" is needed so that a dog who steals items can be commanded to bring them back. Similarly, as discussed with exploratory chewing, teaching the commands "leave it" and "give" are indispensable tools for gaining control with a dog who chews or steals prohibited items (see Sidebar 2).

DOGS: PROBLEM BARKING

Dogs produce a wide range of sounds and use vocal communication frequently to interact with their owners and with other dogs. Barking is perhaps the most common form of vocal signaling used by dogs and occurs in a variety of contexts. Dogs bark to alert their owners to visitors, during greeting, when they are excited, and when they are anxious and stressed. Many owners encourage their dog to bark in certain situations and may even train a "speak" command. So, like most behaviors, barking itself is not a problem. It only becomes so when the barking is repetitive and when it occurs in a context that owners find undesirable. For example, barking to ask to go outside or to warn of intruders are typically encouraged and appreciated by owners. Alternatively, dogs who bark repetitively when a visitor appears or when they wish to come in from outside are considered to be ill-behaved. The similarities between these examples of "desirable" and "undesirable" barking illustrate a common paradox in our relationships with dogs. Desired contexts (i.e. it is acceptable to bark at the door when someone arrives) and undesired contexts (i.e. it is *not* acceptable to bark at the window when someone walks by) can be almost identical from the dog's point of view. It may be unreasonable to expect a dog to understand the distinction between them. Therefore, when diagnosing and treating problem barking in dogs, a complete behavioral history should be collected to identify triggering stimuli and the context in which they occur, the behavior's reinforcement history and the owner's level of consistency when reacting to the dog's vocalizations.

Repetitive Barking

Repetitive barking that occurs in response to people or animals approaching a dog's house or yard can be either territorial in nature or a manifestation of general excitement and arousal. Dogs who are timid or who have not been well-socialized may also bark at visitors or other dogs as an expression of anxiety or fear. Excitable barking that becomes repetitive can occur in dogs who become highly aroused while playing with another dog or person or who are in an anxiety-producing environment. Finally, dogs who are left outside for long periods of time may develop repetitive barking that may be more related to boredom than to territorial behavior. In all of these cases, repetitive barking is commonly labeled as "nuisance" barking, reflecting its effect upon the owners (and possibly upon neighbors) rather than its underlying cause. Therefore, whenever a case of nuisance barking is presented to a behaviorist or trainer, the context in which the behavior is occurring and all triggering stimuli must be identified so that its actual cause can be determined.

Once the underlying cause has been identified, treatment for barking involves managing the context and the triggering stimuli, and counter-conditioning an alternate response. For example, a dog who barks incessantly at the door when visitors arrive is usually hyper-stimulated and may also be showing territorial behavior. This behavior is maintained through both classical and operant learning. Greeting visitors is an unconditioned stimulus that results in excitement and arousal (and barking). The sound or sight of people approaching and the doorbell ringing are neutral stimuli that are paired consistently with the unconditioned stimuli. In other words, the door bell ringing consistently predicts opportunities to see visitors and becomes a conditioned stimulus that triggers great excitement in the dog. Excitable barking is operantly maintained when the dog is allowed to greet immediately after barking at the door (i.e. barking behavior is positively reinforced by social interactions with visitors).

The first step of treatment involves managing the situation to reduce the intensity of both the triggering stimuli and the dog's response. Blocking the dog's access to the door using closed doors or baby gates prevents the reinforcing properties of excitable barking at the door (i.e. barking results in opportunities to greet). Similarly, keeping the dog away from the boundary of his territory also reduces the intensity of the dog's arousal, *provided* counter-conditioning techniques are used to prevent frustration. Simply putting the dog in another room without counter-conditioning another response is a common mistake that causes frustration in the dog and can contribute to an increased level of arousal and exacerbate excitable and territorial behaviors. Therefore, removing the dog from the setting must always be accompanied by counter-conditioning an alternate and incompatible response. The dog is trained to sit or lie down and stay on a mat that is located across the room or located on the opposite side of a baby gate (i.e. not directly at the door). The dog is trained to "go to your mat" in the absence of any triggering stimuli (see target training, Chapter 7, pp. 166–167). The owner gradually shapes the behavior until the dog can be commanded to "go to your mat" from a spot close to the door. As with all operant responses, a very high-value food reinforcer is selected and is reserved for this training. (For some dogs, providing a special food-delivery toy on the mat during the down stay is a strong reinforcer; see Sidebar 1). Basic obedience training is also initiated to teach "sit for greeting" (see pp. 213–214). This can be used for a dog who barks repetitively upon greeting his owner at the door when the owner wishes to still allow the dog to greet. In these cases, teaching a "quiet" command is also helpful as a counter-conditioning exercise (Box 9.2).

Box 9.2 TEACHING EMMA THE "QUIET" COMMAND

Emma is a two-year-old Weimeraner. She lives with two other Weimeraners (Katie and Sally) and with her owner, Jonathan Stevens. Emma is well-behaved in the house and when out on walks and gets along very well with her housemates and with other dogs whom she knows from walks in her neighborhood. With one exception, Jonathan reports few problems with Emma or his other two dogs. The exception is Emma's incessant and very loud excitable barking at dinnertime and when Jonathan is preparing to go for a walk with the dogs. Jonathan lives in a dog-friendly condominium complex and his unit has neighbors above him and to each side. Because Emma is a big dog and her barking is very loud and exuberant, Jonathan has received several complaints from neighbors. He feeds his dogs at 6:30 in the morning, so Emma's barking can be disturbing. Jonathan visits with a local trainer and behaviorist, and undertakes the following program to reduce Emma's excitable barking and to teach her the "quiet" command in the situations when her barking becomes excessive. Here are the steps that he follows:

- *Introducing the "quiet" command:* Jonathan selects situations in which Emma is moderately excited but does not typically bark. He introduces the cue of "quiet" in these situations, simply by saying "quiet" in a pleasant voice and positively reinforcing with a high-value food treat when Emma is calm and not barking. (Note: The purpose of this step is to classically condition the command "quiet" with the presentation of a treat in association with calm and quiet behavior.)
- *Managing the triggering stimuli:* Jonathan notes that being fed with the other dogs increases Emma's excitement at dinnertime and exacerbates her barking. To improve her success rate, he decides to temporarily

feed the dogs separately. He also puts the other two dogs in another room as he prepares for walks to reduce Emma's level of excitement before trips outdoors.
- **Training an alternate behavior:** Jonathan brushes up on Emma's obedience training by training her to sit and stay reliably in the kitchen (where he prepares her food) and near the door (where they prepare for walks). He practices these behaviors separately from walks and dinnertime.
- **Reinforcing "quiet" at dinnertime:** Once Emma reliably offers "quiet" and "sit-stay," Jonathan begins to request "quiet" at dinnertime. As he prepares dinner, he asks Emma to sit and also tells her "quiet." If she sits and is quiet, he immediately reinforces her from the bowl with a piece of her dinner. If Emma remains sitting and does not bark, Jonathan continues to prepare the dinner and to reinforce her with food from the bowl. If Emma barks or jumps up excitedly, Jonathan immediately stops all preparations, becomes motionless (negative punishment). He commands "quiet" and reinforces again when Emma is quiet. Jonathan's rule at dinner becomes: "Dinner preparations continue when Emma is quiet and stop when she is not."
- **Reinforcing "quiet" before walks:** The same training sequence is used prior to walks, Emma is reinforced with treats and praise if she sits and is quiet; all walk preparations abruptly stop if she barks.
- **Reintroducing other dogs:** As Emma progresses and is able to maintain her self-control by sitting and remaining quiet during times of excitement, Jonathan adds in the other dogs and continues to train (he expects a short relapse in the problem barking as excitement increases with the presence of the other dogs).

In cases of territorial barking, the dog's use of the territory must be evaluated. Repetitive territorial barking often develops in dogs who spend an excessive amount of time isolated in a yard or kennel. Dogs who spend many hours in a yard are under-stimulated and experience isolation in the same way as a dog who is left alone indoors for long periods. In these cases, it is uncertain whether the barking is truly territorial in nature or is more related to boredom. Management procedures include reducing the time spent in the yard, introducing interactive toys to the yard for the dog to find during the time that he is outdoors, and increasing exercise and interaction with the owner. The period of time that the dog can typically tolerate without barking should be recorded and the dog must be brought indoors *before* barking begins to avoid inadvertently reinforcing barking behavior. A history of barking that has been followed by being brought indoors (to stop the barking) maintains the barking behavior, effectively reinforcing repetitive barking with the opportunity to escape the yard and to attain social interaction in the home.

Attention-Seeking Barking

Dogs who bark as an invitation to play or during normal greeting can learn to use barking in other situations to elicit their owner's attention. Similar to other forms of attention-seeking behaviors, barking for attention often has a strong reinforcement history. The owner initially interacts with the dog in response to one or two barks and unintentionally reinforces barking. After several repetitions, the dog learns to bark repeatedly to gain this response. (For some dogs, even a verbal reprimand is perceived as a positive interaction and does not effectively punish barking.) When attempting to eliminate attention-seeking barking, the first step is to examine the dog's daily schedule of attention and exercise. In some cases, the dog's needs for exercise and attention are not being met and the dog has learned that the one way to gain attention is through barking. In these situations, treatment involves increasing daily exercise and walks, and ensuring that the dog receives positive interactions with the owner on a regular basis. Basic obedience training is recommended to improve control and to increase positive interactions. For some dogs, simply increasing exercise and providing a regular outlet for mental stimulation in the form of training or play results in a significant reduction in barking for attention.

To address the barking directly, the owner must identify all of the contexts in which the dog barks for attention. Most dogs have an established pattern and, when owners are asked to think about this, they realize that the barking occurs in one or more specific situations. For example, an owner who works at home may report that the dog barks for attention whenever the owner is speaking on the telephone. In an attempt to keep the dog quiet and finish his call, the owner has developed the habit of petting the dog when she barks during phone calls. In another situation, a dog may have learned to bark in anticipation of a daily walk that occurs at 6:00 P.M., and has gradually extended the period of barking such that he begins to bark for his walk as soon as the owner returns home from work at 5:00 P.M. (classical and operant conditioning at work once again). Although extinction is often recommended for attention-seeking behaviors (i.e. ignoring all barking behavior), in reality extinction is very difficult for most owners to use successfully as a behavior modification technique because absolutely no barking behavior can be reinforced. In addition, extinction is always associated with an "extinction burst" comprised of *more* of the problem behavior, which can be very discouraging to owners. An alternate and more successful approach is to prevent barking by preemptively training an alternate behavior. For example, the owner whose dog barks when he is on the phone trains his dog to sit for treats or to lie down with a stuffed bone each time that he picks up the phone. Similarly, the owner of the dog who begins barking for her walk at 5:00 P.M. immediately takes the dog to the yard and plays retrieve for five repetitions upon arriving home from work. The dog is then given a stuffed bone until 6:00 P.M. These approaches meet the dog's needs for activity while at the same time breaking the association between barking and interaction. In all cases, the owner only interacts directly with the dog (petting, praise, getting ready for a walk), if the dog remains quiet (as in extinction), but also preemptively manages the dog's environment and uses counter-conditioning to decouple the association between barking and attention.

Punishment and Problem Barking

Positive punishment such as verbal or physical reprimands are not recommended for problem barking. Aversives that come directly from the owner may suppress the dog's barking temporarily, but it also has the potential to cause fear and avoidance. If the underlying motivation for the barking behavior is anxiety or nervousness, using punishment can increase anxiety and contribute to unwanted responses such as defensive aggression or destructive behaviors. Finally, direct punishment cannot address barking that occurs when the owner is absent. For example, a study of the effectiveness of spraying dogs with a water hose (owner-delivered punishment) to reduce barking at visitors found that while the punishment did stop the dog's barking at the time of the incident, it did not prevent recurrences, and after 90 days the majority of dogs (86 percent) had resumed problem barking.[3]

Anti-Bark Devices: One form of positive punishment that can be effective for certain types of repetitive barking is the use of bark-activated devices. These are intended to interrupt barking with an aversive stimulus such as a loud noise, an electronic shock, or a spray of noxious liquid (citronella). Some products are designed to be mounted on a wall or counter and can be used with dogs who bark at one or two specific sites. More frequently, the bark-activated mechanism is mounted on the dog's collar. These devices administer precisely timed punishment because they are activated by the exact behavior they are intended to suppress – barking. Collars that administer either an electronic shock or citronella spray have been shown to effectively stop barking. Electronic shock collars react to the vibrations of the dog's throat during a bark and deliver a shock of adjustable intensity to the dog's neck. The citronella spray collar includes a small microphone that detects the dog's bark and releases a spray of citronella-scented liquid into the dog's face.

A study comparing these two types of aversive stimuli (electronic shock and citronella spray) found that while the two types of collars were of equally effective, dogs' owners perceived the citronella spray collar as less aversive and more acceptable than the electronic bark collar.[4] This perception may be due to an inability of the owners to appreciate the degree of aversiveness that a noxious smell might have for dogs rather than an actual difference in the dogs' experiences. Additional studies have shown that dogs are more likely to habituate to the citronella spray over time than to other types of aversive stimuli, such as an electronic shock (Sidebar 4). These results suggest that the spray may be perceived as less aversive by dogs. Regardless of the type of collar, anti-bark collars should never be used when the cause of the dog's barking is anxiety or fear, or if the dog has the potential to develop aggression in response to the punishment. Because of their potential for abuse and the risk of increasing rather

Sidebar 4: EVALUATION OF CITRONELLA ANTI-BARK COLLARS

Several studies have been conducted to examine the effectiveness of citronella collars for certain types of nuisance barking. In 2001, Deborah Wells of Queens College in Northern Ireland tested the effectiveness of continual (30 minutes for every day) vs. intermittent (30 minutes every other day) in reducing barking frequencies of dogs who barked at traffic near their yard, while traveling in the car, or at the television. None of the dogs included in the study were demonstrating stress or fear-related barking. All dogs showed dramatic reductions in barking during the first week of wearing the collar. However, over the three-week period, barking gradually increased, albeit to frequencies still lower than those reported prior to treatment. This change suggests habituation to the collar and spray. Dogs who wore the collar every other day showed greater reduction in barking and less tendency to habituate than did dogs who wore the collar every day. These results suggest that the aversive nature of the citronella spray is not severe (i.e. habituation is possible) and that dogs may learn to associate the sensation of the large and bulky collar with punishment for barking (and likewise, its absence signaling no punishment).

Another study, published in 2003, was conducted in veterinary clinics located in the U.S. and Canada. The investigators compared the effectiveness of the citronella collar with a similar but scentless spray collar for reducing barking in healthy dogs who were visiting a clinic for boarding or elective surgery. Although both collars effectively reduced barking, more dogs responded to the citronella collar (76.67 percent) than to the scentless collar (59.6 percent). This study also subjectively evaluated the anxiety level of dogs wearing the collar. The anti-bark collars did not significantly increase or decrease apparent anxiety of the majority of the dogs. However, staff at each of the two clinics reported that the reduction in barking in dogs wearing the collar caused less barking in other dogs in the kennel area, making the environment less disturbing for dogs who were ill or recovering and contributing overall to a less stressful environment.

Because concerns about the stress that might be induced by anti-bark collars is an important consideration for both owners and for the welfare of dogs, a recent study examined behavioral (activity) and physiological (plasma cortisol level) stress responses in kenneled dogs wearing anti-bark collars. Dogs were assigned to wear either a control collar (no punishment), a citronella (lemon) collar, or an electronic collar. Both types of punishing collar effectively reduced barking in response to other dogs walking by. Although both anti-bark collars resulted in transient increases in cortisol levels on the first day of treatment, these levels were still within normal reference range and returned to baseline levels by the second day. The activity level of the dogs was not affected by either type of punishing collar and activity levels did not reflect an increase in anxiety or stress. Contrary to common belief, these results indicate that dogs did not perceive the electronic collar as more aversive than the citronella collar when used to control repetitive barking.

Moffat KS, Landsberg GM, Beaudet R. Effectiveness and comparison of citronella and scentless spray bark collars for the control of barking in a veterinary hospital setting. *J Amer Anim Hosp Assoc,* 29:343–348, 2003.

Steiss JE, Schaffer C, Ahmad HA, Voith V. Evaluation of plasma cortisol levels and behavior in dogs wearing bark control collars. *Appl Anim Behav Sci,* 106:96–106, 2007.

Wells DL. The effectiveness of a citronella spray collar in reducing certain forms of barking in dogs. *Appl Anim Behav Sci,* 73:299–309, 2001.

---

than decreasing problem behaviors, anti-bark collars should only be considered in cases that have been found to be resistant to behavior modification, training and management techniques to reduce repetitive barking.

DOGS: PROBLEM DIGGING

Like most problem behaviors discussed in this chapter, digging is a normal and natural canine behavior. Similar to jumping up to greet and to barking, engaging in digging is inherently rewarding to most dogs (Figure 9.2). Dogs dig in dirt to find a cool place to lie on warm days, to search for bugs or small mammals that live beneath the surface of the lawn, as a solution to boredom, and, less commonly, as an attempt to escape (most dogs will attempt to climb or jump over a fence as opposed to digging out). Dogs who spend many hours in a yard or are under-stimulated and under-exercised learn to entertain themselves by barking (see above) and by being destructive (digging and chewing). When digging is considered to be a problem by the owner, providing for the dog's needs by increasing social interactions, increasing exercise, and reducing the amount of time that the dog spends alone in the yard is the most effective solution. When predation is the primary motivation for digging, ridding the yard of prey species often reduces but does not completely eliminate digging. Dogs that have been selected to have a strong desire to dig, such as terriers and terrier mixes, can often be trained to focus their efforts in a sand pit in one corner of the yard. Toys and bones can be buried in the pit to encourage the dog to dig in that area rather than in the yard. Finally, positive punishment in the form of remote-activated citronella or shock collars or motion-activated sprinkler devices can effectively suppress digging. However, if the dog is digging because of boredom or a lack of stimulation, it is likely that other self-reinforcing and destructive

figure 9.2 Digging is inherently enjoyable for most dogs

behaviors will take its place. For this reason, prevention of digging by reducing opportunities and increasing social interactions and management of the yard are the preferred approaches.

DOGS: OVERLY ACTIVE (HYPERACTIVE)

Although some owners describe their dogs as "hyperactive," most dogs whose energy levels exceed the expectations of their owners are showing normal activity levels for their breed, age, and lifestyle. True hyperactivity as a pathological condition is extremely rare in dogs, and is characterized by excessive activity and chronic restlessness even when adequate exercise and mental stimulation have been provided. Conversely, the majority of dogs whose owners describe them as hyperactive are showing a high level of activity in situations that the owner finds to be problematic. This can be caused by an owner who has unrealistic beliefs about the amount of exercise and attention that the dog requires, and by a lack of basic training and control. Unfortunately, many of the breeds of dogs that are popular today are also breeds that have been selected to be very active and energetic. Golden Retrievers, Labrador Retrievers, Border Collies, Australian Shepherds, and the various "Doodles" all have attributes that make them attractive as family companions but are also breeds that were developed to work as hunting companions or herding dogs, functions that require a temperament that is highly active and athletic. Even though an individual dog may not be used for the breed's intended purpose, owners must be educated regarding the high energy levels of dogs and their needs for regular exercise and activity.

Treatment for overly active dogs involves an initial assessment of the dog's daily opportunities for exercise, mental stimulation, and social interaction. If an increase in exercise is warranted, this can be achieved through longer daily walks,

figure 9.3 Retrieving is an enjoyable activity for many dogs

teaching enjoyable games such as fetch, Frisbee, hide-and-seek, swimming, or arranged play dates with familiar dogs (Figure 9.3). A reputable and well-managed dog day care can provide positive social interactions and exercise for dogs whose owners cannot provide adequate exercise each day. However, owners should be cautioned to carefully observe a facility to ensure that their dog is safe and well-cared-for while enrolled. In cases where the dog must be left alone for a large part of the day, hiring a professional dog walker to exercise the dog during the day is another option. In addition to exercise, dogs who are labeled as overly active often benefit from structured obedience training. Teaching a dog to walk on lead and to respond reliably to the sit, down, come, and stay commands enhances the positive bond that the owner has with the dog and also helps the dog to learn self-control. Some owners and dogs enjoy becoming involved in a canine sport such as Flyball, agility, or Rally, and these activities all provide a positive outlet for a dog's energy.[5]

When unruly behaviors have been unintentionally reinforced by the owner and function as attention-seeking behaviors for the dog, counter-conditioning and counter-commanding are used to train alternate and incompatible responses. As discussed previously, ignoring attention-seeking behaviors in an attempt to extinguish them is recommended by some trainers and behaviorists. However, in reality, extinction alone is rarely successful in practical settings because many of these behaviors are innately reinforcing to the dog and because ignoring the behavior consistently is very difficult for most owners. In addition, owners must be informed that ignoring established attention-seeking behaviors is in essence negative punishment (removing a desirable stimulus to reduce the frequency of an undesirable behavior). As such, this approach can cause a high degree of frustration in the dog, especially if it is attempted without concurrently training alternate and acceptable behaviors. For most dogs, increasing exercise and social stimulation, retraining basic obedience commands, and providing a structured daily routine reduces the excess energy problem while at the same time preserving and enhancing the positive bonds between owner and dog.

CATS: FURNITURE CLAWING

Furniture clawing is an expression of normal marking behavior in cats. In addition to providing a visual and olfactory mark, clawing is also important for claw care (see Chapter 3, p. 50 for a complete discussion of the functions of scratching). Cats often choose to use the sides of couches and chairs as scratching posts because they typically are covered with rough upholstery that is attractive for catching claws and depositing scent. Since cats usually claw mark near resting areas, the couch is a likely marking spot if the cat sleeps on or around this piece of furniture. Ofcourse, regardless of these "natural" aspects of furniture clawing, most cat owners wish to prevent this behavior since clawing is damaging to furniture and other household items.

The best treatment for cats who claw furniture or other undesirable areas is to prevent the behavior by providing the cat with one or more suitable scratching posts (see Chapter 6, pp. 152–154). Optimal scratching posts should be very sturdy and high enough for the cat to stretch out when reaching forward to scratch. The post should be covered with a heavy, rough fabric such as burlap or a similar type of cloth. Scratching posts can be placed near the cat's sleeping area or in an area in which the cat spends a great deal of her time. Placing food tidbits or a toy on the top of the post can encourage the cat to begin to use the post. When a cat has an established furniture-scratching problem, the post can be placed directly in front of the portion of the furniture that the cat is using. The furniture should be covered with tin foil or plastic wrap to prevent the cat from continuing to use that surface. Once the cat is using the post consistently, the owner can gradually move it to an area that is more acceptable. Although the fabric of the post will become worn, it should not be replaced because the cat's scent and visual cues encourage the cat to continue to use the post for scratching.

In most cases, providing an adequate scratching post, taking time to train the cat to use the post, and preventing the use of furniture during training is successful. However, because scratched areas continue to carry a cat's scent and because cats develop preferred scratching surfaces that are often difficult to change, some cats will continue to use furniture as a preferred scratching site. In these cases, management approaches are directed toward reducing damage. Trimming the cat's nails weekly removes the sharp tips of the claws and significantly reduces damage to furniture that is scratched. Commercially available plastic nail caps can be glued onto the cat's nails. These completely prevent damage to furniture, but must be replaced every few weeks. A final solution is declawing (onychectomy). Declawing the cat on the front feet does not stop the cat from marking (and attempting to claw) but rather eliminates the cat's potential to do any damage to furniture. Declawed cats still scent mark, but they are unable to leave any type of visual mark. However, declawing is a painful procedure for the cat and removes an important means of self-defense. This surgery is highly controversial and many cat professionals and veterinarians no longer recommend it. Although still relatively common in the United States and Canada, declawing surgeries for pet cats is prohibited in the United Kingdom, Germany, Switzerland, and Finland. Therefore, as a solution for furniture clawing, onychectomy should only be considered as a final resort after all other training and management approaches have failed.

CATS: JUMPING UP ON COUNTERS/FURNITURE

Most cats enjoy exploring tabletops and counters and prefer to rest in elevated places. These behaviors are self-reinforcing in that cats prefer spots where they are protected and can view their territory. In addition, exploring tabletops and counters can provide positive consequences in the form of appealing odors and occasional food tidbits that are present. Some cats develop a habit of walking along desks and workspaces to attain the owner's attention or to play with available "toys" such as paper clips and pens. (Speaking from experience, it is rather difficult to write when there is a cat walking across one's computer keyboard.)

Treatment for "counter crawling" in cats first requires that all foods and potential "toys" are removed from countertops and desks. For some owners, the easiest approach is to block the cat's access to these areas when the cat cannot be closely supervised. Second, several alternate and equally appealing elevated rest and play areas must be provided. The owner should select several acceptable areas around the house such as windowsills or the back of a couch or chair. At least one of these spots should provide a view of the outdoors. Kitty condos are excellent alternatives as most include elevated platforms for climbing and perching, scratching surfaces, and cubby holes for hiding (Figure 9.4). The cat is encouraged to use these areas by providing petting and attention, or tossing treats to the cat whenever she approaches or uses acceptable spots. Kitty condos often have areas where dangling toys can be attached to entice the cat to play. When the cat attempts to use an unacceptable spot, the owner should simply remove her and place her in one of the acceptable areas. Over time, the presence of the cat's scent and the lack of opportunities for positive reinforcement on counters and desks condition the cat to prefer the alternate spots.

figure 9.4 Cats enjoy the elevated resting spots provided by multi-level "kitty condos"

Many cats learn to stay off counters when their owners are present, but continue to use these surfaces whenever the owner is absent. This usually occurs because the owner has used some form of punishment to stop the cat from jumping onto counters. Punishments such as yelling or swatting the cat are obviously associated with the owner and also carry the risk of causing the cat to fear and avoid the owner. Although the use of a squirt bottle or loud mechanical noise such as banging a metal pan are often recommended and promoted as "environmental" (i.e. not associated with the owner), cats *do* learn that the person is the source of these aversive stimuli and that the presence of the owner is predictive of being punished. For these reasons, any form of punishment that is delivered by the owner will only reduce the behavior when the owner is present, and always carries the risk of frightening the cat.

Remote punishers are devices that emit a loud noise or blast of air whenever the cat attempts to jump onto a prohibited counter. Examples include tin cans containing marbles that will rattle and startle the cat when he jumps up, or the presence of cellophane paper or tin foil. Motion-detection products designed specifically as remote punishers for cats are also available. These are placed onto counters and emit a noise or blast of air when activated by motion. While remote punishers are effective when present, cats do learn that the presence of the device is associated with the punishment and so will often continue to use the spot if the remote punisher is not present. Therefore, it is important to continue to reinforce the appropriate places for the cat to rest and play when remote punishment is used.

CATS: NOCTURNAL ACTIVITY

Although domestication has significantly altered the cat's social behavior, cats are still by nature more nocturnal (active at night) or crepuscular (active at dawn and dusk) than diurnal (active during the daylight hours). However, because most human households operate on a diurnal schedule, cats are expected to adapt to daily activity patterns that are somewhat unnatural to them. Given this evolutionary background, it is not surprising that some cats disturb their owners during the night by playing and getting into other forms of mischief. Nocturnal activity is most common in kittens and adolescent cats who have high activity levels. For most cats, this problem abates when they reach 12 to 18 months of age. However, some cats continue nocturnal activity, even as adults. Undesirable nocturnal behaviors include chasing around the house, knocking over objects or damaging furnishings, exuberant play sessions between cats in multiple-cat homes, and attention-seeking behaviors directed towards the sleeping owner. In extreme cases, the cat will even ambush the owner during the night, attacking feet, hands, or faces. For obvious reasons, although being active at night may be "natural" for cats, when owners are losing sleep, their home is being damaged, or they are being injured, these behaviors are problematic.

   As with any other behavior that reflects a need for play and attention, treatment involves ensuring that the cat is receiving adequate exercise and interaction opportunities with the owner during the daylight hours. Because a cat is more likely to sleep through the night if activity and interaction occur in the evening before bed, the owner should provide nightly play sessions approximately 1 to 2 hours before bedtime. Playing immediately before bedtime is not recommended because the cat may still be aroused and feeling playful if the session ends too close to bedtime. A set of novel toys should be selected and used only for the evening play sessions. Toys that the cat can chase and pounce on are preferred by cats (Box 9.3). Management solutions include confining the cat to one part of the house, locking the cat out of the bedroom at night, or setting up remote punishers in areas of the home that are being damaged.

Box 9.3 BOOTS AND HIS TOYS

Boots is an eight-month-old, neutered, male Maine Coon Cat. His owner, Sandy Miles, has had Boots since he was 10 weeks old. Boots is a very active and outgoing cat, typical of the Maine Coon breed. He is very affectionate towards Sandy and also towards Sandy's older cat, Barney. Barney and Boots play often during the day while Sandy is at work and frequently sleep together on one of the kitty condo platforms in their home. The problem is that Boots is also very active in the middle of the night and attempts to engage both Sandy and Barney in play. Both cats sleep on the bed with Sandy, and Boots will wake at around 3:00 A.M. every night. He walks back and forth across the bed, meowing loudly. If ignored, he will alternatively bat at and pounce upon Barney or Sandy (neither of whom is interested in playing).

To reduce Boots's nighttime festivities, Sandy begins a program of active play and attention for Boots. She selects from the following types of cat toys and rotates these each evening to maintain Boots' interest. On evenings that she is not at home, she provides Boots with suspended toys or a food delivery toy. Here are some of the toys and environmental enrichment devices that she selected:

- **Batting toys:** Many of the play behaviors that cats show are part of normal predatory behaviors. Batting toys can be balled-up paper, an empty toilet paper tube, or a small stuffed toy. There are even cat toys available that are designed to vibrate, move, and squeak to simulate prey. Most cats will bat, chase, and pounce upon "prey-like" toys. Like many cats, Boots also enjoys "eviscerating" his toys by lying on his back, manipulating the toy with his front paws and raking the toy with his back feet.
- **Suspended toys:** Sandy selects a synthetic bird toy that is suspended from an elastic cord. The toy bounces erratically when Boots jumps and bats at it. This toy is hung from door handles or sometimes on Boot's climbing structure. For safety, Sandy only offers suspended toys when she is home to ensure that Boots does not become tangled in the cord or attempt to ingest part of the toy.
- **Interactive toys:** A variety of toys are designed for play between cats and their owners. Examples include laser lights, dangling toys suspended from a pole, and retrieving toys such as small balls or rings. Boots's favorite is a ping-pong ball! He especially enjoys it when Sandy places the ball in the bathtub, where he can bat it around and up the side of the tub. Ping-pong balls can also provide hours of entertainment for a cat who loves to chase and who is trained to retrieve.
- **Toys for hiding in:** Some cats love to explore and hide in new spaces. An inexpensive hiding toy for cats includes different sized cardboard boxes and paper (not plastic) bags.
- **Climbing structures:** Kitty condo structures that have several platform levels, scratching posts, and cubbyholes for hiding provide opportunities for jumping, stretching and climbing. Boots and Barney use their structure to play ambush games with each other and for rest time.

CATS: EXCESSIVE VOCALIZATION

Cats communicate vocally with their owners and with other cats. The "meow" sound in particular is used for interactions with people (and is very infrequently used to communicate with other cats). Cats are capable of producing a wide variety of meow types in different contexts and with apparently different meanings. In fact, many owners become quite proficient at interpreting their cats' vocalizations and responding in a manner that provides for their cats' needs. Therefore, it is not surprising that some cats develop attention-seeking vocalizations that are excessive. Many cases of "too much meowing" occur early in the morning when the cat is requesting food, attention, or play from the owner. Other stimuli for prolonged vocalizations include territorial behaviors, such as when a cat detects an unfamiliar cat outside of the home or is experiencing discomfort or pain, and as a sign of cognitive dysfunction in a geriatric pet. Although excessive vocalization can be reported in cats of any breed, it is more common in the oriental breeds such as Siamese and Burmese, presumably because these breeds have a tendency to use vocal communication more frequently than other breeds.

Because excessive vocalizations may be related to pain or illness, a complete veterinary check-up is warranted whenever there has been a sudden change in a cat's frequency of vocalizations. Persistent calling while in the litter box is suggestive of lower urinary tract disease, and a cat who shows a change in sleep-rest patterns and wakes the owner with persistent meowing may be developing age-related cognitive dysfunction. In these cases, medical treatment is often effective. When the cat's vocalizations are determined to be attention-seeking behaviors, the owner's response to the meows must be investigated. For example, if the owner has interpreted all meows as requests for food, this reinforces meowing for attention and also establishes food as the only type of reward to which the cat responds (which can lead to overweight conditions). Similarly, if each meow leads to petting and interaction, vocalizing for attention may have a strong positive reinforcement history of which the owner is often not even aware.

Treatment includes providing proactive positive interaction *before* meowing occurs. The owner first lists the times and situations in which the cat typically meows for attention. Play sessions and interactive toys are provided to preempt these attempts and redirect the cat to other activities. Similar to attention-seeking behaviors in dogs, although extinction is often recommended for attention-seeking vocalizations, the ability of owners to consistently ignore all meows and the unrelenting persistence of many cats makes this an impractical solution for most owners. Preventing the behavior through differential reinforcement of alternative behaviors is most effective and has the benefit of preserving the bond between owner and cat (Box 9.4). Finally, although some experts recommend using an aversive stimulus such as spraying with a water pistol or compressed air canister when the cat meows for attention, these solutions are not recommended because of the risk of causing fear and because they do not address the underlying cause of the behavior – a need for attention and affection.

Box 9.4 TIGGER'S PERSISTENT MEOWING PROBLEM

Tigger is a two-year-old domestic longhair cat, owned by Mike Powers. Mike has had Tigger since he was nine weeks old and reports that Tigger is very affectionate and playful. Mike provides a variety of toys for Tigger, which are located in a small basket kept in his living room. In addition, Tigger enjoys cuddling on Mike's lap when he is watching television or working on his home computer. At night, Tigger sleeps curled up at the end of Mike's bed.

Two months ago, Mike changed jobs and now has a much longer commute to and from work. As a result he must leave the house at 7:00 A.M. and often does not return home until 7:00 in the evening. Recently, Tigger has started to meow persistently; starting at about 4:00 in the morning and continuing until Mike gets up to feed him and play with him. Mike feels badly because he realizes that Tigger is lonely and under-stimulated because he is forced to spend more hours alone. However, Mike is losing sleep over this problem and needs a solution. He recognizes Tigger's needs for attention and exercise and so makes the following changes to Tigger's daily schedule and to his interactions with Tigger:

- *Increased play and exercise in the evening:* Mike realizes that he has reduced the amount of time that he plays with Tigger because of the added work hours and the stress of his new job. Knowing that playing with Tigger is an enjoyable (and stress-reducing) activity for him as well as for Tigger, he recommits to regular evening play sessions. Because he realizes that Tigger needs an outlet for his energy level, he includes several new toys in their play and also institutes a few "toy management" rules (see below).
- *Modified feeding schedule:* Because there is a possibility that Tigger is waking Mike because he would like to be fed, Mike begins to provide a small meal to Tigger shortly before bedtime.
- *Provision of interactive toys:* Mike selects several new toys for Tigger that include a laser toy and a "fishing rod" toy that Mike holds and manipulates during play sessions. He also finds a small stuffed mouse that Tigger loves and teaches Tigger to retrieve the toy, using small cat treats as a reinforcer. In addition, he purchases several types of interactive toys that can be attached to doorknobs or that move erratically and hides these around the house for Tigger to find during the day while he is at work.
- *Toy management:* Knowing that cats enjoy novelty and that a toy will maintain Tigger's interest longer if it is sporadically available, Mike begins to rotate Tigger's toys, providing one or two "new" toys each day. In addition, all of the interactive toys that Tigger and Mike play with together are put away, brought out for play sessions only.
- *Extinction of meowing:* To break the nightly cycle of "mid-night meows," Mike stops responding to any of Tigger's meowing in the middle of the night. Because Mike's time is limited with Tigger, he does not wish to shut Tigger out of the bedroom. As an alternative, he purchases a set of ear plugs and commits to one week of ignoring all meows. Tigger attempts to meow several times the first and second nights, but eventually goes back to sleep. Ensuring that Tigger's needs for companionship and exercise are met and adding extinction solves the problem and soon both Tigger and Mike area adapted to their new daily routine.

CATS: PLANT EATING AND PICA

Eating Houseplants

Plant eating is not that unusual, probably because cats who live outdoors commonly consume grass and plant eating may simply be the indoor cat's version of this behavior. However, plant eating can have serious consequences if the plant that is consumed is toxic. For this reason, cat owners should grow only non-toxic plants in their homes and, if their cat goes outdoors, in their gardens. Cats can be prevented from consuming houseplants by hanging plants or not allowing the cat access to rooms that have plants on the floor. If the cat has a definite preference for certain plants, applying a noxious-tasting substance to the leaves provides effective environmental punishment. Similar to digging in dogs, providing a small flowerpot that is planted with grass or catnip and training the cat to use only this pot as a "grazing area" can provide an alternate and acceptable outlet for this behavior.

Pica (Eating Non-Nutritional Items)

The term *pica* refers to the ingestion of non-nutritional items. In cats, pica most commonly manifests as plant eating (see above) or fabric eating (also referred to as "wool chewing"). Cats use their molar teeth to chew on pieces of cloth. A cat who shows this behavior apparently seeks out opportunities to chew on cloth and is capable of producing large gaping holes in clothing or fabric in a manner of minutes. There seems to be a breed-specific predilection for this behavior, as it is reported more frequently in Siamese and Burmese cats. While it is has been hypothesized that fabric chewing in cats is related to early weaning and represents a neotenized behavior pattern, there are not definitive data to prove this. It is more probable that fabric eating is a form of obsessive-compulsive behavior characterized by stereotypical chewing. Because some cats appear to chew fabric when they are hungry, increasing the number of meals that the cat is offered each day or feeding free-choice may reduce the behavior. For other cats, growing a safe plant or patch of grass in the home specifically for the cat to consume may redirect the cat from potentially dangerous clothing. Finally, because fabric chewing can be related to chronic stress in the cat's life or a form of obsessive-compulsive disorder, some cats respond favorably to pharmacological treatment. A veterinary behaviorist should be consulted for recommendations.

REVIEW QUESTIONS

1. Describe the steps used to teach a dog to sit for greeting as an alternate behavior to jumping up to greet.
2. What are the most common causes of destructive chewing in dogs?
3. Describe the two primary types of anti-bark collars and the advantages and disadvantages of each.
4. What types of training and household management techniques can owners use to keep their cat off countertops?
5. Discuss ways in which owners can prevent or reduce overly exuberant nocturnal activity in their cats.

REFERENCES AND FURTHER READING

Adamelli S, Marinelli L, Normando S, Bono G. **Owner and cat features influence the quality of life of the cat.** *Appl Anim Behav Sci*, 94:89–98, 2005.

Adams GJ, Clarke WT. **The prevalence of behavioural problems in domestic dogs: A survey of 105 dog owners.** *Australian Vet Pract*, 19:135–137, 1989.

American Association of Feline Practitioners. **Feline Behavior Guidelines.** AAFP, 43 pp., 2004.

Beaver BV. **Effectiveness of products in eliminating cat urine odors from carpet.** *J Amer Vet Med Assoc*, 194:1589–1591, 1989.

Beaver BV. **House soiling in cats: A retrospective study of 120 cases.** *J Amer Anim Hosp Assoc,* 25:631–636, 1989.

Beaver BV. **Feline behavioral problems other than house soiling.** *J Amer Anim Hosp Assoc*, 25:465–469, 1989.

Beaver BV. **Owner complaints about canine behavior.** *J Amer Vet Med Assoc*, 204:1953–1955, 1994.

Bennett PC, Rohlf VI. **Owner-companion dog interactions: Relationships between demographic variables, potentially problematic behaviors, training engagement and shared activities.** *Appl Anim Behav Sci*, 102:65–84, 2007.

Byrne RW. **Animal communication: What makes a dog able to understand its master?** *Curr Biol*, 13:R347–R348, 2003.

Call J, Brauer J, Kaminski J, Tomasello M. **Domestic dogs (*Canis familiaris*) are sensitive to the attentional state of humans.** *J Compar Psych*, 117:257–263, 2003.

Clark GI, Boyer WN. **The effects of dog obedience training and behavioural counseling upon the human-canine relationship.** *Appl Anim Behav Sci,* 37:147–159, 1993.

Coppinger RP, Feinstein M. **Why dogs bark.** *Smithsonian Magazine,* January:119–129, 1991.

Curtis TM. **Preventing behavior problems in cats.** *Compend Contin Educ Pract Vet*, Feb:119–121, 2006.

Donaldson J. **Culture Clash: A Revolutionary New Way of understanding the Relationship between Humans and Domestic Dogs**. James and Kenneth Publishers, Oakland, CA, 221 pp., 1996.

Hart BL, Hart LA. **Canine and Feline Behavioral Therapy.** Lea and Febiger, Philadelphia, PA, 275 pp., 1985.

Haug LI, Beaver BV, Longnecker MT. **Comparison of dogs' reactions to four different head collars.** *Appl Anim Behav Sci*, 80:1–9, 2002.

Heidenberger E. **Housing conditions and behavioral problems of indoor cats as assessed by their owners.** *Appl Anim Behav Sci*, 52:345–364, 1997.

Hennessy MB, Davis HN, Williams MT, Mellot C, Douglas CW. **Plasma cortisol levels of dogs at a county animal shelter.** *Physiol Behav*, 62:481–490, 1997.

Hetts S, Estep DQ. **Behavior management: Preventing elimination and destructive behavior problems.** *Vet Forum*, November:60–61, 1994.

Hilby EF, Rooney NJ, Bradshaw JWS. **Dog training methods: Their use, effectiveness and interaction with behaviour and welfare.** *Anim Welfare*, 13:63–69, 2004.

Horwitz D, Mills D, Heath S. **BSAVA Manual of Canine and Feline Behavioural Medicine.** British Small Animal Veterinary Association, Gloucester, UK, 288 pp., 2002.

Jagoe A, Serpell J. **Owner characteristics and interactions and the prevalence of canine behaviour problems.** *Appl Anim Behav Sci,* 47:31–42, 1996.

Jongman EC. **Adaptation of domestic cats to confinement.** *J Vet Behav*, 2:193–196, 2007.

Kobelt AJ, Hemsworth PH, Barnett JL, Coleman GJ. **A survey of dog ownership in suburban Australia – Conditions and behavior problems.** *Appl Anim Behav Sci,* 82:137–148, 2003.

Kobelt AJ, Hemsworth PH, Barnett JL, Coleman GJ, Butler KL. **The behaviour of Labrador retrievers in suburban backyards: The relationship between the backyard environment and dog behavior.** *Appl Anim Behav Sci*, 106:70–84, 2007.

Landsberg G, Hunthausen W, Ackerman L. **Handbook of Behaviour Problems in the Dog and Cat.** Butterworth-Heinemann, Oxford, UK, 211 pp., 2004.

Ledger RA. **Owner and dog characteristics: their effects on the success of the owner-dog relationship. Part 1: Owner attachment and ownership success.** *Vet Internat*, 11:2–10, 1999.

Ledger RA. **Owner and dog characteristics: their effects on the success of the owner-dog relationship. Part 2: Owner expectations and ownership success.** *Vet Internat,* 12:8–18, 2000.

Lindsay SR. **Handbook of Applied Dog Behavior and Training; Volume 2: Etiology and Assessment of Behavior Problems.** Iowa State University Press, Ames, IA, 328 pp., 2001.

Loveridge G. **Environmentally enriched housing for dogs.** *Appl Anim Behav Sci,* 59:101–113, 1998.

Marinelli L, Adamelli S, Normando S, Bono G. **Quality of life of the pet dog: Influence of owner and dog's characteristics.** *Appl Anim Behav Sci,* 108:143–156, 2007.

Miklósi A, Polgárdi R, Topál J, Csányi V. **Intentional behaviors in dog-human communication: an experimental analysis of "showing" behaviour in the dog.** *Anim Cognition,* 3:159–168, 2000.

Miklósi A, Polgárdi R, Topál J, Csányi V. **Use of experimenter-given cues in dogs.** *Anim Cognition,* 1:113–121, 1998.

Miller J. **The domestic cat: Perspective on the nature and diversity of cats.** *J Amer Anim Hosp Assoc,* 208:498–501, 1996.

Miller DD, Staats SR, Partlo C. **Factors associated with the decision to surrender a pet to an animal shelter.** *J Amer Anim Hosp Assoc,* 209:738–742, 1996.

Moffat KS, Landsberg GM, Beaudet R. **Effectiveness and comparison of citronella and scentless spray bark collars for the control of barking in a veterinary hospital setting.** *J Amer Anim Hosp Assoc,* 29:343–348, 2003.

Morgan M, Houpt KA. **Feline behavior problems: The influence of declawing.** *Anthrozoos,* 3:50–53, 1989.

Odendaal JSJ. **An ethological approach to the problem of dogs digging holes.** *Appl Anim Behav Sci,* 52:299–305, 1996.

Overall K. **Clinical Behavioral Medicine for Small Animals.** Mosby, St. Louis, MO, 544 pp., 1997.

Pal SK. **Urine marking by free-ranging dogs (*Canis familiaris*) in relation to sex, season, place and posture.** *Appl Anim Behav Sci,* 80:45–59, 2003.

Podberscek AL. **Positive and negative aspects of our relationship with companion animals.** *Vet Res Comm,* 30 (suppl):21–27, 2006.

Polsky RH. **Electronic collars: Are they worth the risks?** *J Amer Anim Hosp Assoc,* 30:463–468, 1994.

Rochlitz I. **A review of the housing requirements of domestic cats (*Felis silvestris catus*) kept in the home.** *Appl Anim Behav Sci,* 93:97–109, 2005.

Sales G, Hubrecht R, Peyvandi A. **Noise in dog kenneling: Is barking a welfare problem for dogs?** *Appl Anim Behav Sci,* 52:321–329, 1997.

Schalke E, Stichnoth J, Ott S, Jones-Baade R. **Clinical signs caused by the use of electric training collars on dogs in everyday life situations.** *Appl Anim Behav Sci,* 105:369–380, 2007.

Schilder MBH, van der Borg JAM. **Training dogs with help of the shock collar: Short and long term behavioral effects.** *Appl Anim Behav Sci,* 85:319–334, 2004.

Shyan-Norwalt MR. **Caregiver perceptions of what indoor cats do "for fun."** *J Appl Anim Welfare Sci,* 8:199–209, 2005.

Steiss JE, Schaffer C, Ahmad HA, Voith V. **Evaluation of plasma cortisol levels and behavior in dogs wearing bark control collars.** *Appl Anim Behav Sci*, 106:96–106, 2007.

Takeuchi Y, Ogata N, Houpt KA, Scarlett JM. **Differences in background and outcome of three behavior problems of dogs.** *Appl Anim Behav Sci*, 70:297–308, 2001.

Topál J, Miklósi A, Csányi V. **Dog-human relationship affects problem solving behaviour in dogs.** *Anthrozoos*, 10:214–224, 1997.

Turner DC. **Treating canine and feline behaviour problems and advising clients.** *Appl Anim Behav Sci*, 52:199–204, 1997.

Voith VL, Borchelt PL (editors). **Readings in Companion Animal Behavior.** Veterinary Learning Systems, Trenton, NJ, 276 pp., 1996.

Wells DL. **The effectiveness of a citronella spray collar in reducing certain forms of barking in dogs.** *Appl Anim Behav Sci*, 73:299–309, 2001.

Wells DL. **A review of environmental enrichment for kenneled dogs, *Canis familiaris*.** *Appl Anim Behav Sci*, 85:307–317, 2004.

Wells DL, Hepper PG. **Prevalence of behaviour problems reported by owners of dogs purchased from an animal rescue shelter.** *Appl Anim Behav Sci*, 69:55–65, 2000.

Yeon SC, Erb HN, Houpt KA. **A retrospective study of canine house soiling: Diagnosis and treatment.** *J Amer Anim Hosp Assoc*, 35:101–106, 1999.

Footnotes

[1] Bennett PC, Rohlf VI. Owner-companion dog interactions: Relationships between demographic variables, potentially problematic behaviors, training engagement and shared activities. *Appl Anim Behav Sci*, 102:65–84, 2007.

[2] Heidelberger E. Housing conditions and behavioral problems of indoor cats as assessed by their owners. *Appl Anim Behav Sci*, 52:345–364, 1997.

[3] Pageat P, Tessier Y. Disruptive stimulus: Definition and application in behavior therapy. In: *Proc of the First Internat Conf Vet Behav Med*, UFAW, Potters' Bar, UK, p. 187, 1997.

[4] Juarbe-Diax SV, Houp KA. Comparison of two anti-barking collars for treatment of nuisance barking. *J Amer Anim Hosp Assoc*, 5:231–235, 1996.

[5] Marder A, Reid P. Canine behavior problems: Behavior modification, obedience, and agility training. *Compend Continuing Ed Pract Vet*, 18:975–983, 1996.

CHAPTER 10

Separation, Fear, and Anxiety Problems in Dogs and Cats

Separation anxiety and fear-related behaviors make up a substantial proportion of the behavior problem cases referred to behaviorists and trainers. Separation anxiety in dogs is a prevalent problem, comprising up to 70 percent of cases in some clinics.[1,2] In addition, a substantial number of dogs show some behavioral signs of separation stress at some point in their life, even though their owners do not seek professional help for the problem.[3] Fears and general anxiety problems are also relatively common. A survey of pet owners found that almost 40 percent of dogs were fearful of loud noises, while 22 percent dogs were fearful of unfamiliar adults, 33 percent were fearful of children, and 14 percent were fearful of other dogs.[4] More than half of owners who adopt dogs from shelters report problems associated with fearful behaviors during the first year following adoption.[5] Although separation anxiety is relatively uncommon, shyness and fear-related behavior problems are seen in many cats. Some cat owners successfully manage these fears in their cats, but untreated fear-related behaviors can adversely affect a cat's welfare and quality of life.

NORMAL VS. PROBLEMATIC ANXIETY AND FEAR REACTIONS

Fear has adaptive functions and is a natural and normal part of an animal's behavioral repertoire. In all species, fear responses to certain types of stimuli are innate and are important for survival. The tendency to run away from painful experiences or from unfamiliar animals or situations has distinctive survival value and was selected for during both the dog and the cat's evolutionary history. In addition, a fear of unfamiliar animals or people is inherent for most wild animal species. However, for the domestic dog and cat, the genetic changes associated with domestication have attenuated these reactions. When pets live with families, socialization to new people, places, and other animals during early development and throughout life also contributes to reductions in timidity and fear.

Although socialization can often prevent fear-related problems from developing in many pets, it is not the only factor that influences fear-related behaviors. Fear reactions can become problematic when negative associations are made with harmless stimuli or when a dog or cat's fear occurs at intensity or frequency that it impacts the animal's safety, quality of life, or relationship with the owner and other family members.

Similar to fear responses, separation stress is a normal behavior pattern. It is expressed most intensely in puppies and kittens. Distress crying and increased activity (even hyperactivity) are normal infant responses to separation from an attachment figure, most typically the mother. Young puppies and kittens use these behaviors to communicate hunger, chilling, or loneliness. The mother reacts by returning to the litter and tending to her young. During weaning, the gradual introduction of repeated periods of separation from the mother causes initial distress that decreases with time and maturation. Mild levels of anxiety are also expected when puppies and kittens are adopted into new homes. As newly adopted pets begin to bond with their new family, primary social attachments are shifted to human caretakers and possibly to other pets in the family.

Minor distress upon separation from attachment figures and demonstrations of joy and elation during reunions are expected in a social animal. Many dogs prefer to be in the same room as their owners, whine or pace briefly when they are left home alone, and become very excited when their owners return. A dog who prefers to go for a ride with her owner in the car and who enjoys spending most of her time at the owner's feet is *not* showing abnormal attachment behavior. Similarly, although cats vary dramatically in their degree of sociability, a cat who cuddles in his owner's lap and meows for attention is not showing an abnormal level of attachment. Conversely, a pet who reacts to being isolated or confined with an extreme level of anxiety that is out of context with the situation is showing an abnormal response to separation. Severe separation reactions, especially when they are retained beyond puppy or kittenhood, are considered to be maladaptive behaviors because they cause prolonged agitation and stress and place the animal at risk of injury. For example, dogs who break teeth or injure themselves jumping out of windows or cats who over-groom and cause skin lesions in response to isolation are demonstrating excessive and pathological levels of anxiety.

SEPARATION ANXIETY

The terms *separation anxiety* or *separation stress* are used to describe problem behaviors that dogs (and, less frequently, cats) show in response to the distress of isolation from their owners and occasionally from other pets. Separation stress is characterized by one or more problem behaviors that include agitation (pacing, jumping up, panting), distress vocalizations (howling, barking, whining), destructive behaviors (chewing, digging), hyperactivity (excessive greeting behavior, constant attention seeking), and house soiling (urination or defecation). Physiological signs may also occur, such as excessive salivation, self-mutilation, and vomiting. Because dogs and cats can be destructive, vocalize, and house soil for reasons that are unrelated to separation stress, it is important to differentiate between the various underlying causes for an accurate diagnosis of separation stress (Sidebar 1).

Sidebar 1 | DIAGNOSING SEPARATION ANXIETY

The behavioral signs of separation anxiety reflect generalized but severe distress and anxiety in response to isolation. These signs include excessive destructiveness, house soiling, and vocalizations. However, each of these behaviors can have underlying causes that are unrelated to separation stress. The following set of questions and answers identify significant features that differentiate separation anxiety from other underlying causes of these behavioral signs.

QUESTION	RESPONSE: SEPARATION STRESS
When does the dog/cat show destructive behavior?	**Only when isolated** from the owner or primary attachment figures
What types of items are destroyed?	Items that are often **touched or used by the owner,** such as clothing, eyeglasses, TV remotes, seat cushions
Where are chewing/digging behaviors directed?	Usually **near areas of entry and exit** such as interior and exterior doors, door trim and window ledges

QUESTION	RESPONSE: SEPARATION STRESS
Does confining the dog reduce the problem?	**No. The dog does not tolerate crating** or confinement well. The dog will destroy bedding or toys left in the crate and shows **repeated and often frantic attempts to escape the crate**
When does the dog/cat house soil?	**Only when isolated** from the owner or primary attachment figures (the dog/cat is otherwise reliably house-trained)
Where does the dog/cat house soil?	**In same room that destruction occurs;** may be near doorway (Cats: On the owner's bed)
When does the pet vocalize?	As the owner is preparing to leave (whining, meowing) and during **isolation (barking, howling)**
When does the pet demonstrate anxious or nervous behaviors?	Whenever the owner is **preparing to leave** and engages in typical pre-departure cues such as putting on coat, picking up keys, opening garage door

Risk Factors and Predisposing Temperament Traits

Dogs: Dogs who have a history of rehoming or were adopted from shelters have been reported to show a higher incidence of separation anxiety and fear-related behaviors than dogs obtained from other sources. For example, one study reported that 26 percent of dogs diagnosed at a behavior clinic with separation anxiety were adopted from a shelter, compared with only 8 percent of the general population of dogs.[6] Another study that examined almost 500 cases found that dogs originating from shelters were significantly more likely to demonstrate signs of anxiety and separation stress than dogs who were obtained from purebred breeders, friends, pet shops, or advertisements.[7] What is not obvious from these studies is the underlying cause of this association. Are dogs relinquished because of existing problems associated with separation, or does the environment of the shelter and the experience of abandonment predispose a dog to future episodes of separation stress? Other factors that may influence a dog's risk of developing separation-related problems include gender (males are more often affected than females), multiple rehomings, an abrupt change in the dog's living environment or daily routine, and having some form of noise phobia (thunderstorms, fireworks).

There also appears to be a relationship between general anxiety, fear, and separation stress. Dogs who exhibit separation anxiety tend to have a nervous or anxious temperament in general and often show behaviors that indicate a high degree of dependency and attachment to their owner. They may greet the owner in a hyper-excitable state using prolonged and exaggerated greeting rituals. They are more likely to follow the owner around the home and to show distress if isolated in another room. For example, the owner may report that his dog follows him into the bathroom or will not willingly go outside alone for elimination. These dogs are often intolerant of confinement and attempts to use a crate are met with increased anxiety and destructiveness. Some dogs with separation anxiety have been described as having "barrier frustration," a form of claustrophobia in which the dog cannot tolerate confinement to small areas or separation behind a baby gate or other type of barrier. Interestingly, many of these dogs will tolerate confinement to the owner's car, but cannot tolerate isolation in any other situation. The differences in these two responses can possibly be explained through classical conditioning. While confinement to a crate has consistently predicted prolonged periods of isolation (and therefore, stress and anxiety), dogs are typically left in cars for very short periods while their owner runs an errand. Therefore, confinement in the car may have been classically conditioned to predict short and tolerable periods of separation.

Although much of the information about over-attachment and separation anxiety comes from case studies, one well-controlled study found that dogs who were diagnosed with separation anxiety were three times more likely to follow their owners excessively, four times more likely to show excessively excited greeting behaviors, and more than four times more likely to be anxious prior to their owner's departure than were dogs without separation anxiety.[8] Conversely, a recent study that administered a standard test of attachment to dogs with and without separation anxiety found that dogs with the disorder did *not* have attachment scores that were higher than the control dogs.[9] However, this study used a small set of reactions to separation in a novel (laboratory) setting and so may not have actually elicited the attachment behaviors that dogs demonstrate in their normal home environment. Finally, it is often assumed that the owners of pets with separation anxiety have intentionally or unintentionally reinforced over-dependence in their pets by "spoiling" their dog. However, studies of the behavior of dogs with separation anxiety do not support this belief. Owners of dogs with separation anxiety are *not* more likely to allow their dog to sleep on the bed, to feed them treats from the table, or to provide multiple type of toys and other treats than are owners of dogs that do not have problems with separation.[10]

Finally, it is possible that the selection for juvenile traits in companion animals predisposes certain individuals and perhaps certain breeds to experiencing over-attachment and dysfunctional reactions to separation. For example, it has been theorized that breeds of dogs that were originally developed for high activity and for working closely with humans may be more predisposed to separation problems than breeds that were developed to function more independently.[11] Examples include the herding breeds and certain types of sporting and working breeds. However, studies that have examined breed predispositions to separation stress in dogs have reported inconsistent associations with breeds and do not even show agreement on purebred versus mixed breed dog predispositions.[12]

Cats: Although separation anxiety is less well documented (and probably less common) in cats than in dogs, clinical reports in recent years indicate that it is a genuine problem. Risk factors are similar to those in dogs, and include adoption from a shelter or frequent rehoming, having a dependent and "clingy" temperament, and living with a single owner. Cats who experience a sudden change in living situation or daily schedule that requires long periods of time alone are also at increased risk for separation anxiety.

BEHAVIORAL SIGNS AND DIAGNOSIS OF SEPARATION ANXIETY

The most obvious feature of separation anxiety is that stress-related behaviors are shown *only* when the human caretaker is not present. Typically, if the owner has had the opportunity to observe the dog, these behaviors commence either prior to or within a few minutes following the owner's departure. Although it is commonly believed that dogs are destructive or vocal for only the first 30 to 60 minutes of isolation, studies of the behavior patterns of dogs with separation stress have found that while the initial 30 to 60 minutes of separation was most stressful to the dogs, many are active and destructive repeatedly throughout periods of isolation.[13]

Signs of stress or anxiety are often triggered in affected dogs in response to the owner's "pre-departure cues." These are classically conditioned stimuli that have predictably preceded isolation. Significant pre-departure cues may include picking up car keys, locking doors, or putting on a coat. Owners often describe their dogs as upset and overly solicitous as they prepare to leave the home. The dog may show physical signs of stress such as trembling, salivation, and pacing and may attempt to block the owner's access to the door. Cats are often described as being "clingy" prior to departure, requesting petting and frequently vocalizing. In dogs, vocalizations that are related to separation, as opposed to watch dog barking, play behaviors, or boredom, may be triggered by pre-departure cues. Whining is typically reported during pre-departure, while repetitive barking and howling develop later, during isolation. Stress-related vocalizations of dogs are high-pitched and strained, distinct from those made during territorial defense or play.

The importance of pre-departure cues and the classically conditioned associations that develop cannot be overemphasized. Dogs who are extremely distressed by isolation focus on any cue that consistently predicts the owner's impending departure. For example, if an owner's normal morning preparations include rising in response to an alarm, showering, making coffee, dressing, and leaving the house, the dog may gradually back-chain each association to the point that signs of agitation and distress begin when the owner's alarm clock rings. Dogs with severe separation anxiety often have a host of pre-departure cues to which they react. During treatment, all of these must be identified and isolated so that they can be decoupled from departure during counter-conditioning treatment (see p. 253).

Although signs of anxiety and vocalizations often begin as the owner prepares to depart, destructive behaviors and house soiling occur exclusively in the owner's absence. When dogs with separation anxiety are destructive, they usually direct their effort toward a point of exit from the home or room in which the dog is kept, or toward an item that carries a strong scent of the owner. Dogs typically scratch and dig at the base of doors, doorknobs, or window dressings, or may chew couch cushions or items of clothing. When separation anxiety is the cause of elimination in the home, the dog is fully house-trained when the owner is home and the elimination occurs only during isolation, regardless of whether the dog had eliminated outside prior to the owner's departure. Cats, for reasons that are not completely understood, often target the owner's bed for urination and sometimes defecation when the owner is absent. This behavior can be differentiated from inappropriate elimination due to litter-box aversions or surface preferences because it only occurs when the cat is left alone. In addition, most cats with separation anxiety only show problem behaviors when they are left alone for extended periods of several days, rather than during the owner's normal workday routine.

TREATMENT OF SEPARATION ANXIETY (DOGS)

The primary goal when treating separation anxiety is to reduce the dog's level of anxiety and improve her level of security and confidence when isolated. Training and behavior modification programs for separation stress include using successive approximation (shaping), counter-conditioning, desensitization, and management techniques. In some cases, administering anti-anxiety medication is helpful as adjunctive therapy (see pp. 253–258).

Reducing Dependency on Owner

Many dogs with separation anxiety demonstrate signs of hyper-attachment to their owners. These dogs are typically described as having an anxious or nervous temperament and are hyper-vigilant about maintaining close proximity to the owner. They may follow the owner from room to room and show signs of distress when separated from the owner or confined, even when the owner is still at home. Teaching a basic control exercise such as the down-stay can be used to provide opportunities for reinforcing relaxed postures and behaviors. The dog is first trained to lie down and stay on command, with the owner standing close. Positive reinforcement is used to shape increasing durations of time that the dog remains in a down position (see Chapter 7, pp. 163–164 for complete instructions). Negative reinforcement should not be used at any time when teaching this exercise, because the goal is to train the dog to offer a reliable down-stay while feeling relaxed and comfortable. High-value food treats and quiet praise are used as positive reinforcers. After durations of 1 to 2 minutes with the owner remaining next to the dog have been reliably trained, the owner begins to introduce short distances away from the dog. This can be in the form of "yo-yo" training, as described in Chapter 7 (p. 169). Gradual shaping allows the owner to move across the room, sit at a desk several feet away from the dog, or move back and forth doing chores around the home. The owner frequently returns to the dog to calmly reinforce the stay and relaxed behaviors using verbal praise, gentle petting, and food treats. If the dog shows distress at any point during this training, the owner must back up and work in closer proximity. This rule is very important because repeated episodes in which the dog experiences stress or anxiety while engaged in the down-stay will condition the down-stay as an aversive stimulus as opposed to the desired outcome of the down-stay as something pleasurable. Some owners choose to teach the dog a "go to your bed" command for this portion of the training. This is very helpful when distance is expanded to separation in another room, as the bed can then be placed in different rooms.

Once the dog has been trained to lie down and stay on command and tolerates short separations from the owner, the owner uses this command periodically throughout the day when home with the dog. The objective is always to pair separations from the owner with feelings of pleasure and relaxation (Sidebar 2). Each practice down-stay should be positively reinforced and all relaxed and calm behaviors are rewarded with attention and food treats. Under no circumstances should the down stay exercise be used as punishment. Finally, it is often helpful for owners to practice down-stay exercises after the dog has gone for a walk or had a period of activity and exercise and is tired.

Finally, it is important for owners to understand that this part of the behavior modification program is not intended to weaken the bond that they have with their dog. Rather, the training functions to increase the dog's feelings of security and comfort, and to decrease feelings of distress when separated from the owner. Although some behaviorists also suggest ignoring the dog whenever attention-seeking behaviors are shown, this is not recommended because the use of negative punishment can cause frustration and could increase the dog's level of anxiety rather than decrease it. Moreover, many

Sidebar 2: SHAPING DOWN-STAY TO REDUCE DEPENDENCY

Dependency Responses

Close to Owner ⟶ **Pleasure/Relaxation/Comfort**
(unconditioned stimulus)   (unconditioned response)

Separated from Owner ⟶ **Anxiety/Fear/Frustration**
(unconditioned stimulus)   (unconditioned response)

Successive Approximation (Shaping) Down-Stay

Step 1 – Teach Down-Stay Using Positive Reinforcement

Down-Stay (near owner) ⟶ **High-Value Treats** ⟶ **Pleasure/Relaxation**
(operant response)   (positive reinforcement)   (unconditioned response)

Step 2 – Shape Duration

Down Stay (increasing time) ⟶ **High-Value Treats** ⟶ **Pleasure/Relaxation**
(operant response)   (positive reinforcement)   (unconditioned response)

Step 3 – Shape Distance (Separation)

Down Stay (increasing distance) ⟶ **High-Value Toy** ⟶ **Pleasure/Relaxation**
(operant response)   (positive reinforcement)   (unconditioned response)

Step 4 – Gradually Shape Both Duration (Time) and Distance (Separation) from Dog in Various Rooms and Situations

---

owners justifiably find such advice unacceptable and are unwilling to comply with it. An alternate approach that can be used for dogs who demonstrate excessive attention-seeking is to teach an alternate and incompatible behavior that can be used preemptively to redirect the dog (see Chapter 9, pp. 213–214 for a complete discussion).

Counter-Conditioning Pre-Departure Cues

When destructive behaviors are caused by anxiety or frustration, alleviating these emotions should prevent behaviors that are destructive to the owner's home or dangerous to the dog's safety. Therefore, counter-conditioning is central to almost all components of a program designed to treat separation stress. These techniques are designed to change the dog's emotional response to impending isolation from anxiety and fear to emotions of calmness and relaxation. Pre-departure cues lend themselves to counter-conditioning techniques because they can be easily separated into small cues to be presented to the dog in situations that are decoupled from actual departures. Common pre-departure cues that dogs react to include picking up keys, putting on a coat, gathering together work items such as a briefcase or purse, or opening the garage door with a remote controller. At the start of the training, the owner makes a list of all of the pre-departure signals that the dog reacts to, listed in order of increasing negative effect upon the dog. The cue that has the *weakest* predictive properties is selected *first* for counter-conditioning training. In other words, if the dog strongly reacts to the owner putting on her coat but shows a moderate or inconsistent response to the owner making coffee, the latter cue is counter-conditioned first.

A specifically designated high-value toy or chew bone is selected by the owner. This may be a hard bone that can be stuffed with moist treats or peanut butter or a food-delivery toy that the dog enjoys. A toy or bone that can occupy the dog for 30 minutes or more will be most effective. Initial training involves simply presenting this toy to the dog while he is relaxed and happy (Sidebar 3). The owner allows the dog to chew on the toy for several minutes and then removes the toy while the dog is still very interested in it. The toy is then put away, out of reach of the dog. This step is repeated on several occasions to ensure that the dog continues to enjoy the toy and learns to accept the removal of the toy. Decoupled pre-departure cues are then added, beginning with the lowest intensity cue. The sequence of presentation is very important – the pre-departure cue is given immediately *before* presenting the special toy (i.e. the pre-departure cue predicts an opportunity to chew on the highly desirable toy). In addition, the special toy is now *only* provided when the owner is engaging in pre-departure cues. The owner gradually desensitizes the dog to the first pre-departure cue by repeatedly exposing the dog to the cue throughout the day, providing the special toy for several minutes and never following the cue with departure. For example the owner may pick up the keys, counter-condition the dog using the special toy, and then continue about her business around the house (Sidebar 3). After several minutes the toy is removed. This sequence can be repeated several times a day and with different pre-departure cues. These counter-conditioning exercises function to reduce the dog's anxiety in anticipation of the guardian's absence and also to build a positive association between the special toy and isolation cues. When the dog has become counter-conditioned to all pre-departure cues, a program of graduated departures can be introduced.

Sidebar 3: COUNTER-CONDITIONING PRE-DEPARTURE CUES WITH A HIGH VALUE TOY

Conditioned Pre-Departure Cue Example (Keys)

Picking up Keys → **Owner Departs (Isolation)** → **Anxiety/Fear**

(conditioned stimulus)   (unconditioned stimulus)   (unconditioned response)

Behavior Modification Program

Introduce High-Value Toy

Special High-Value Toy → **Pleasure/Relaxation**

(unconditioned stimulus)   (unconditioned stimulus)

Counter-Condition Pre-Departure Cue

Picking up Keys → **Special High-Value Toy** → **Pleasure/Relaxation**

(conditioned stimulus)   (unconditioned stimulus)   (unconditioned response)

**Repeat with All Pre-Departure Cues
(Decoupled from Owner Departing and Isolation)**

Desensitization to Isolation

After pre-departure cues have been counter-conditioned and the dog has been conditioned to calmly accept periods of separation using the down stay exercises, graduated departures can be introduced. A pre-selected high-value toy is used to counter-condition and desensitize the dog to isolation in a separate room while the owner is at home. A daily schedule providing graduated durations of separation is created. Each practice session is paired with the high-value toy that was used to condition the dog to pre-departure cues. As with the previous phase of the treatment program, this toy is only provided to the dog during the training sessions. At the start of this phase of the program, the dog is left alone for very short periods and the owner always returns to the dog *before* the dog has lost interest in the toy. This rule ensures that the dog is not left alone long enough to induce a stress response. As the program progresses, the length of time that the dog is left alone must always be shorter than the time within which the dog becomes anxious. If the pre-departure cues have been well-conditioned, and the dog has been trained to offer relaxed down-stays apart from the owner throughout the home, most dogs will tolerate absences of several minutes. The owner offers the special toy, leaves the dog in the room for a few minutes, returns, reinforces calm behavior with quiet and calm praise, removes the special toy and releases the dog from the room. This sequence is repeated several times throughout the day. The special toy is always put away after releasing the dog and is only presented again at the start of another practice session.

If pre-departure cues have been successfully decoupled from predicting the owner's departure, the presence of the special toy becomes a "safety cue" for the dog. It predicts isolation periods in which the owner is gone for very short time and no anxiety occurred. This is why it is imperative that the special toy is *not* provided at any other time, especially in circumstances during the treatment program when the owner is required to be away for an extended period and there is a high probability that the dog will experience anxiety. The toy should never be given to the dog when the owner leaves for a period longer than the dog is capable of tolerating because the item will then lose its value as a safety cue if it becomes associated with anxiety.

As the dog becomes acclimated to the room and begins to anticipate the desired toy (safety cue), the caretaker can gradually progress to increased durations, using a sequence of variable time periods (for example, 10 minutes, 4 minutes, 12 minutes, 1 minute). Over time, periods of 30 to 45 minutes are conditioned. In some cases, it is helpful to provide two special toys as the duration away increases. Once a dog tolerates 45 minutes of isolation with the owner still at home, the owner can begin to add the pre-departure cues and, eventually, to actually leave the house for short periods of time. When these new cues are added, however, the time period should once again be decreased to several minutes and then gradually increased to 30 minutes or more. Again, a schedule of gradually increasing time periods can be devised, but should be adjusted during treatment to reflect the dog's response to each level of separation.

Throughout desensitization, the dog's response always determines the rate at which training proceeds. Because dogs differ significantly in their abilities to tolerate isolation and to counter-condition to the safety cue, the owner uses the dog's response during the previous session to determine whether the dog is ready to tolerate increased duration in the current session. A written time schedule is useful as a guide, but the dog's ability to tolerate each level of separation should be the ultimate criteria to use when determining progression to a new level. The owner increases the duration of separation when the dog exhibits no pre-departure anxiety to the current level and the dog does not appear to be stressed or show exaggerated greeting behaviors when the owner returns. The most important factor is that the dog remains non-anxious. If the schedule is increased too rapidly and anxiety results, the problem is often exacerbated and the usefulness of the safety cue is lost.

Management Approaches

A variety of management practices can be used in conjunction with behavior modification when working with dogs with separation anxiety. Dogs who experience a regular and consistent schedule of exercise, training, and attention tend to have lower anxiety levels and are more likely to relax and rest during times when they are not interacting with their owners. In addition, a dog who learns to expect attention and exercise at certain times every day is more manageable and may tolerate isolation better as a result of learning to anticipate regular periods of exercise and rest. Daily obedience training should be incorporated into these routines as training can also help to increase a dog's confidence and security.

When destruction is serious and there is potential for injury to the dog, confinement may be necessary to prevent further destruction and to protect the dog. Crating is often the first type of confinement that owners attempt. Unfortunately, if a crate has been used previously, this is often unsuccessful if the crate has become classically conditioned to predict isolation and anxiety. As a result, many dogs with severe separation anxiety are crate-intolerant and may destroy the crate and injure themselves in attempts to escape. In cases in which the crate is not an option but confinement is necessary, the owner may be able to use a small room, such as a kitchen or playroom, or a secure outdoor kennel. The pivotal criterion for this space is that it is an area to which the dog has not yet developed an aversive conditioned response. A stuffed bone or food-delivery toy can be provided, but because confinement has the potential to be stressful, it should not be the same toy that is being developed as a safety cue during the counter-conditioning portion of the behavior modification program. Finally, many dogs who are seriously affected cannot tolerate any form of confinement at home. In these cases, some options that can be used when the owner must be away for more than a short period include enrolling the dog at a doggy daycare, boarding the dog at a kennel, or hiring a pet sitter to come in to stay with the dog.

Adjunctive Drug Therapy

In some cases, drug therapy is a helpful adjunctive therapy for separation anxiety. Drugs in a class called tricyclic antidepressants are often prescribed because they reduce anxiety and can augment the effects of a behavior modification program. The two that are most commonly used are amitriptyline and clomipramine. Of these two, only clomipramine has been specifically labeled for treating canine separation anxiety and has been shown in controlled trials to be an effective addition to behavior therapy.[14] The results of these studies found that dogs treated with both the drug and behavior therapy improved more rapidly than did dogs who were treated with behavior therapy alone. However, another study reported that clomipramine reduced overall activity level and suppressed dogs' attention-seeking behaviors but did not significantly improve response to behavioral therapy.[15] A second class of drugs, the benzodiazepines, is sometimes prescribed for immediate control in severely affected dogs during times that the owners are forced to leave the dog alone. Of these drugs, alprazolam and clorazepate are most frequently used. These medications can help owners to reduce extreme panic in a dog when they have no choice but to leave the dog alone. The medication is given 1 to 2 hours before the owner leaves and is intended to reduce anxiety for a single episode. Benzodiazepines can be used concurrently with the tricyclic antidepressants, and so are helpful in preventing relapses during a behavior modification program. Other drugs that are occasionally used include the selective serotonin reuptake inhibitors (fluoxetine or paroxetine), the monoamine oxidase B inhibitor selegiline hydrochloride, and phenothiazines (Table 10.1).

The inclusion of drug therapy can be helpful in reducing anxiety during the initial stages of behavior therapy. The majority of dogs are slowly weaned off of these medications as their behavior improves. In all cases, drug therapy should

table 10.1 Medications Used as Adjunctive Therapy for Separation Anxiety

CLASS OF DRUG	EXAMPLES	MODE OF ACTION	SIDE EFFECTS	COMMENTS
Tricyclic Antidepressants	Clomipramine Amitriptyline	Neurotransmitter effects; reduces arousal, anxiety	Increased drinking and urination, GI upset, urinary or fecal retention, glaucoma	Should not be prescribed in pets being treated with monoamine oxidase inhibitors
Monoamine Oxidase Inhibitors	Selegiline	Inhibits panic or phobic reactions, reduces anxiety	Gastrointestinal upset, restlessness and hyperactivity	May take as long as 6 weeks to show any benefits
Benzodiazepines	Diazepam Alprazolam	Anti-anxiety; reduced state of arousal; fast acting	Lethargy, sedation, ataxia, increased appetite, liver toxicity (rare)	Can be used to for immediate control of anxiety during severe episodes
Selective Serotonin Reuptake Inhibitors	Paroxetine Fluoxetine	Enhance availability and effects of serotonin	Loss of appetite, diarrhea, vomiting	May take several weeks to affect behavior, Used effectively for urine spraying

only be used under the supervision of a veterinary behaviorist and should be administered in conjunction with a behavior modification program. Finally, in addition to prescribed drugs such as the tricyclic antidepressants, a hormonal derivative called Dog Appeasing Pheromone (DAP) has been tested as an adjunctive medical therapy for dogs with separation anxiety and may help to reduce anxiety in some dogs (Sidebar 4).

Sidebar 4: THE EFFECTS OF DOG APPEASING PHEROMONE (DAP) ON SEPARATION ANXIETY IN DOGS

Shortly after giving birth, the sebaceous glands located around the female dog's mammary glands secrete pheromones that are known to have a calming effect upon her puppies. Dog Appeasing Pheromone (DAP) is a synthetic and commercially available form of these pheromones. The aerosolized compounds are administered using either a heated electric diffuser that mounts on a wall or from a small container mounted on the dog's collar. Studies of DAP have suggested that it can be an effective adjunctive therapy for reducing fear of loud noises or riding in the car, and for calming dogs who are living in a shelter environment. Recently, a group of researchers examined the potential for DAP to act as an adjunctive therapy for dogs with separation anxiety.

A group of 67 dogs with diagnosed separation anxiety was recruited through veterinary behavior centers. Commonly reported behavioral signs in the dogs were destructiveness, excessive vocalizations, and house soiling. Dogs were also reported to show signs of hyper-attachment to their owners. Following recruitment, the dogs were randomly assigned to two groups. The standard reference group received a behavior modification program to treat separation anxiety plus clomipramine. The test group received the same behavior modification program plus DAP. Clomipramine was administered twice daily as an oral capsule and DAP was administered using a diffuser installed in an area that the dog spent most of the day or was confined when the owner was absent. Placebo diffusers and capsules were supplied to dogs in the clomipramine and DAP groups, respectively. Dogs were treated for 4 to 5 weeks.

Owners and their veterinarians evaluated the effects of treatment on both the primary and secondary behavioral signs of separation anxiety. Undesirable behaviors decreased in both groups of dogs. Overall assessments by owners indicated no significant difference between clomipramine and DAP as adjunctive therapies. In addition, more dogs in the clomipramine group than in the DAP group experienced undesirable side effects. Gastrointestinal upsets (vomiting, diarrhea, reduced appetite) occurred in dogs who received clomipramine but not in dogs who were exposed to the DAP. Owners of dogs in the DAP group showed a significantly greater rate of compliance than did owners of dogs in the clomipramine group. Results of this study indicate that the efficacy of treatment with DAP was not inferior to including clomipramine with a behavior modification program for separation anxiety. However, because there was no negative control group included in this study, improvement cannot be attributed directly to either clomipramine or DAP and may have been due to behavior modification alone in both groups of dogs. If it is effective, DAP has the practical benefits of fewer undesirable side effects, ease of administration, and higher owner compliance than the drug clomipramine.

Gaultier E, Bonnafous L, Bougrat L, Laront C, Pageat P. Comparison of the efficacy of a synthetic dog-appeasing pheromone with clomipramine for the treatment of separation-related disorders in dogs. *Vet Rec*, 156:533–538, 2005.

Ineffective Treatments for Separation Anxiety

Punishment is ineffective as an approach to reduce the destruction associated with separation anxiety and will quite likely increase a dog's anxiety, further exacerbating the problem. Although many owners are aware that they should never punish their dog "after the fact," for example, when they return home to find a mess, it is unfortunate that some trainers and behaviorists continue to promote the use of remote punishers such as bark collars or "setting the dog up" by hiding and waiting until destruction or vocalizations occur and then reprimanding the dog (i.e. "catching him in the act"). It cannot be emphasized strongly enough that these approaches are ineffective and inhumane for dogs experiencing anxiety or fear and should *never* be used. Punishment appears to work because it interrupts and suppresses behavior at the time it is administered. As a result, the owner may erroneously believe that this approach is effective. To the contrary, while shocking a stressed dog for barking or harshly reprimanding an anxious dog for destructive behaviors are aversive enough to inhibit these behaviors at the time that the punishment occurs, this should not be confused with preventing anxiety-related behaviors in the future. Because they add to the anxiety and fear of a dog who is already highly stressed, aversive stimuli (shocking, yelling, hitting) are abusive and cruel and should never be promoted or condoned. Other approaches that some owners attempt but which are generally ineffective (though not damaging or abusive) include getting a second dog to keep the affected dog company, crating without including a behavior modification program, and attempting to use a safety cue such as leaving the radio on or providing a stuffed toy without taking the necessary steps to classically condition the cue.

PREVENTION OF SEPARATION STRESS

As discussed previously, separation distress is a normal component of all puppies' behavioral repertoire. As they grow and develop, puppies usually become habituated to periods of isolation through regular, short periods of time during which they are kept in a separate room or in a crate. Socialization to a variety of people and experiences, coupled with regular, short periods of isolation, gradually teaches puppies to readily accept new experiences and to be comfortable when separated from their human social partners. A "safety cue" toy can be introduced early in life, such as a hollow bone or hard rubber toy stuffed with treats. This special toy is provided whenever the puppy is isolated, pairing the toy with short periods of "alone time." Because the early introduction of this cue prevents its association with severe anxiety, it can quickly become classically conditioned to predict pleasurable emotions of relaxation and eating special treats. As the dog matures, regular periods of isolation coupled with a consistent daily routine of exercise, attention, play, and handling are essential for the prevention of separation-related problems during adulthood.

There is some evidence that dogs who respond to commands consistently or who have been enrolled in obedience training classes may be less likely to develop separation problems. For example, a study of dogs and their owners compared the effects of training and human interaction on the human-dog relationship over an eight-week period.[16] Groups of dogs who had either completed an eight-week obedience class or who had spent a comparable time period in positive interactions with their owner were reported to have fewer problems with separation stress compared with dogs who received neither type of interaction. Other studies of the dog-owner relationship have supported these results, showing that dogs who either responded reliably to obedience commands or who attended obedience classes were less likely to show separation-related behavior problems.[17,18] Obedience training may lower the risk of separation stress in dogs by improving the overall quality of the dog-human relationship, since owners of dogs who are obedience trained report fewer behavior problems of all types. Additional benefits that training might have could be related to the level of control and graduated separations that are inherent in teaching a dog to sit or lie down and stay as the trainer moves away or to another room. Most training programs introduce these exercises through shaping, and it is possible that the gradual introduction of separation from the owner, along with positive reinforcement of calm behavior (staying) serves to attenuate over-attachment behaviors. Alternatively, the association between obedience training and fewer problems may be that owners who choose to obedience train their dogs are less likely to perceive their dogs' behavior as problematic in the first place.

A CAUTIONARY NOTE: SEPARATION ANXIETY OR SOMETHING ELSE?

Separation anxiety has received a great deal of attention in recent years, in part because of enhanced understanding of this problem, but also because of the development of effective treatments that include behavior modification programs and pharmacological therapies (see Sidebar 5 and Box 10.1). This attention has allowed many trainers, behaviorists and veterinarians to help owners to improve the quality of their dogs' lives and, in many cases, to keep the dog in the home. However, this increased awareness, especially on the part of owners, may also lead to misdiagnoses. Signs of separation anxiety are similar to the behavioral signs of under-stimulated and under-exercised dogs whose social and physical needs are not being met. For example, a standard component of separation anxiety behavior modification programs is to increase the dog's daily exercise and institute a program of regular obedience training. While increasing exercise and mental stimulation may reduce frustration or stress experienced by the dog when left alone because the dog is tired, it is also possible that some of the dogs who respond positively to these changes are actually *not* suffering from separation stress. In the absence of other presenting signs that are strongly suggestive of anxiety and of hyper-attachment, the parsimonious explanation for destructive behaviors or repetitive barking are instead boredom and a lack of mental stimulation. Similarly, dogs who have not been thoroughly house-trained may eliminate when alone but at no other time because they have learned that elimination in the presence of the owner leads to punishment, not necessarily because they are suffering from separation anxiety (see Chapter 8, pp. 188–189).

IS BEHAVIOR MODIFICATION EFFECTIVE IN TREATING SEPARATION ANXIETY? | Sidebar 5

Treatment programs designed for dogs with separation anxiety include counter-conditioning, systematic desensitization, and management techniques. Even when owners are committed, these programs can be labor-intensive, time consuming, and complicated. According to one study, most dogs with separation-related behavior problems do not receive formalized treatment and are either managed by their owners, rehomed, or abandoned. Therefore, behavioral therapy programs that are known to effectively reduce a dog's anxiety and destructive behaviors are needed for owners who cannot effectively manage this problem. Two studies have recently examined the effectiveness of behavior modification programs in treating separation anxiety.

The first study, conducted by Blackwell's group, recorded the success rate of a standard behavior modification program. Owners of 34 dogs with separation anxiety were given a standard set of behavior therapy instructions to follow for 12 weeks. The control group included 16 dogs whose owners did not request help. At the end of the study period, problem behaviors were reduced in 81 percent of the treated group. In contrast, the untreated control group showed little change and no significant improvement in any problem behaviors. The investigators also compared the treated group's responses to 30 dogs who had been treated with a behavior program that was specifically tailored to each dog's specific problems and to the lifestyle of the owner. Of those 30 dogs, all owners (100 percent) reported improvements in their dogs' behavior after 12 weeks of treatment.

Another study, conducted by Takeuchi and associates, examined treatment outcome and level of owner compliance with a personally designed separation anxiety behavior therapy program. Owners of 52 dogs treated for separation anxiety at a behavior clinic were interviewed six months or longer after beginning treatment. Problem behaviors had improved in 62 percent of the treated dogs and 38 percent had no change, had worsened, or were no longer in the home. Owners who had been given a list of five or fewer instructions were more likely to comply and to report success than were owners who were given a longer list of instructions. Clients were most likely to comply with directives that required little time and were not complicated. For example, owners followed instructions for not punishing the dog for separation-related misbehavior, for providing a special toy as a safety cue, and for increasing the amount of exercise that the dog received. Conversely, instructions with lowest compliance were those for counter-conditioning and decoupling pre-departure cues and for gradual desensitization to isolation.

Together, these results suggest that a small set of easy-to-implement treatment instructions has the greatest likelihood of owner compliance and success. When possible, treatment instructions should be specifically tailored to the dog's behaviors and the owner's lifestyle. Although a gradual program of desensitization to gradually increasing periods of time alone is the theoretical "gold standard" of behavior modification, a treatment can only be effective if it is used. Therefore, it may be more effective to concentrate on the treatment and management techniques that have known success and with which owners are most likely to comply.

Blackwell E, Casey RA, Bradshaw JWS. Controlled trial of behavioral therapy for separation-related disorders in dogs. *Vet Rec*, 158:551–554, 2006.

Bradshaw JWS, McPherson JA, Casey RA, Larter IS. Aetiology of separation-related behaviour in domestic dogs. *Vet Rec*, 151:43–46, 2002.

Takeuchi Y, Houpt KA, Scarlett JM. Evaluation of treatments for separation anxiety in dogs. *J Amer Vet Med Assoc*, 217:342–345, 2000.

Box 10.1   PEACHES THE BORDER COLLIE

Background: Peaches is a three-year-old, spayed, female Border Collie, owned by Anne Price. Anne has completed several obedience classes with Peaches. She relates that Peaches was very easy to train and is well-behaved at home and when out in public. Peaches is a very active dog and Anne enjoys long walks with her every evening and runs with Peaches 4 or 5 mornings a week. Peaches is very affectionate ("clingy almost") with all people, but she is especially attached to Anne. Peaches is also friendly to other dogs and has several dog friends who she plays with on weekends.

Description of problem: Recently, Peaches has started to be destructive when Anne is not at home. At first, Anne would come home and find a piece of laundry on Peaches's bed. Since Peaches was not chewing the item up, Anne did not worry about it. Then, gradually, Peaches began to chew on the items she took to her bed, and last week, Anne found two shoes and a winter hat on the bed, both destroyed. Anne also reports that a neighbor has heard Peaches howling, sometimes for several hours at a time while Anne is at work. Yesterday, Anne found deep scratch marks at the base of her back door. Anne lives alone and Peaches is free in her home when she is at work. Until recently, Peaches had always been fine with this. However, Anne changed jobs about nine months ago and now has a 45-minute commute to work. Previously, Anne worked within a 10-minute drive from work and also worked from home two or three days a week. Her new position does not allow her to work from home at all. Together, these changes have resulted in significantly longer workdays and longer durations alone for Peaches.

Additional background: A coworker informed Anne that Peaches is angry with Anne for being left alone for so many hours and is "getting back" at Anne by chewing up things that are important to Anne. The coworker has told her that she should "set up" Peaches by leaving several items out for Peaches to chew, pretend to leave, wait a few minutes and then come running back into the house to harshly reprimand Peaches if she has taken any of the items that were left out. She told Anne that she must startle Peaches by yelling and swatting her, so as to make this correction work the very first time. Anne has not yet done this, because she feels badly that she may have caused this problem and because she believes that Peaches is distressed when she is away. Anne states that she absolutely adores Peaches, and that while she does not see this to be a terrible problem, she is worried that Peaches is unhappy or stressed, and that the problem seems to be gradually worsening.

Diagnosis and treatment: Separation stress and possibly boredom-induced destructive behavior as a result of recent changes in daily routine, time with owner, and duration of time alone. Behavior modification program: Counter-condition a safety cue (Anne chooses a hollow bone stuffed with peanut butter); use down-stay training in the home to increase Peaches's confidence when separated from Anne; counter-conditioning and desensitization of pre-departure cues and time alone; use doggy daycare 2 to 3 times per week; hire pet sitter to come in at noon on long days to play with and walk Peaches; shift daily routine to provide exercise regularly in the evenings.

Outcome: Using behavior modification and management, Anne saw a reduction in Peaches's anxiety level within one week and complete resolution of signs within one month. (Anne did *not* take her co-worker's unsolicited advice.)

These distinctions are more important today, given the recent use (and documented benefits) of clomipramine and other drug therapies for dogs with actual separation stress. While these drugs, when coupled with a complete behavior modification program, are beneficial to the animals who need them, the attraction of a "quick fix" to frustrated owners may lead to medicating dogs in lieu of providing for their basic needs. Because informed consent by the animal is impossible and because medications do have the potential for abuse, veterinarians who prescribe medications for the amelioration of anxiety in dogs and cats must always include a careful diagnostic protocol and encourage compliance with a behavior modification program before prescribing any adjunctive medication for use in a behavior modification programs.

TREATMENT PROTOCOL FOR CATS WITH SEPARATION ANXIETY

Treatment for separation anxiety in cats follows the same general principles as treatment for dogs, with several distinctions. Because the normal feeding behavior of cats does not include prolonged chewing and gnawing, stuffed bones are not an appropriate selection for a safety cue. However, cats do enjoy batting around toys that move erratically and many cats play with food-delivery toys that are specifically designed for cats. These products can be classically conditioned to predict opportunities for positive experiences in the same way as those used with dogs. Although opportunities for increased exercise and stimulation are more limited for cats than for dogs, owners can compensate by providing interactive toys and increasing play. Rotating toys is especially important for cats, as novelty has been shown to be an important factor in maintaining interest in object play in cats. Because the majority of cats with separation stress eliminate in a specific spot (often the owner's bed), the targeted area should be made inaccessible to the cat during treatment. While training a reliable down-stay is not possible with cats, other methods can be used to reduce a cat's degree of dependency on the owner. These include confining the cat to another room with interactive toys for short periods of time while the owner is home and providing petting and attention on a predetermined schedule, rather than always in response to the cat's attention-seeking behaviors (Box 10.2).

Box 10.2   OREO'S SEPARATION PROBLEM

Background: Oreo is a beautiful, black-and-white, neutered, domestic longhair male cat owned by Rob Murphy. Rob adopted Oreo from his local shelter six months ago. Oreo is extremely affectionate and bonded to Rob immediately. At home, Oreo follows Rob around his apartment, sits in his lap and never lets Rob out of his sight. He is a very playful cat and enjoys batting at his "mouse toy" and retrieving it when Rob tosses it for him. Rob finds Oreo's affectionate temperament and playful behaviors endearing and does not consider them to be a problem. In fact, no problems with Oreo occurred until Rob had to leave for an overnight business trip, one month ago.

Description of problem: Rob provided sufficient food and water and a clean litter box for Oreo. He packed for his trip and left home at his usual time on a Tuesday morning. He returned late in the evening on Wednesday. Upon returning, Oreo met Rob at the door, showing his typical excitement and affection. Rob did notice that Oreo seemed even more "clingy" than usual and was meowing persistently. When Rob went into the bedroom, he found that Oreo had urinated several times on his pillow and in the center of his bed. Rob attributed this to Oreo's first time alone and did not worry about it too much. However, two weeks later, when Rob again had to travel, he noticed that Oreo became very agitated as he took out his luggage and packed for his trip. He followed Rob as he prepared to leave, pacing in front of Rob and meowing constantly. Because he was going to be away for three days, Rob had arranged for a friend to check in on Oreo each evening. When the friend visited on the second day, Oreo was happy to see him but had again urinated on Rob's bed. (Rob relates that Oreo is completely reliable about using his litter box when Rob is at home and maintains his normal workday routine.)

Diagnosis: Separation anxiety in response to change in routine and isolation lasting overnight.

Treatment: Rob purchased a variety of interactive and food-delivery toys for Oreo. After playing with each toy and teaching Oreo to use the food-delivery toys, Rob selected four toys that were Oreo's favorites. He began presenting these to Oreo while he was at home, with Oreo confined to another room. Rob practiced these play sessions several times a day whenever he was at home with Oreo. Rob also encouraged Oreo to play in other rooms by hiding his favorite toys in the kitchen or study while Rob was in another room to reinforce Oreo's confidence and to reduce his attention-seeking and "clingy" behaviors. Rob also purchased a cat bed for Oreo. He placed the bed in the living room and reinforced Oreo with attention and food treats whenever he was resting on the bed. For the first trip away, Rob hired a pet sitter to come in three times a day. He closed the door to his bedroom and hid a variety of food-delivery and interactive toys around the apartment. Oreo's new bed was placed in his favorite room of the apartment, the living room. When the house sitter came to visit, she was instructed to pick up several of the toys and replace them with new toys that Rob had left for this purpose.

Outcome: After one additional urination episode (the house sitter forgot to keep the bedroom door closed), Oreo had no further house-soiling problems while Rob was away. Rob continues to use the interactive toys and the house sitter when he has to be away overnight on business to provide Oreo with attention and affection when he is away.

FEAR-RELATED BEHAVIOR PROBLEMS

Fear-related behavior problems in dogs and cats vary significantly in the type of stimulus that elicits fear and the intensity of the pet's reaction. For example, dogs and cats who show generalized fear to a wide range of situations are often referred to as being shy or timid. In other cases, a dog or cat may develop very specific and intense fears toward a certain type of person or a particular location or situation. An example is the dog who becomes so agitated and frantic upon hearing fireworks that he is in danger of injuring himself in his attempts to escape, or a cat who is terrified of a visiting houseguest and hides under a bed, refusing to come out to eat or use the litter box. While behavior modification is often helpful in reducing fear-related behaviors, the long-term prognosis can be highly variable and depends upon the age of onset, the duration and the intensity of the fearful response, the ability to control the eliciting stimuli, and the dog or cat's basic personality type.

figure 10.1 Fearful (nervous) dog

When confronted with fearful stimuli, animals have one or a combination of three possible reactions available to them; freezing, fleeing, or fighting. In dogs, freezing responses include a lowered head and crouched body posture and hiding or attempting to stay close to the owner. The dog's eyes are wide, ears are back, and the mouth is either closed with retracted lips or opened and panting (Figure 10.1). Cats will reduce their body size by lying down with feet tucked underneath the body, and may lower the head. Fleeing behaviors include all attempts to run away or avoid the stimulus. Dogs who react aggressively usually will demonstrate growling, snapping at air, or biting. Defensive threats may be followed by an attempt to flee. Most fearful animals only become aggressive if their attempts to flee have been impeded or if they have learned to used aggression preemptively to avoid a frightening experience.

Common Fears in Dogs and Cats

Fear-related behaviors can be divided into four general categories: fear of new places or situations, fear of unfamiliar people, fear of unfamiliar dogs/cats, and noise phobias. Some pets exhibit only one specific type of fear, while others may show several apparently unrelated fear responses or phobias. Because many cats live exclusively indoors and tend to be very attached to territory, fear behaviors in cats are most commonly expressed when they are removed from their normal living spaces (such as a trip in the car, or a move to a new home), or when a new person or cat enters into the cat's established territory.

Fear of New Situations/Places: Fear of new places and experiences (**agoraphobia**) is seen most commonly in dogs who live in kennels or who rarely leave their homes, and in house cats who live exclusively indoors. For example, a dog whose only trips away from home are infrequent visits to the veterinarian's office, the grooming shop, or a boarding kennel may learn that any trips away from home lead to unpleasant experiences. Similarly, dogs from large breeding kennels may have had little or no experiences away from their kennel setting. Cats who live exclusively indoors, who have not been well-socialized to visitors or to different types of people, and who only leave the home for trips to the veterinarian's office frequently are fearful in all new settings. In both dogs and cats, age has a significant effect upon the prognosis with these types of fear reactions, with older adult pets being less responsive to treatment than young adolescents. In some situations, an animal may become agoraphobic as a result of an unpleasant or traumatic experience. In one case study, a dog became unwilling to leave the house to go for her regular (and previously enjoyable) walks around the neighborhood after having been attacked and badly bitten by a neighbor's loose dog. While the ability of dogs and cats to eventually overcome situational fears varies greatly, many will retain a certain level of timidity for their entire lives.

Fear of Unfamiliar People or Animals: Dogs and cats who are timid or fearful when meeting new people often have not been adequately socialized to different types and ages of people. A typical example is a dog who is owned by a quiet elderly couple and who shows nervous or fearful behavior when approached by young children. Similarly, the fear of other animals can develop as a result of a lack of socialization or continued exposure during adulthood. Some dogs and cats have not learned normal intra-species communication patterns and may be unable to either send or perceive normal canine or feline communication signals. When approached by another animal of the same species, these inabilities lead to fearful responses and avoidance. In addition, traumatic events such as being attacked by another animal or stranger can cause the development of a fearful response.

Noise Phobias: Noise phobias make up a large proportion of fear-related problems in dogs. The most common stimuli are thunder, gunshots, and fireworks. While some noise phobias can be managed successfully with a behavior modification program, the prognosis varies greatly depending upon the individual, the duration of the phobia, the ability to control exposure to the stimulus, and success in finding an effective artificial stimulus to use during the exposure exercises. Fear of thunderstorms is more common in dogs than in cats. This type of phobia is unusual in that it typically develops in mature adults, and gradually increases in intensity as the dog ages. Dogs with thunderstorm phobias usually display a gradient of fearful behavior that is directly proportional to the intensity of the storm. Owners report that the dog begins to pace nervously and stays close as the weather becomes dark and there are signs of an approaching storm. One of the problems in treating thunderstorm phobia is the number of stimuli to which the dog may be reacting. These often include wind, rain, changes in atmospheric pressure and ionization, lightening, and odors. Because the easiest stimulus to replicate is auditory (i.e. thunder), this is what caretakers usually focus on and is the cue that is used in desensitization programs.

The Role of Avoidance Learning in Maintaining Fear-Related Behaviors

One of the primary difficulties in solving fears and phobias in dogs and cats is that the behaviors that allow the pet to escape or avoid the fearful situation are strongly reinforced each time that they are successful. This is called avoidance learning and it occurs whenever a behavior in which an animal engages allows the animal to avoid exposure to an aversive stimulus. In the case of fear, when a fearful experience causes the pet to attempt to escape (flee), and the animal is successful in doing so, the escape behaviors are immediately reinforced. For example, a dog who was once frightened by a young neighbor boy banging two metal garbage cans together learned to run into the garage whenever the boy approached. The association between the boy and the loud noise (classical conditioning) caused the dog to continue to show this behavior even when the boy did not make any loud noises. Running into the garage essentially *prevented* the dog's exposure to noise, regardless of whether the boy had any intention of being noisy. Because the dog runs away each time that he sees the boy, he (the dog) never has the opportunity to learn that the boy is harmless and is not going to make noise. The behavior of running away was negatively reinforced each time that it occurred because the dog was able to avoid exposure to an unpleasant stimulus (the boy, who was a predictor of a frightening noise). A similar example is the cat who hides under the bed and refuses to come out each time that a visitor arrives. Because hiding allows the cat to avoid exposure to the aversive stimulus (an unfamiliar person) and results in an abatement of unpleasant feelings of fear, the cat preemptively hides each time that she hears someone at the door and so never has the opportunity to habituate to unfamiliar and non-harmful visitors.

A common error is often made regarding the relationship between the emotion of fear and the behaviors that an animal engages in to escape a fearful stimulus. As stated previously, the behaviors that a dog or cat learns will "work" for avoiding situations that cause them to experience fear are reinforced whenever they are successful. It is the *behaviors* that have been reinforced, *not* the fear. In other words, the sequence that is operantly learned, running away or hiding, function to "turn off" the aversive stimulus (operant learning). The mistake is made when owners are incorrectly advised to neither comfort nor come to their pet's aid when their pet is frightened, because doing so will presumably "reinforce the pet's fear." Such advice is incorrect and inhumane, as it is analogous to telling a parent to refuse aid to her child when the toddler is screaming in fright at the sight of a clown (which are admittedly quite scary). Certainly very few parents would inform the clown that they are going to ignore their terrified child so as not to "reinforce his fear!"

Unfortunately, such advice is given frequently to pet owners and its prevalence reflects confusion between operant and classical learning situations. It is important to remember that fear is a basic emotional response that is involuntary and has important biological functions. Pets (like people) do not *choose* to be fearful; it is a basic reaction to situations that the animal perceives to be threatening or dangerous and leads to escape (or defense) behaviors. It is false to state that a dog or cat (or person) chooses or willingly *decides* to experience fear. However, this is exactly what is implied when owners are advised to not comfort or aid an animal who is panicking in fear because of the belief that such actions will "reinforce the pet's fear." The caring (and effective) approach to dealing with a pet's fearful response is to calmly and quietly come to the pet's aid and remove him from the fear-inducing situation. Nothing can be done at that point in time to alter the pet's response because fear has already been triggered and it is the responsibility of the owner to protect the dog or cat from excessively stressful and frightening situations. Simply ignoring the pet's fear in a misguided attempt to change behavior is counterintuitive to most owners who love their pets, and is definitely not going to help the pet to overcome his fear.

Treatment of Fear-Related Behavior Problems

The first step in the treatment of fear-related behavior problems is to completely identify the attributes of all stimuli that elicit fear responses in the pet. The owner should list all possible situations in which the dog or cat has shown fear or timidity in the past. Effective treatment approaches include management to control exposure to eliciting stimuli, and counter-conditioning and desensitization to change the pet's emotional responses to triggering stimuli.

The most important component of behavior modification programs for treating fear-related behaviors in both dogs and cats is counter-conditioning. The ultimate goal of counter-conditioning is to change the pet's immediate response to the particular stimulus or situation from one of fear (and its attending behaviors of flight or freeze) to one of calmness or even pleasure. In order to accomplish this, the owner *must* be able to introduce the stimulus at an intensity at which the pet does *not* react. For example, a dog who is fearful of unfamiliar people while out on walks can be exposed to people who are walking far enough away to cause an alerting response, but which does not trigger fear and avoidance behaviors. Similarly, a cat who is frightened by visitors coming in the door can be confined to a room that is furthest away from the door or in which the cat is known to feel comfortable. When these controls are in place, the owner can begin to counter-condition using high-value food treats, praise and petting. In the dog's case, as soon as the dog notices a person while out walking (but does not show fear), the owner quietly says the dog's name (or uses a safety cue such as "look") and provides very high-value treats and praise (calm, quiet tone of voice). As the person moves away or the owner and dog pass the person (still at a safe distance), the treats and praise stop.

This classical conditioning sequence is practiced until the dog reacts to the sight of another person by looking to the owner for treats and praise (i.e. no fear is triggered and the dog's emotional state is calm and positive). This response signifies that classical conditioning has been accomplished. The dog now views the sight of a person at a distance as a reliable predictor of high-value treats and praise from the owner. This pleasurable stimulus-to-stimulus association changes that dog's emotional response to the stranger from one of apprehension and fear to one of pleasure. In the cat example, the owner stands at the room's doorway, or next to the wire crate if the cat is confined to a crate as the visitor arrives. The sound of the door or visitor arriving is paired with the provision of high-value treats. Similarly, when the visitor retreats, food treats and praise stop. This sequence is practiced until the cat anticipates treats and attention when she hears the door or a visitor.

The most important rule of counter-conditioning fear-related behaviors is that situations in which fear is triggered should be carefully avoided. A common mistake that is often made with this program is that owners will observe that the pet is not showing anxiety or fear when presented with a low level of the stimulus (i.e. person far away) and then intensify the stimulus too rapidly, for example by bringing the person closer or allowing the visitor to enter the room. Often, this results in fear or nervous behaviors in the pet. If the owner allows this to occur, the positive associations that were conditioned previously may well be lost as the dog or cat now has reason to believe that getting treats when seeing a visitor is predictive of that visitor coming too close. In cases in which it is difficult to control the stimulus, the owner should simply stop treats and remove the pet from the situation before fear is triggered. This cannot be emphasized strongly enough because, even if unintentional, each episode in which fear is triggered is more than just a step back – it may reduce the effectiveness of future counter-conditioning attempts.

Because it is impossible to completely control a dog or cat's experiences and living situation, if the situation does arise in which a person approaches too close or a visitor comes into the home and attempts to pet the cat and the pet reacts fearfully, the owner should react neutrally ("come on, sweetie, let's turn around and walk the other way" or "let's put you in another room"), removing the pet from the situation as quickly as possible. In these cases, treats should not be provided because there is no utility in providing treats to a frightened animal.

The owner must understand that the provisions of treats are always contingent upon the pet seeing the trigger, *not* reacting in fear, and receiving treats. As stated previously, the classically conditioned stimulus-stimulus relationship that is desired is stranger/visitor predicts treats/petting (and their associated pleasant emotional responses).

Some owners may find that counter-conditioning alone is sufficient to reduce their pets' fear and to control fear-related behaviors. However, adding a program of systematic desensitization to counter-conditioning can allow further improvement. When the dog or cat simply notices the stimulus (for example, a person walking on the opposite side of the street or a visitor entering the home), and shows reduced fear and calm behavior, the owner can begin to desensitize to increased intensities of the stimulus. The owner can develop a gradient of these stimuli that progresses from the least fearful to the most fearful in the dog's perception. For example, the dog who is frightened by unfamiliar people can be counter-conditioned as people are allowed to approach more and more closely. A sit-stay should be trained so that the dog can learn to offer a behavior that can be positively reinforced (and which is incompatible with moving away). If the dog shows no fear at each step, and relaxed responses are practiced on several occasions, the stimulus intensity can be increased (i.e. a person moves closer). Again, the objective is to always introduce a stimulus that stays within the pet's comfort level and that does not elicit fear, allowing the counter-conditioning of relaxation and neutral (i.e. non-fearful) behavior. While some owners are able to progress all the way to the highest intensity stimulus, many reach a certain point and the dog or cat is capable of going no further. In these cases, a lessening of fear is often achieved, but the problem is not completely resolved.

The use of desensitization and counter-conditioning can be a bit more complicated when treating noise phobias, because these problems require that a suitable artificial stimulus is found. Recordings of thunderstorm noises are available for this purpose, but because most dogs and cats who are thunder-phobic react to multiple stimuli such as changes in air pressure, blowing wind, and temperature changes, the use of these recordings are frequently not successful in completely eliminating fear and anxiety. In some cases, drug therapy may be advantageous during the treatment program if exposure to the eliciting stimuli cannot be prevented (for example, during a real storm). Synthetic pheromones may also be helpful as adjunctive therapy for treating various types of fear-related behaviors in cats. Pharmacological therapy for behavior problems should always be considered carefully and must always be under the direct supervision of a veterinary behaviorist.

Some cases of fear-related behaviors are resistant to treatment because the degree of fear that the dog or cat shows is extreme, the problem has persisted for many months or years, or the pet has a strong genetic predisposition to shyness. Additionally, some owners are unable or unwilling to commit to a complete behavior modification program with their pet. In such cases, managing the pet's environment to prevent exposure to fearful stimuli can be the best solution. Using the examples discussed previously, walking the dog only in areas where unfamiliar people are not likely to approach and confining the cat when visitors arrive are acceptable solutions. However, the caretakers should be aware that management approaches mask the underlying problem and that fears do not usually abate over time without treatment.

Treatments That Are Not Recommended: "Flooding" involves exposing the animal to the fear-inducing stimulus and only removing the stimulus when the subject shows no fear (see Chapter 5, pp. 130–131). For success, it is imperative that the dog, the environment, and the stimulus are well-controlled. When used, the initial stages of flooding involve presenting mild forms of the stimulus, if that is possible. The dog or cat is prevented from escaping and reinforcements (food treats, praise, and petting) are given only when the subject stops showing fear-related behaviors and becomes calm. Once exposure to flooding has started, the trainer must stay committed until the pet becomes calm. Since it is not known how adversely an animal will react, there is always the risk of unintentionally making the pet more fearful when flooding is used. When successful, flooding is an expedient method for solving fear problems and involves much less time commitment on the part of the caretaker. However, there are serious disadvantages to this method. When the pet is very fearful, she may injure herself in attempting to escape, become aggressive, or become so highly stressed that calmness can never be achieved. It is also important that an owner (or trainer) not misinterpret exhaustion or learned helplessness with calm behavior. An added danger is that animals responds to flooding by generalizing their fear to other associated stimuli that are presented during the flooding process, such as the setting, the trainer, or the equipment that is used for restraint. In general, fears that have been exhibited for a long duration do not respond well to flooding, and may actually be exacerbated by this technique. Because of the many risks associated with flooding and because of the many uncontrolled factors involved, flooding is not recommended as an approach to treating fear-related behaviors. Similarly, punishment is not an effective treatment approach for fear-related behavior problems because aversive stimuli serve only to intensify fear and escape behaviors. In addition to being inhumane, using punishment with a fearful dog or cat is often an underlying cause of the development of fear aggression.

REVIEW QUESTIONS

1. Compare and contrast behavioral signs of separation stress/anxiety in dogs and in cats.
2. What are "pre-departure cues" and why are they an important consideration when treating separation anxiety?
3. Define systematic desensitization and counter-conditioning.
4. Describe the role of avoidance behaviors in maintaining fearful responses in dogs and cats.
5. List management approaches that can be used to control separation anxiety and fears in dog and cats.

REFERENCES AND FURTHER READING

Appleby D, Pluijmakers J. **Separation anxiety in dogs: The function of homeostasis in its development and treatment.** *Vet Clin North Amer: Small Anim Pract*, 33:324–344, 2003.

Bert B, Harms S, Langen B, Fink H. **Clomipramine and selegilne: Do they influence impulse control?** *J Vet Pharmacol Therap*, 29:41–47, 2006.

Bradshaw JWS, McPherson JA, Casey RA, Larter IS. **Aetiology of separation-related behaviour in domestic dogs.** *Vet Rec*, 151:43–46, 2002.

Clark GI, Boyer WN. **The effects of dog obedience training and behavioral counseling upon the human-canine relationship.** *Appl Anim Behav Sci*, 37:147–159, 1993.

Flannigan G, Dodman NH. **Risk factors and behaviors associated with separation anxiety in dogs.** *J Amer Vet Med Assoc*, 219:460–466, 2001.

Frank D, Minero M, Cannas S, Palestrini C. **Puppy behaviors when left home alone: A pilot study.** *Appl Anim Behav Sci*, 104:61–70, 2007.

Goddard ME, Beilharz RG. **Factor analysis of fearfulness in potential guide dogs.** *Appl Anim Behav Sci*, 12:253–265, 1984.

Griffith CA, Steigerwald ES, Buffington T. **Effects of a synthetic facial pheromone on behavior in cats.** *J Amer Vet Med Assoc*, 217:1154–1156, 2000.

Horwitz DF. **Diagnosis and treatment of canine separation anxiety and the use of Clomipramine hydrocholoride (Clomicalm).** *J Amer Anim Hosp Assoc*, 36:107–109, 2000.

Jagoe A, Serpell J. **Owner characteristics and interactions and the prevalence of canine behaviour problems.** *Appl Anim Behav Sci*, 47:31–42, 1996.

King T, Hemsworth PH, Coleman GJ. **Fear of novel and startling stimuli in domestic dogs.** *Appl Anim Behav Sci*, 82:45–64, 2003.

King JN, Simpson BS, Overall KL, and others. **Treatment of separation anxiety in dogs with clomipramine: Results from a prospective, randomized, double-blind, placebo-controlled, parallel-group, multi-center clinical trial.** *Appl Anim Behav Sci*, 67:255–275, 2000.

King JN, Overall KL, Appleby BS, and others. **Results of a follow-up investigation to a clinical trial testing the efficacy of clomipramine in the treatment of separation anxiety in dogs.** *Appl Anim Behav Sci*, 89:233–242, 2004.

Kronen PW, Ludders JW, Erb HN, Moon PF, Gleed RD. **A synthetic fraction of feline facial pheromones calms but does not reduce struggling in cats before venous catheterization.** *Vet Anesthesia Analgesia*, 33:258–265, 2006.

Landsberg G. **The distribution of canine behavior cases at three behavior referral practices.** *Vet Med*, 86:1011–1018, 1991.

Lern M. **Behavior modification and phamarcotherapy for separation anxiety in a 2-year-old pointer cross.** *Can Vet J*, 43:220–222, 2002.

Levine ED, Ramos D, Mills DS. **A prospective study of two self-help CD based desensitization and counter-conditioning programmers with the use of Dog Appeasing Pheromone for the treatment of firework fears in dogs (*Canis familiaris*).** *Appl Anim Behav Sci*, 105:311–329, 2007.

Lund JD, Jorgensen MC. **Behaviour patterns and time course of activity in dogs with separation problems.** *Appl Anim Behav Sci*, 63:219–236, 1999.

McCrave EA. **Diagnostic criteria for separation anxiety in the dog.** *Vet Clinic North Amer, Small Anim Pract*, 21:247–255, 1991.

Mills D. **Management of noise fears and phobias in pets.** *In Practice*, 27L248–255, 2005.

Parthasarathy V, Crowell-Davis SL. **Relationship between attachment to owners and separation anxiety in pet dogs (*Canis lupus familiaris*).** *J Vet Behav: Clin Appl Res*, 1:109–120, 2006.

Podberscek AL, Hsu Y, Serpell JA. **Evaluation of clomipramine as an adjunct to behavioral therapy in the treatment of separation-related problems in dogs.** *Vet Rec*, 145:365–369, 1999.

Prato-Previde E, Custance DM, Spiezio C, Sabatine F. **Is the dog-human relationship an attachment bond? An observational study using Ainsworth's strange situation.** *Behav*, 140:225–254, 2003.

Schwartz S. **Separation anxiety syndrome in cats: 136 cases (1991 – 2000).** *J Amer Vet Med Assoc*, 220:1028–1033, 2002.

Schwartz S. **Separation anxiety syndrome in dogs and cats.** *J Amer Vet Med Assoc*, 222:1526–1532, 2003.

Shore ER, Riley ML, Douglas DK. **Pet owners' behaviors and attachment to yard versus house dogs.** *Anthrozoos*, 19:325–334, 2006.

Takeuchi Y, Ogata N, Houpt KA, Scarlett JM. **Differences in background and outcome of three behavior problems of dogs.** *Appl Anim Behav Sci*, 70:297–308, 2001.

Tod E, Brander D, Waran N. **Efficacy of dog appeasing pheromone in reducing stress and fear related behavior in shelter dogs.** *Appl Anim Behav Sci*, 93:295–38, 2005.

Tuber DS, Hothersall D, Peters MF. **Treatment of fears and phobias in dogs.** *Vet Clinics North Amer, Small Anim Pract*, 12:607–623, 1982.

Voith VL, Ganster D. **Separation anxiety: Review of 42 cases. Abstract.** *Appl Anim Behav Sci*, 37:84–85, 1993.

Voith VL, Borchelt PL. **Separation anxiety in dogs.** In: *The Domestic Dog: Its Evolution, Behavior, and Interactions with People*, Cambridge University Press, Cambridge, UK, pp. 124–139, 1995.

Voith VL, Goodloe L, Chapman B, Marder AR. **Comparison of dogs presented for behavior problems by source of dog.** Paper presented at AVMA Annual Meeting, Seattle, WA, July 18, 1993.

Voith VL, Wright JC, Danneman JP. **Is there a relationship between canine behavior problems and spoiling activities, anthropomorphism, and obedience training?** *Appl Anim Behav Sci*, 34:263–272, 1992.

Wells DL, Hepper PG. **Prevalence of behaviour problems reported by owners of dogs purchased from an animal rescue shelter.** *Appl Anim Behav Sci*, 69:55–65, 2000.

Footnotes

[1] McCrave EA. Diagnostic criteria for separation anxiety in the dog. *Vet Clin North Amer, Small Anim Pract*, 21:247–255, 1991.

[2] Landsberg G. The distribution of canine behavior cases at three behavior referral practices. *Vet Med*, 86:1011–1018, 1991.

[3] Bradshaw JWS, McPherson JA, Casey RA, Larter IS. Aetiology of separation-related behaviour in domestic dogs. *Vet Rec*, 151:43–46, 2002.

[4] Voith V, Borchelt PL. Fears and phobias in companion animals. In: *Readings in Companion Animal Behavior*, Veterinary Learning Systems, Trenton, New Jersey, pp. 140–152, 1996.

[5] Wells DL, Hepper PG. Prevalence of behaviour problems reported by owners of dogs purchased from an animal rescue shelter. *Appl Anim Behav Sci*, 69:55–65, 2000.

[6] Voith VL, Ganster D. Separation anxiety: Review of 42 cases. *Appl Anim Behav Sci*, 37:84–85, 1993.

[7] Voith VL, Goodloe L, Chapman B, Marder AR. Comparison of dogs presented for behavior problems by source of dog. Paper presented at AVMA Annual Meeting, Seattle, WA, July 18, 1993.

[8] Flannigan G, Dodman NH. Risk factors and behaviors associated with separation anxiety in dogs. *J Amer Vet Med Assoc,* 219:460–466, 2001.

[9] Parthasarathy V, Crowell-Davis SL. Relationship between attachment to owners and separation anxiety in pet dogs (*Canis lupus familiaris*). *J Vet Behav: Clin Appl Res,* 1:109–120, 2006.

[10] Voith VL, Wright JC, Danneman JP. Is there a relationship between canine behavior problems and spoiling activities, anthropomorphism, and obedience training? *Appl Anim Behav Sci,* 34:263–272, 1992.

[11] Niego M, Sternberg S, Zawistowsky S. Applied comparative psychology and the care of companion animals: 1. Coping with problem behavior in canines. *Humane Innov Altern Anim Exper,* 4:162–164, 1990.

[12] Voith VL, Borchelt PL. Separation anxiety in dogs. In: *The Domestic Dog: Its Evolution, Behavior, and Interactions with People,* Cambridge University Press, Cambridge, UK, pp. 124–139, 1995.

[13] Lund JD, Jorgensen MC. Behaviour patterns and time course of activity in dogs with separation problems. *Appl Anim Behav Sci,* 63:219–236, 1999.

[14] King JN, Simpson BS, Overall KL, and others. Treatment of separation anxiety in dogs with clomipramine: Results from a prospective, randomized, double-blind, placebo-controlled, parallel-group, multi-center clinical trial. *Appl Anim Behav Sci,* 67:255–275, 2000.

[15] Podberscek AL, Hsu Y, Serpell JA. Evaluation of clomipramine as an adjunct to behavioral therapy in the treatment of separation-related problems in dogs. *Vet Rec,* 145:365–369, 1999.

[16] Clark GI, Boyer WN. The effects of dog obedience training and behavioral counseling upon the human-canine relationship. *Appl Anim Behav Sci,* 37:147–159, 1993.

[17] Jagoe A, Serpell J. Owner characteristics and interactions and the prevalence of canine behaviour problems. *Appl Anim Behav Sci,* 47:31–42, 1996.

[18] Takeuchi Y, Ogata N, Houpt KA, Scarlett JM. Differences in background and outcome of three behavior problems of dogs. *Appl Anim Behav Sci,* 70:297–308, 2001.

CHAPTER 11

Problem Aggression in Dogs and Cats

Aggression problems are the most commonly reported behavior problems in dogs and the second-most commonly reported problems in cats.[1,2] Between 3 and 5 million people in the United States are bitten by dogs each year and more than 2 million of these bites result in serious injury.[3,4] Children are bitten more often than adults and are particularly susceptible to serious injury or fatality from dog bites (Sidebar 1). Aggression between dogs is also a serious problem. This may occur between dogs who are strangers, or between dogs who live together in the same home. Problem aggression in cats includes biting and scratching that is directed toward people or toward other cats. Aggression between cats varies from occasional hissing or scuffling between housemates to serious physical attacks against any cat that is encountered. The most commonly reported form of inter-cat aggression occurs when a new cat is introduced into a home. Feline aggression that is directed toward people may include the owners, visitors, or both and has a number of underlying causes.

Sidebar 1: PREVENTING DOG BITES – TEACHING CHILDREN HOW TO BEHAVE WITH DOGS

The majority of dog bites that cause serious injury and hospitalization are inflicted upon children, and the most commonly bitten group are children under five years of age. These bites are often serious, with 80 percent of bites to young children involving the face, head, or neck. Contrary to popular belief, most of these bites are inflicted either by the family dog or by a dog belonging to friends, not by unfamiliar dogs or strays. Several studies of interactions between young children and dogs in homes have shown that children often initiate contact and frequently interact inappropriately with dogs. Behaviors such as pulling the dog's tail and ears, lying across the dog, hitting the dog, and attempting to ride the dog are commonly reported. Other studies examining the circumstances of dog bites to young children found that children under the age of five often incur injury after provoking the dog through unsafe behaviors such as disturbing the dog while he was eating, teasing the dog, getting between a mother dog and her puppies, or attempting to lift the dog.

Given the high proportion of young children who are bitten by familiar dogs within home or yards and evidence of unsafe and inappropriate behaviors by children, community dog-bite prevention efforts are being designed to target young children and focus on teaching children how to behave appropriately toward dogs. Programs vary in the methods that they use; some include demonstrations with a live and well-trained dog, others use photographs and illustrations of dogs, and one innovative program has children interact with a computer-generated cartoon dog. Regardless of the teaching approach that is used, most programs teach details of the following general concepts:

- *Gentleness and respect:* Children are taught (it does not come naturally to them) to approach dogs calmly and quietly, to pet gently and appropriately, and to never hit, tease, pull tails, lie upon, or attempt to ride a dog.
- *Do not disturb:* Children should never approach nor interfere with **any** dog who is eating, sleeping, chewing on a bone or toy, confined in a car, tied out or loose in a yard.
- *Listen to the dog:* Most programs include basic information describing canine body postures and communication signals that indicate both safe (friendly) and not safe (fearful, aggressive) dogs.
- *Always ask first:* Although most bites involve known dogs, children must still be taught to never approach an unfamiliar dog without first asking the permission of the owner. (Loose dogs should never be approached; the "Be a Tree" technique is used when an unsupervised dog approaches the child.)

Finally, because the number of children who are bitten by their own dog is substantial, education of parents is an important component of bite prevention programs. Fiona Wilson's study found that many parents were unaware of the dangers that their dog could pose to their children and the majority of parents allowed unsupervised interactions between very young children and the family dog. Despite a substantial number of previous dog bites, the parents also believed that their dog would not bite if disturbed while sleeping, eating, guarding a toy, or when challenged by the child. These results indicate that parent education about the previously described behaviors is as important as is direct intervention with the children when designing bite-prevention programs.

De Keuster T, De Cock I, Moons CPH. Dog bite prevention: How a blue dog can help. *FECAVA Symp*, pp. 136–138, 2005.

Jalango MR. On behalf of children: When teaching about pets, be certain to address safety issues. *Early Child Educ J*, 33:289–292, 2006.

Wilson F, Dwyer F, Bennett PC. Prevention of dog bites: Evaluation of a brief educational intervention program for preschool children. *J Commun Psychol*, 31:75–86, 2003.

Whenever problem aggression is reported in a dog or cat, the target of the aggression and its severity, frequency, and potential for causing harm must be immediately evaluated. In some cases, it is necessary to clearly identify the problem as aggression and to distinguish the pet's behavior from play or predatory behaviors. Aggressive behavior in both dogs and cats can be classified according to the target(s) and with respect to underlying motivation. As with all forms of behavior problems, proper treatment and management relies upon correctly determining the cause and triggering stimuli. An added concern with problem aggression is the safety of people and other pets and the potential for harm during treatment. Treatment approaches for aggression must always include management procedures for keeping others safe and preventing any further bites or aggressive episodes. Because of the seriousness of aggression problems (especially in dogs), this chapter is intended to provide an overview of the types of aggression problems seen in dogs and cats and methods for preventing and treating problems, but does *not* provide complete training for working with dangerously aggressive dogs or cats. (See Appendix 1 and Appendix 2 for additional readings and for information regarding professional certification for dealing with severe behavior problems.)

OVERVIEW OF AGGRESSION

The simple definition of *aggression* is any canine or feline behavior that inflicts or intends to inflict harm or injury to another individual (human or non-human). However, this definition is very broad and would also include predation and possibly some forms of rough or uninhibited play. A more precise definition classifies aggression as a type of agonistic behavior characterized by arousal, an intent to cause harm or injury, and evidence that the target individual reacts aversively either by attempting to avoid the aggressor or by responding in defense. *Agonistic behaviors* refer to the set of behaviors that any animal uses to resolve conflicts with another individual. Although *agonistic behaviors* are often used synonymously with *aggression,* they include all of the behaviors that can be used to solve conflicts, such as include flight and submission. Therefore, aggression is one type of agonistic behavior that an animal might show during a conflict. The term *arousal* implies change to an emotional state that accompanies the animal's intent to cause harm. For example, a dog who is playing expresses an affective (emotional) state that is affiliative and does not have the intent to adversely affect her playmate. Conversely, the emotional state of an aggressive dog or cat differs both in its expression and its intent. It involves a set of facial expressions and associated body postures that communicate threat or impending attack to the target victim. This more narrow definition of aggression therefore includes threats (growling, lunging, or hissing) that may not actually result in contact and excludes predation or play behaviors that can cause harm but which do not involve the same emotional state.

Although problem aggression that is directed toward owners, unfamiliar people, or other animals is a serious behavior problem, it is important for owners to understand that as a species-specific behavior pattern, aggression is not maladaptive. Aggressive behaviors have been selected within species for the advantages that these behaviors confer to individuals involved in conflict. Indeed, if aggressive behaviors were completely maladaptive, they would not exist. Competition between group members and between species is part of survival and natural selection. Successful defensive behaviors that are aggressive (fight) prevent an animal from becoming a meal for a predator or from losing its territory to competing animals of the same species. However, just because a trait confers advantages in certain circumstances, this does not mean that trait is never harmful to other individuals or that it is not maladaptive when expressed in inappropriate circumstances.

Because aggressive behaviors are a part of every dog's and cat's behavioral makeup, individuals cannot (and should not) be classified as being either "aggressive" or "non-aggressive." Every dog and cat is capable of showing aggressive behavior. In addition, simply labeling an animal as "aggressive" provides no information about the context in which the aggression occurs or the factors that trigger the behavior. Nor is such a label helpful in developing an effective treatment program. Moreover, the acceptability of the behavior is typically determined by evaluating the appropriateness of the situation in which the animal demonstrates aggression, as well as the severity of the aggressive response. Problem aggression can be classified according to several schemes. These include dividing aggression according to the target (i.e. the owner, strangers, or visitors to the home, familiar dogs/cats or unfamiliar dogs/cats), the type of threat behaviors that the dog shows (offensive or defensive), or determining if the behavior is learned or inherited. The most widely used method for diagnosing and treating problem aggression relies upon determining the function that the aggression has for the dog or cat. This functional classification approach takes into account the context in which the aggression occurs as well as the types of behavior patterns and body postures that the dog or cat exhibits.

For dogs and cats who live within human social groups, aggressive behaviors that are classified as problem aggression are typically responses that are either directed at inappropriate targets (owners, other pets), are triggered in situations that are not acceptable, or are expressed at a level of intensity that is dangerous. When the stimulus that triggers the behavior seems to be abnormally subtle or unreasonable or when the dog's aggressive threshold is unusually low, the behavior may be classified as abnormal. However, in most cases of problem aggression, the pet is showing a normal response to the situation or has learned through prior experience to escalate his aggressive response. Just as with other problem behaviors, owners must understand that the line that they may draw between acceptable and not acceptable behavior may seem clear to them, but may not be quite so obvious to their dog or cat. For example, an owner may encourage his two-year-old Rottweiler-mix to bark at the door whenever someone approaches, in the belief that this will promote desirable protective behaviors. However, when the same dog rushes and nips at someone coming into the house, suddenly the dog's behavior is classified as a problem. Similarly, an owner who teaches her kitten to chase her feet in play may suddenly decide that this behavior is aggressive play when the same cat, as an adult, begins to ambush family members and inflict painful bites on their ankles.

PROBLEM AGGRESSION IN DOGS

While aggression is a normal and adaptive behavior in dogs, actual attacks and physical injury are rare and conflicts between animals are either avoided altogether or resolved through highly ritualized species-specific communication signals (see Chapter 3 for a complete discussion). When these signals are correctly sent, received, and interpreted by both participants, aggressive displays and bites are either avoided altogether or are inhibited. Still, harmful bites *do* occur and have several possible causes. Because aggression exists on a continuum and because one dog may show several types of aggression in different contexts, classifying aggression can be difficult. An individual dog's propensity to show aggression can be affected by many factors, including genetics, neurochemical factors, the living environment, and prior learning experiences. Although there is not complete agreement on functional classifications, the most common schemes identify primary types of canine aggression as dominance-related, possessive, territorial, defensive (fear-related), or pain-induced. Within these categories, human-directed aggression includes dogs who aggress toward their owner or toward visitors or unfamiliar people, and dog-directed aggression includes dogs who fight with other dogs in their home (also called *social aggression*) and dogs who are aggressive only toward unfamiliar dogs.

Dominance-Related Aggression

Dominance-related aggression is an offensive form of aggression that is seen during competitive interactions over the control of valued resources or in response to a perception of challenges to the dog's social status. Aggression may be directed toward humans or toward other dogs (intra-species), but is typically displayed toward individuals with whom the dog has an established social relationship (family members or other dogs in the home). Dominance-related aggression can range from mild posturing and growling to direct attacks and uninhibited biting. It is usually reported to begin at the age of social maturity, between one and three years of age, and is most common in intact males. However, females and neutered dogs of both sexes are capable of developing dominance aggression. There is some evidence that spaying may result in an increase in dominance aggression in females who were already showing signs of aggression.[5] However, these data do not indicate that spaying *causes* dominance aggression and do not suggest that keeping a female pet intact will prevent dominance-related aggression.

Dominant Behavior: It is very important to make the distinction between a dog who is confident or assertive and uses dominant signals in certain contexts, and a dog who is showing dominance-related aggression. As discussed in Chapter 3 (pp. 72–75), the popularization and overuse of the concept of dominance in dogs has caused a great deal of confusion. Dominant facial expressions and body postures comprise a normal and necessary set of social signals that dogs use to communicate confidence and to maintain control over valued resources. There are many dogs who have confident, "leader-type" temperaments but who do not demonstrate problem aggression toward their owner or other dogs. In fact, this type of temperament is often purposefully selected for in certain breeds or for certain types of work.[6] These dogs greet their owners and other dogs with confident body postures and rarely show deference or appeasing body postures. In a multiple-dog home, they are more likely to preempt toys and choice sleeping areas, although even the control of resources is often flexible and context-specific among dogs. Many confident or assertive dogs do not become aggressive and maintain peaceful and affectionate relationships with their owner and with other dogs in their home. For these dogs, early and consistent training that reinforces deferential behaviors and teaches control exercises prevents pushy and attention-seeking behaviors. Training confident dogs to readily give valued resources to their owner and to comply with basic obedience commands (sit, down, stay, come) can avert opportunities for resource guarding or an escalation of assertive behaviors to aggression.

Human-Directed Dominance Aggression: Dominance aggression that is directed toward the owner or other human family members typically occurs in situations that involve a challenge to the dog's social status or to his control over a valued resource. Typical provoking situations include disturbing the dog while he is sleeping, attempting to restrain the dog for grooming, standing over or abruptly hugging the dog, or physically reprimanding the dog. Attempts to take toys, bones, food or other possessions away from the dog may also trigger an aggressive response. (Note: Although possessive aggression is often seen in dogs with dominance-related aggression, possessive threats alone are *not* diagnostic of dominance-related aggression. See "possessive aggression," pp. 289–290). Because dominance-related aggression is highly individualized, owners are often very proficient at predicting the circumstances that will evoke an aggressive response in their dog. Some owners describe the dog as "becoming very still, or frozen" immediately before aggressing. It is possible that these signs are warning signals that the owner has either misinterpreted or does not react to quickly enough in order to prevent aggression.

Learned behavior is an important component of dominance-related aggression. A dog who first shows an aggressive response to a situation that he perceives to be threatening usually warns with a growl or snarl. Most owners, being quick learners, jump back or move quickly away. This consequence, though certainly appropriate and wise from the owner's viewpoint, serves to reinforce the dog's aggression. For example, the dog learns that growling when his owner picks up his feet to cut nails causes his owner to stop handling his feet, a desirable outcome for the dog. Similarly, if the owner stops approaching when the dog growls while holding a stolen sock, the dog learns to use a growl to maintain control over the valued resource. Similarly and somewhat paradoxically, the use of punishment with dominant aggressive dogs also results in an escalation of aggression. When an owner attempts to punish a dog for dominance-related aggression, the dog will almost invariably intensify his aggressive response. For example, a dog who growls when asked to move is using normal canine offensive signals to communicate an unwillingness to comply and a threat of aggression. If the owner responds to the dog's growl with physical punishment, the dog learns that growling did not have the desired outcome (being allowed to stay where he was). The next time that the dog is asked to move, he escalates the growl to snarling and eventually to biting. The owner must understand that the dog is *already* challenging the owner with the growl and has no reason not to escalate this behavior if challenged. After several repetitions of this scenario, the dog will preemptively bite when the owner attempts to move him. If previous punishment has had the effect of suppressing growls, the dog may now show no warning signals at all and will immediately bite when challenged. Therefore, the use of punishment is neither effective nor safe as a treatment approach when dealing with dominance-related aggression (see treatment next page).

Dominance-Related Aggression between Dogs: As discussed in Chapter 3, dominant and subordinate interactions serve principally to maintain peace and support social cohesion. Dogs use deference and appeasement signals to communicate to a confident or threatening dog that they are not a threat. Contrary to the highly popularized (and incorrect) version of dominance, social status within a group of dogs is normally maintained by deference signals, not through aggressive displays. When dominant signals are used, they are often subtle and non-aggressive and serve to maintain access to desired resources such as food, toys, or resting spots. Even in these contexts, social relationships are often fluid and highly context-specific. Although problems certainly occur and can be very serious, many dogs living in multiple-dog homes are able to coexist peacefully with no or minimal agonistic interactions and are often strongly bonded to one another (see Chapter 3, pp. 72–75).

Just as the terms *dominance* and *dominance aggression* are often misapplied when referring to interactions between humans and dogs, so too are they frequently misused when describing conflicts between dogs. The underlying cause of aggression between dogs who live in the same home is most often either fear or anxiety, resource guarding (possessive aggression), or redirected aggression during periods of arousal or excitement. The most common situations in which familiar dogs fight are when competition occurs over food, toys, or access to the owner, and during occasions that stimulate high arousal or excitement. Although status may be involved in these situations, it is incorrect to assume that status conflicts are the underlying cause whenever dogs fight. True dominance-related aggression is relatively rare and occurs when an assertive dog (often the younger) *repeatedly* attempts to control the behavior and movement of another dog in the home through offensive aggressive displays. These challenges occur when dogs are competing for interactions with the owner, when entering or leaving the home, and over the use of toys, food, and resting places. Most or all of these contexts must be present for a diagnosis of dominance-related aggression. In addition, the aggressing dog will have an offensive threat body posture characterized by holding the tail high, with a stiff, forward-directed body posture. The dog may attempt to place his head over the shoulders of the other dog. Dominance-related aggression between dogs often begins around the onset of sexual maturity in intact animals (six to nine months) and at social maturity in neutered animals (one to three years). Conflict can also occur when a new dog is introduced into a home or when an older dog becomes infirm and a younger, more assertive dog shows increasingly dominant behaviors.

Treatment of Dominance-Related Aggression: The goals of treatment for dominance-related aggression are to modify the dog's behavior and manage his environment to prevent aggressive threats and to ensure the safety of all people and other dogs. The treatment program must be tailored specifically to address the context and intensity of the dog's aggression and to the lifestyle and capabilities of the owner. In cases of extreme aggression when the dog's response is out of proportion to the context or when the triggers for aggression are highly unpredictable, the owners must consider their ability to keep others safe and the likelihood of a successful treatment. Every dog is unique, and owners differ in their motivation and ability to modify the dog's environment and implement a consistent training program.

The first step in a treatment plan for dominance-related aggression is to modify the dog's daily schedule and living environment to avoid all situations and stimuli that have the potential to provoke an offensive threat. For example, if the dog growls when asked to get off the owner's bed, he is no longer allowed access to the bedroom. If handling or grooming elicits threats, these procedures are discontinued until they can be reintroduced using counter-conditioning and desensitization exercises (see next page). Regardless of whether the dog resource-guards, all toys, chew bones, and food bowls are picked up and placed out of the dog's reach. Access to the outdoors is controlled by leashing the dog before approaching the door and teaching a sit-stay or wait command prior to going outdoors. Finally, the dog is attached to a long lead when free in the home and is never allowed off-lead when outdoors.

A structured program of training is introduced. Because most basic obedience classes that are available to pet owners cannot adequately address dominance aggression problems, instruction for this training should come from a trainer or behaviorist who provides individualized, in-home consultations for aggression problems. Many dogs who develop aggression problems have had little or no previous obedience training. Therefore, training a reliable sit, down, stay, and come is one of the most important components of a behavior modification program for aggression. The down-stay exercise is especially useful with dominance-related aggression because using this exercise allows the owner to shape and positively reinforce deferential body positions and calm and relaxed behaviors. It is imperative that only non-confrontational techniques that use positive reinforcement are used to teach these exercises (see Chapter 7, pp. 164–170 for complete instructions). In addition to the sit-stay and down-stay exercises, the dog must also be trained to come when called, to wait at the door, and to "leave it" and "give" (Chapter 9, pp. 218–219). The dog is required to offer a sit-stay or down-stay prior to gaining access to any valued resources such as meals, toys, walks outside, and interactions with the caretaker. The most important rule of a treatment program for dominance-related aggression is that the dog is placed in non-confrontational situations in which all food, playtime, attention, and other valued resources are available as reinforcements for calm, non-threatening, and well-controlled behavior. (Some trainers and behaviorists refer to this concept as the "nothing in life is free" or "NILIF" procedure; Sidebar 2.) The owner must also be taught to recognize and positively reinforce all deferential behaviors when they are offered by the dog.

Sidebar 2: GENERAL GUIDELINES FOR "NOTHING IN LIFE IS FREE" PROCEDURE (NILIF)

The general premise of behavior modification programs for dogs who are demonstrating pushy behaviors or dominance-related threats is to change the dog's relationship with the owner from one of conflict and challenge to one in which the owner has control over all resources and in which the dog readily offers deferential behaviors for reinforcement. This includes removing the dog's access to all valued resources, teaching and practicing obedience exercises (sit, down, stay, come), and managing the dog's environment to prevent opportunities for the dog to practice bullying behaviors. The basic steps of this program include the following:

- **Equipment:** A halter-type collar should be used for all training. The dog should wear the collar and an attached leash or long line while free in the home and while loose in a yard.

- **Obedience training:** Control obedience commands of *sit, down, stay, come, give,* and *leave it* should be taught and practiced several times daily in various rooms of the home. This training is lifelong, not just during treatment for behavior problems.

- **Prevent conflict:** All situations that the dog may perceive to be a challenge are avoided. The dog is trained to come when called from different parts of the home, and from resting areas using the attached line (the owner should never call the dog from closer than the end of a six-foot line). Any non-compliance from the dog is ignored; the owner turns and walks away, saying nothing.

- **Toy management:** All toys and chew bones are picked up. These are provided to the dog as positive reinforcers for calm and deferential behaviors (response to *sit, down, stay, wait, come when called*).

- **Feeding time training:** The dog is trained to sit and stay prior to all meals. After putting the food bowl down, the dog is allowed to eat undisturbed. When the dog is finished, the owner calls the dog away from the bowl, reinforces with a food treat, and takes the dog outside to eliminate. The food bowl is picked up and put away.

- **Space management:** The dog's access to desirable sleeping areas and to any spots that were protected is blocked for several weeks. When the dog has been trained to reliably respond to obedience commands, the dog is allowed limited access and is trained to come away from these areas on command (the owner always uses a long line for *come when called*, again, always staying six feet or more away from the dog). All correct responses are positively reinforced.

- **Positively reinforce all deferential and calm behaviors:** The *sit* or *down* command is used prior to all activities and interactions with the dog (i.e. sit before petting; down before going outside; sit before receiving dinner). Positive reinforcement (treats, praise, and petting) is provided each time that the dog complies with a command. The dog is ignored if he does not comply (the owner simply walks away).

Finally, any undesirable behaviors that the dog offers are never reinforced. For example, if the dog constantly seeks attention by staring at or pawing the owner, the owner simply gets up and walks away. Several minutes later, the owner may call the dog to him, request a sit-stay or down-stay, reinforce the dog with a food treat and then provide calm and positive interactions (petting, praise). Similarly, if the dog mouths when the owner attempts to put on a collar and lead, the owner ends the interaction by walking away. Another attempt should include the sequence of coming when called, followed by sit, and then the collar and lead are put on the dog. The owner should never allow himself to be placed into a struggle for control with the dog (i.e. confrontations are always avoided). Finally, counter-conditioning and systematic desensitization are used to re-introduce handling and other circumstances that were previously triggers for the dog. For example, handling is reintroduced during down-stays, during which gentle touching is reinforced. The brush can be reintroduced and brushing paired with food treats (classical conditioning; see Chapter 4, pp. 82–87). Every situation in which the dog has shown threatening behavior must be identified and reintroduced using counter-conditioning and desensitization. If there are some contexts that present a very high risk of aggression or that are not important to the owner to train, management must then be directed toward preventing the dog from being in those situations.

Treatment for Dominance-Related Inter-Dog Aggression: When aggression occurs between two dogs who live together, treatment includes the same principles. The multi-dog home must be managed in a manner that prevents all aggressive displays between dogs and positively reinforces appropriate and non-aggressive behaviors in a variety of situations and settings. If one or more of the dogs are intact males, **castration** is recommended. This usually results in a decrease in the level of aggression in most males, but not all dogs respond noticeably. The primary rule that owners must follow is to correctly identify the dog in the interaction whose behavior is most appropriate and non-aggressive and reinforce that dog. The message that the owner works to communicate is that pushy and obnoxious behaviors such as stealing toys or bones, shouldering dogs out of the way, or preventing access to the owner for attention will not be reinforced or tolerated (Box 11.1). Conversely, calm, non-threatening behaviors and compliance with control commands (sit, down, stay, come) are consistently and positively reinforced. For example, both dogs are taught to sit calmly for petting and interactions. These pleasant interactions are provided to both dogs when they comply and withheld if one or both do not. The owner does not reprimand the dogs if they do not comply with the sit request; she simply gets up and walks away from them. Management techniques involve the removal of valued toys and bones to prevent resource-guarding conflicts between dogs (see possessive aggression for a complete discussion). Owners must understand that treating inter-dog aggression within a home is *not* about imposing some preconceived idea of rank or attempting to identify an "alpha" dog and support that dog. To the contrary, treatment is aimed toward reinforcing appropriate deferential and non-aggressive behaviors that are normal and typical of peaceful social groups.

Box 11.1 TEACHING OLIVER AND BAXTER TO SHARE

Description of problem: Sally Sloan is the owner of two neutered male Labrador retrievers, Oliver (two years old) and Baxter (four years old). The dogs have lived together for 18 months (since Sally adopted Oliver). Although they generally get along well and play together often in the yard and in the house, Sally reports that there has been increasing tension between the two dogs. Oliver has started to steal any toy or bone that Baxter might be chewing on, is playing very roughly with Baxter, and attempts to block his access to Sally when she is petting both dogs. (Sally says that she has never intervened when Oliver steals or is pushy because a friend at work told her it is best to let dogs work these issues out for themselves.) Unfortunately, although Baxter allows Oliver to take his bones, he is becoming increasingly anxious and "worried looking" when he has a toy or bone because Oliver will seek him out to steal his possession. Sally states that yesterday, Baxter was so anxious that he ran under the kitchen table with his toy and growled when Oliver attempted to take it away. Sally also notices that Baxter is less and less willing to play with Oliver because Oliver "body slams" Baxter and plays so roughly that Baxter becomes frightened. She says that Baxter increasingly hides under her desk to avoid Oliver, and avoids coming over for petting and attention if Oliver is near Sally. Finally, although both dogs completed an introductory obedience class, Sally admits that she has not kept up with their obedience training.

Diagnosis and treatment: Oliver is possessive and showing pushy behaviors that are probably related to status (dominance-related aggression). In response, Baxter is reacting with fear that may be escalating to defensive aggression. A behavior modification program and training program focuses on teaching Oliver to behave appropriately with Baxter, to decrease Baxter's anxiety, and to prevent aggressive interactions between the two dogs. Here are the rules that Sally follows with her dogs:

- **Obedience training:** Sally is instructed to reinstitute obedience training with her dogs, concentrating on practicing deferential behaviors such as sit, down, and stay. She also teaches "leave it" and "give." Positive reinforcement (food treats, praise, and petting) is used for all training.

- **Reinforcing deferential behaviors:** Both dogs are required to sit/stay before meals, before going outside, after being called to come, and prior to all interactions with Sally. If they do not comply, Sally simply turns and walks away and tries again several minutes later. When they do comply, she first reinforces the dog who obeys most rapidly and who is more calm and controlled. She reinforces both dogs for complying (i.e. she does *not* attempt to reinforce a presumed status order between the two dogs). If Oliver attempts to block Baxter's access to Sally she simply stands up and walks out of the room, calls both dogs to her, tells them to sit, and reinforces *both* for sitting before petting.
- **Management of valued resources:** Oliver must be taught that he is not allowed to steal toys, bones, or food from Baxter. All toys and bones are picked up and Sally provides these resources to both dogs only under supervision. When chewing on bones, the dogs are always separated by at least six feet. If Oliver attempts to take Baxter's possession, he is immediately interrupted with "leave it" and returned to his own bone. To ensure that Oliver complies, Sally initially attaches a lead to Oliver for this training.
- **Rough play:** Rough play is stopped using negative punishment. Sally uses the "settle" command to interrupt all rough play. Both dogs then do a short down-stay until they are calm. They learn that rough play leads to a loss of play opportunities.
- **Feeding:** Dogs are either fed in separate rooms or are trained to eat only out of their own bowl and are not allowed to approach each other's bowls while eating.

Outcome: After several weeks, Sally reports that Baxter and Oliver are playing gently together and that Oliver is better behaved, both with Baxter and – an added benefit – with Sally. She plans to continue with their training and has enrolled both dogs in another manners-training class so that she can keep practicing control exercises with her two dogs.

Refractory Cases: Unfortunately, not all cases of dominance-related aggression can be treated successfully. The degree of danger to people and dogs in the home, and the limitations of the owner, must be considered. Some dominant-aggressive dogs are unresponsive to treatment, even when a detailed training program is used and when the owners are dedicated and consistent. In other cases, the owners may be unwilling or unable to commit to a complete behavior modification program, or to live in a situation in which the dog may continue to be a danger to the family or to others. In such cases, options such as carefully placing the dog in another home or euthanasia must be considered. Some owners may wish to consider using pharmacological therapy in conjunction with the behavior modification program. A veterinary behaviorist should always be consulted if considering adjunctive drug therapy.

Possessive Aggression

Possessive aggression is often used synonymously with the term "resource guarding" and refers to aggressive displays that occur during competition for valued resources. Highly valued resources for dogs include toys, chew bones, their food bowl, and access to the owner. In addition, some dogs fixate on unusual items such as pieces of tissue, articles of clothing, or even the television remote control. Guarding these unusual items usually signifies learned behavior associated with a history of object-stealing by the dog. When an owner has reacted by chasing or harshly reprimanding the dog for stealing something, the dog may learn to respond by guarding the stolen item.

Possessive aggression may be offensive or defensive in nature. Dogs who show offensive possessive aggression are confident dogs who may tease their owner (or another dog) with the item and demonstrate offensive threat postures when challenged. Conversely, dogs who are fearful or anxious when they are attempting to maintain control of a possession often run and hide with the item, and will show defensive threat postures when confronted. Dogs who demonstrate defensive possessive aggression usually have a history of being physically punished for stealing or guarding forbidden items. In some of these cases, the motivation for aggression may have components of both offensive and defensive threat. Some authors refer to this as "conflict" aggression, because dogs who are anxious or nervous (because they anticipate punishment) may show ambiguous signals when they aggress. For example, a dog who wants to maintain control over a possession but has been punished or teased in the past may show components of both offensive and defensive aggression in his behavior when guarding. Finally, food bowl aggression can be classified as a subcategory of possessive aggression in dogs who guard their food bowl but show no other form of aggressive behavior. It can occur in dogs who have been compelled to guard their food from other pets or have learned to distrust humans who have repeatedly approached (and possibly removed) their food bowl.

Treatment for Possessive Aggression: First, all objects that the dog guards or steals must be removed and kept out of the dog's reach. The owner supervises any and all future access that the dog has to these objects. Basic obedience commands are taught or reviewed (sit/down/stay/come) and the dog is also trained to relinquish possessions using the "give" command and to refrain from picking up objects using the "leave it" command (see Chapter 9, pp. 218–219). Both "give" and "leave it" must be shaped, starting with items that are considered by the dog to be of low value rather than items that the dog has a history of guarding. As discussed previously, operant conditioning using positive reinforcement and successive approximation are used to teach these exercises safely and successfully. When dogs are possessive over food bowls or toys with other dogs in the family, but not with people, both management of the home environment and training are used to prevent stealing and guarding behaviors between dogs (Box 11.1).

Food bowl aggression is often treated as a special case because it frequently occurs independently of other forms of aggression. In addition, food bowl aggression can rapidly escalate given the dog's repeated opportunities to practice this form of resource guarding during mealtimes each day. Therefore, when food bowl aggression is reported, the owners need an immediate plan to prevent additional opportunities for the dog to aggress while still allowing the owner to feed the dog. (Note: Dogs who aggress over high-value food treats such as table scraps or other human foods are treated using the same protocol as that used for other forms of possessive aggression.) Some dogs demonstrate very mild food bowl guarding behaviors. These dogs may begin to wolf their food when someone approaches or stiffen slightly as they continue to eat. More serious and potentially dangerous levels of aggression occur when the dog freezes over the bowl, stops eating, growls, and bites or attempts to bite when a family member approaches.

Similar to other types of possessive aggression, dogs who guard their food bowls may be showing either offensive or defensive threat postures. Defensive aggression often develops as a result of an owner frequently removing the dog's food bowl while he is eating, with the misguided intent of *preventing* food bowl aggression. The use of this inappropriate technique teaches the dog that a person approaching his bowl predicts something aversive, such as the chance of losing his meal or being teased while he is eating. Unfortunately, this sets the stage for defensive food bowl guarding behaviors. Similarly well-intentioned but wrong advice is to sit with one's hand in the dog's bowl while the dog is eating. (As an analogous example, not many people would appreciate having another person's hand in their plate during dinnertime and most would react adversely, not positively, to such an experience.) For the dog, this often serves to increase, not decrease, anxiety about their food bowl. Finally, a history of being physically punished for mild food guarding behaviors can cause aggression to escalate and lead to more severe defensive threats.

Management techniques are the simplest approach for treating mild food bowl guarding. The dog is meal-fed two or three meals a day in a secure room with a closed door or gate and is trained to sit/stay prior to getting his meal. After being released to eat, the dog is allowed to eat his meal without interruption. When the dog has finished eating, he is released from the room and taken outdoors for elimination. While the dog is outdoors, the bowl is picked up and put away. When food guarding is directed toward other dogs in the home, the same procedure is followed. Each dog is taken outdoors separately to prevent excitable or redirected aggressive displays immediately after eating. For many households, using these management techniques to avoid triggers and prevent aggressive displays are adequate for reducing the dog's anxiety near the food bowl and for ensuring safety of the people and other dogs in the home. For more severe cases or when owners do not wish to comply with management approaches, additional training using counter-conditioning and desensitization can be used to reduce food bowl guarding behaviors (Sidebar 3).

Sidebar 3

COUNTER-CONDITIONING AND DESENSITIZATION PROGRAM FOR DOGS SHOWING FOOD BOWL AGGRESSION

Counter-conditioning is used to change a dog's emotional response to a person approaching her food bowl from anxiety to pleasure and anticipation. Once this change has been initiated, systematic desensitization is added to the program to acclimate the dog to gradually closer proximities to people. The two components for a successful and safe program to change a dog's response to her food bowl include both counter-conditioning and desensitization. **Counter-conditioning:** The dog learns that the presence of a person in the room during meal time predicts positive experiences (more food) as opposed to aversive experiences (taking food away). **Desensitization:** The dog learns (gradually) that the proximity of a person does not pose a threat and continues to predict pleasurable experiences.

Training Program (Note: Dogs who guard their food bowls must be meal-fed and not allowed to free-feed):

- **Select a safe distance:** Prior to the first training meal, the owner selects a spot in the kitchen (or room where the dog is fed) that is far enough from the dog's feeding spot to not evoke any reaction from the dog (i.e. the dog must continue to eat in a relaxed and calm manner; no freezing, wolfing food, or growling). The owner places a chair at that spot.

- **Dividing dinner:** The dog's ration of food for the meal is divided in half. One half is placed in his bowl and the second half is poured into a plastic container. The owner has the dog sit-stay prior to receiving his bowl. After placing the bowl on the floor, the owner, carrying the container with the food, walks to the chair and sits down. The owner says nothing as the dog eats his meal.

- **Come to me for more:** As the owner sees that the dog is finishing, he stays seated, and calls the dog to him. When the dog comes, the owner requests a sit and positively reinforces come/sit with one or two pieces of food from the container.

- **Return to bowl:** The owner then calmly stands, walks back to the dog's bowl, and pours the second half of the meal into the bowl, keeping out one or two small pieces of food. The owner then returns to the chair and remains seated until the dog finishes eating.

- **One more "come when called":** When the dog has finished eating, the owner again calls the dog to him, reinforces with the food treats and then leaves the room (with the dog). The dog is taken outside to eliminate and the bowl is picked up while the dog is outdoors or out of the room.

- **Practice for three days:** This sequence is repeated for at least three days (six meals). The dog must be showing relaxed and happy behaviors upon returning to the owner and when the owner places the food in the bowl, before proceeding. (This is of utmost importance as the sequence must be practiced for adequate repetitions that the dog views the owner approaching the bowl as something positive, not as an aversive.)

- **Reducing distance (the moving chair):** When the dog's behavior indicates excitement and pleasure upon seeing the owner rise from the chair to place the second part of the meal into her bowl, the owner can begin to place his chair several inches closer to the dog's bowl each day.
- **Final caution:** When the owner can sit quietly approximately 1 or 2 feet away from the dog and the dog is not guarding her food, the training is successful. To be safe and to show dogs the respect they deserve, owners should not touch, pet, or otherwise harass their dog during mealtime.

Territorial Aggression

When humans or animals approach their home or yard, many dogs show orienting behaviors and bark but do not become aggressive. This is normal behavior and only becomes a problem when the dog becomes uncontrollable or aggressive. Areas that dogs are most likely to consider their home territory include the owner's house, yard, car, and possibly other areas where the dog is frequently walked or confined. For confident dogs, territory may be perceived as a valuable resource that must be protected and competitively maintained. Territorial threats are offensive in nature for these dogs. Similar to possessive aggression, offensive territorial aggression may occur in dogs who show dominance-related aggression. For fearful or nervous dogs, territorial aggression is a learned behavior that has proven to be effective in driving off intruders and reducing feelings of anxiety and fear. This second type of territorial defense is often seen in poorly socialized dogs who have been confined to a small area or tied to a doghouse. Territorial aggression is most prevalent in dogs who are kept tied outside for long periods of time, are confined to a small yard, or who have been encouraged by their owners to guard entryways or yards.

Treatment for Territorial Aggression: First, the dog's access to guarded areas is restricted to prevent exposure to all eliciting stimuli. This is accomplished by simply bringing the dog into the house or changing the way in which the dog is managed within the home. Within the house, any doors or windows that the dog guards are off-limits to the dog. This is an important component of treatment because there is the potential for an aggressive response on each occasion that the dog has unsupervised access to guarded areas. A unique characteristic of territorial behavior is that it is always immediately reinforced when the intended target moves away. For example, a dog who barks and growls at the mailman as he approaches the house each day to deliver mail sees the mailman turn and leave. As the mailman departs, the dog's agitation decreases, thus reinforcing the barking and growling behaviors.

Counter-conditioning and desensitization are effective techniques for use with territorial behaviors. The dog is first trained to offer a behavior that is incompatible with charging the door (or fence) and barking. A typical response that is trained is "come when called" and sit. For dogs who rush doors, some owners prefer to teach the dog to go to a bed or mat in another room. Target training is used to teach this exercise (see Chapter 7, pp. 166–167). A reliable response to this command must be trained in the *absence* of triggers for a territorial response before attempting to desensitize the dog to eliciting stimuli such as approach of visitors to the house. In other words, the owner repeatedly calls the dog away from the door or window or from the fence in the yard when the dog is calm and not aroused. In addition, practicing the response in the absence of triggers must continue to be used throughout the dog's life to prevent the owner's "coming when called" command from becoming a predictor for people or animals approaching the door. When the dog's response is reliable, a desensitization program is introduced, starting with the approach of family members or friends to the guarded area. As the dog begins to associate the presence of visitors with responding to the counter-command and receiving positive reinforcement for that response, the stimuli can be gradually intensified. As with all types of desensitization programs the intensity of the stimulus (e.g. visitors entering the home or approaching the yard) must never be introduced at a level that elicits an aggressive response.

A second component to training involves counter-conditioning the dog's emotional response to greeting visitors. The arrival of the owners or visitors at the doorway is paired with food treats or even the dog's dinner. As soon as someone arrives, the dog is asked to come and sit (or to go to his mat), and the owner provides very high-value treats to both reinforce the "come when called" *and* to counter-condition the dog's emotional response to a visitor. The goal is that the stimulus of someone approaching becomes conditioned to predict pleasant experiences (treats, petting). Initially, the owner (not the visitor) provides all of the reinforcement for quiet and controlled behavior. As the dog's behavior improves, the owner and the visitor can provide treats. However, a very common mistake that owners make with this training is to attempt to have visitor provide treats to a dog who is already aroused, barking, and possibly even aggressive. This is at the very least ineffective and potentially dangerous. A visitor should only be asked to provide treats once training and behavior modification have changed the dog's response to the visitor's presence to calm and friendly behaviors. Just as with other types of counter-conditioning, if a threat or aggressive episode is triggered by mistake, the owner must consider that training session a loss, remove the dog from the setting, and try again another time. In some cases of territorial aggression, complete success may be achieved and the dog becomes calm and friendly toward visitors. In many cases, however, especially if the dog has a very strong reinforcement history for territorial aggression, the goal may be simply to decrease the dog's aggression or to teach the dog to move to another room whenever visitors arrive.

Fear-Related Aggression

Fear is a basic emotional state in which an animal responds to an aversive stimulus by either attempting to escape (flight) or by becoming defensive (fight). In all animals, fear and its attendant behaviors serve to protect the individual from potentially harmful stimuli. In dogs, the most frequently reported causes of fear-related aggression occur when a dog is approached by an unfamiliar person or dog. Body postures and facial expressions of fear include lowered body position and head, tucked tail, wide eyes (whale eye), retracted ears, and panting. When a fearful dog becomes defensively aggressive, he will retract his lips to show his teeth in a snarl (Figure 11.1). Although there are some components of this posture that are similar to submissive body postures, a major difference is that dogs who are fearful or defensive have widened eyes (not narrowed), usually gaze at the person or dog who is approaching (or show a "whale eye" as they lean away), and show piloerection. In contrast, dogs who are showing submission have narrowed, "squinty" eyes, do not have piloerection, and are initiating contact and interaction (i.e. they want to greet). Because some submissive and friendly dogs show a "greeting grin" that displays their teeth and resembles a snarl, it is not uncommon for people to misinterpret these signs.

figure 11.1 Defensive threat facial expression

Common underlying causes of defensive aggression include inadequate socialization to people, other dogs, or novel situations, or prior harmful or abusive experiences. When dogs are initially exposed to a fear-inducing situation, most attempt to flee or freeze in place. Aggression is typically used only if the dog perceives himself to be unable to escape from the aversive stimulus. For some dogs this perception may occur while on a leash or when tied out or confined to a yard. Because an aggressive response usually has the immediate effect of terminating the impending interaction, the aggressive behaviors are immediately reinforced. In other words, a dog who growls and air snaps at an approaching stranger or at a dog running toward her learns that an aggressive display stops the approach of the unfamiliar dog or person and prevents further feelings of fear by causing the stranger to move away. After several repetitions of this experience, a dog who is fearful starts to show aggressive body postures and vocalizations *preemptively,* before the person or dog is in close proximity. For example, an owner may relate that his dog barks aggressively whenever she sees an unleashed dog approaching and will growl or bite as the dog rushes up to greet. This occurs because the dog has learned that these behaviors and body postures are effective in preventing exposure to the frightening stimulus. Because the dog's fear is allayed, the dog never has the opportunity to learn that the approaching dog may actually be harmless. Therefore, each aggressive episode continues to reinforce and strengthen future fear-related aggressive responses.

Treatment for Fear-Related Aggression: Treatment programs for fear-related aggression use the same approach that is used to treat fears and phobias (see Chapter 10, pp. 269–272). Counter-conditioning is the primary tool that is used and, if possible, a desensitization program is added. The first phase of behavior modification involves counter-conditioning the dog to sit and look to the owner for a food treat on command. The owner may use a command such as "look," and the dog is consistently reinforced with food treats for all friendly and calm behaviors. It is also recommended that the dog become accustomed to wearing a head-halter training collar to allow better control of the dog's head and to aid with keeping the dog focused on the owner during training. All of the initial training for "sit and look" take place in non-distracting environments where the dog feels calm and comfortable. When the dog has been trained to maintain a sit-stay and to look toward the owner in these non-threatening situations, a standard program of counter-conditioning is started (see Chapter 10, pp. 269–272 for a complete description of counter-conditioning for fear-related behavior problems).

When aggression is a component of a dog's fear response, it is of utmost importance that the desensitization proceeds at a rate at which the dog never feels compelled to react aggressively. Long-term prognosis for dogs with fear aggression depends upon the length of time that the dog has exhibited the behavior, the intensity of the aggressive response, and the commitment and ability of the owner to prevent aggressive episodes during the treatment period. Most dogs show improvement and many begin to accept and enjoy interactions with new people. However, many nervous dogs continue to have episodes of fear or aggression throughout their lives. In all cases, careful household management and preventing the dog from being exposed to triggering stimuli that cannot be controlled are an essential component of the treatment plan.

Fear-Related Aggression toward Unfamiliar Dogs: Aggression between unfamiliar dogs occurs most commonly when one dog is on-lead during a walk and is rapidly approached by another dog who is running loose, or between two off-lead dogs who are loose in a yard or dog park. Owners of dogs who are defensively aggressive toward strange dogs often report that their dog is typically either friendly or aloof toward other dogs when off-lead, but becomes immediately nervous and shows preemptive threat postures when on-lead or when forced to greet in a small confined area. Aggressive responses in dogs who are on-lead toward uncontrolled dogs are so common that it has been erroneously labeled "leash aggression." This form of aggression can be exacerbated by an owner who has punished the dog for growling or aggressing when another dog appears. Although the use of strong aversives such as collar corrections, yelling and hitting can suppress a dog's aggressive response at the time it is administered, it does *not* change the dog's motivation for aggressing and actually makes the problem worse. Having experienced punishment from the owner upon seeing another dog, the dog is now fearful of *both* the approaching dog and the impending punishment from the owner.

Owners *must* be educated to understand the difference between temporarily suppressing a behavior and changing a dog's response to an unpleasant stimulus (i.e. solving the problem). In the case of fear-related behaviors toward other dogs, dogs who are on-lead feel vulnerable because they learn through experience that they cannot move freely away from the approaching dog and are in essence "cornered" by a dog who is making them nervous. Punishing a dog in such a situation will clearly not teach the dog to be calm and friendly to approaching dogs and certainly will not change her perception of other dogs as a threat. Unfortunately, because harsh physical punishment will suppress a behavior as it is occurring, owners can experience a false sense of success. They must be educated that excessive punishment suppresses all behavior, and that this is *not* to be confused with a dog learning not to respond to an aversive stimulus. Usually, all that one has to do is ask how many times the dog has been punished for aggressing toward other dogs. If the answer is multiple times (and the dog is still aggressing when other dogs approach), then this is clearly information that the use of punishment is not effectively changing the dog's behavior or emotional response.

Treatment for Fear-Related Aggression toward Unfamiliar Dogs: Dogs who are defensively aggressive toward unfamiliar dogs may have a history of poor socialization or having had a traumatic or unpleasant experience with a strange dog. When the problem is associated with being walked on lead, once the problem has occurred several times, the owner understandably becomes tense and nervous whenever a strange dog approaches their dog. These emotions are communicated directly to the dog and cause a vicious cycle of increased nervousness and defensive behaviors. Treatment involves using counter-conditioning to change the dog's emotional response to the sight of an approaching dog while at the same time protecting the dog from forced contact with an unfamiliar dog. Counter-conditioning a connection between "dog and high-value treats" is used to change the dog's reaction to the sight of another dog. The dog is then gradually desensitized to dogs who approach more closely (Box 11.2). Although many dogs who are defensively aggressive toward unfamiliar dogs can be conditioned to react calmly and to "look to their owner" without aggressing when they see another dog, most never become comfortable with being rushed by unfamiliar dogs. Just like people, many dogs value their personal space and for the owner to have an expectation of comfort when an unfamiliar dog forces contact is unreasonable.

Box 11.2 TEACHING RANGER TO TOLERATE UNFAMILIAR DOGS

Background: Ranger is a neutered male Shepherd/Lab mix, approximately six years old. He was recently adopted by Peter Johnson and his partner Mark Ellis. Ranger has had little or no prior training and arrived at the shelter as a neglect case when he was found tied out to a dog house with inadequate shelter and water. Peter and Mark are a volunteer foster home and initially took in Ranger to foster because he was showing increasing levels of aggression toward other dogs when he was taken outside for exercise. The shelter decided to first try fostering to remove Ranger from the kennel setting. Peter and Mark fell in love with Ranger and permanently adopted him after three weeks. Ranger is very affectionate and friendly. He has never shown any aggression toward either Mark of Peter, nor toward any visitors to their home. He also happily greets people while out walking or in the park. Ranger does not guard his toys or his food bowl, is house-trained, and other than lacking some basic manners, is a delight to live with. Peter and Mark have one other dog in their home, Lucy. Because they knew Ranger's history at the shelter, they very slowly and carefully introduced the two dogs, keeping them completely separated for a week and taking them for walks together several times a day with each person handling one dog. Once Ranger was consistently relaxed around Lucy, they carefully introduced supervised "free time" between the two dogs. The two dogs are now completely comfortable with each other, sleep in the same room, and play often.

Description of problem: Ranger continues to show aggression toward dogs he does not know when out walking. If Ranger sees another dog approaching, he begins to bark loudly. If the dog continues to approach, even if the dog seems friendly, Ranger's barking escalates and he will lunge and snap at the dog if the dog rushes up to him. Peter has noticed that if a dog approaches very slowly and calmly, Ranger is less likely to be aggressive and may greet in a friendly manner. He also notices that if the second dog is off-lead and uncontrolled, Ranger's behavior is much worse than if the other dog is on-lead.

Treatment: Peter and Mark begin the following training and behavior modification program to reduce Ranger's anxiety when he sees other dogs and to prevent aggressive responses:

- ***Training an alternate response:*** Ranger is fitted with a halter-type collar and is trained to respond to the command "look" by turning to face his owner and offering a sit for treats. Roger and Peter also learn to step to the side of the walkway and to position themselves between Ranger and an approaching dog. (These exercises are practiced often and repeatedly when no dogs are approaching to prevent Ranger from associating the command "look" as a predictor of an approaching dog.)
- ***Counter-conditioning:*** When Ranger sees another dog at a distance while out walking, Roger immediately cues him with "look, sit" (spoken in a relaxed and happy voice), and provides multiple very-high-value treats as soon as Ranger turns to face him. It is of utmost importance that the cue and response occur *before* Ranger shows any signs of arousal or agitation. When the dog retreats or passes by, the treats stop and Ranger continues his walk.
- ***Prevention:*** If the situation arises that a dog approaches too quickly or Roger and Mark do not respond quickly enough and Ranger begins to react, they immediately turn 180 degrees and walk away in the opposite direction. No attempt is made to counter-condition because once Ranger is anxious, that session is lost as an opportunity to change his emotional response. The best that can be done is to reduce Ranger's anxiety as quickly as possible and try again next time.
- ***Management:*** Roger and Mark walk Ranger only in areas where other owners are responsible and keep their dogs on-lead. They do not take Ranger to a dog park (ever), and are careful when walking in the park so as to avoid having dogs rush up and jump on Ranger.

Outcome: Ranger's level of arousal and anxiety decrease almost immediately when he has an alternate behavior that keeps him feeling safe. After one month of training, he will sit quietly off to the side of the path when another dog and owner pass by, is relaxed, and is taking treats. Although Ranger will probably never become friendly to unknown dogs, he is no longer showing aggression when he sees them. Peter and Mark understand that they must manage this to prevent unwanted interactions indefinitely.

It is also very important to point out that a dog who rushes other dogs, leaps upon them wildly in greeting, or physically assaults another dog in exuberance is *not* demonstrating normal or "friendly" dog greeting behaviors. Rather, these dogs are pushy, obnoxious dogs who have not learned proper canine communication signals. Normal canine greeting behaviors include sniffing of the face or groin region (often using a side presentation), and do *not* include one dog leaping upon the other (Figure 11.2 and Figure 11.3). Greeting sniffs and examinations are typically followed either by an invitation to play or by simply ending the interaction and moving on. Owners who present a rambunctious dog without manners as "oh, my dog is just friendly" are at fault for allowing their dog to harass other dogs and for not training their dog for control around other people and dogs. However, because one cannot control the behavior of other people and because people will continue to disobey leash laws, it is the responsibility of responsible owners to keep their dogs safe and to avoid unpleasant encounters with loose and uncontrolled dogs. A solution for owners when they are confronted with a loose dog who is not under control is to immediately turn and walk in the

figure 11.2 Two unfamiliar dogs greeting – face sniffing

figure 11.3 Two unfamiliar dogs greeting – side presentation and sniffing

opposite direction, completely ignoring the approaching dog. (If necessary, turning and throwing a handful of treats toward the dog can effectively stall a loose dog's approach.) This prevents an unpleasant encounter for the fearful dog and hopefully also sends a message (negative punishment!) to the owner of the uncontrolled dog that they are behaving inconsiderately and need to control their dog. In cases when it is impossible to get away from a loose dog, owners can use a "safety sit-stay" in which they train their dog to sit and stay behind them, while they block the oncoming dog's access to the dog. Finally, it should go without saying that dogs who are defensively aggressive should *never* be taken to dog parks.

Ineffective Treatments for Aggression Problems in Dogs

Physical punishment has been shown to be ineffective in the treatment of all types of aggression and, in many circumstances, only serves to exacerbate the problem. Pain has long been recognized as an elicitor of aggression in animals who are both offensively and defensively motivated. Similarly, using physical punishment when a dog shows aggression toward another dog functions to escalate the ongoing conflict between the two dogs and can result in redirected aggression toward the owner. Although owners must of course break up a fight, and yelling loudly is often the best way to do this, breaking up a fight with punishment should not be confused with solving the problem or changing the relationship between the two dogs. Finally, punishing a dog who is defensively aggressive only adds to the dog's fear and lack of trust – certainly not an outcome that is compatible with reduced aggression. Indeed, many cases of severe fear-related aggression are *caused* by the use of excessive punishment when the dog initially showed a defensive threat posture. Case studies of dominant-aggressive dogs also report that attempts to discipline the dog were a common cause of escalated aggression and bites to the owner or others in the home. Therefore, if the dog's owner has been using any form of physical reprimands, dominance rolls, or direct challenges to the dog, these procedures should be immediately discontinued. They are ineffective, inhumane, and unsafe.

Another common recommendation for inter-dog aggression within homes is for the owner to support the "dominant" dog. It should be evident by now that this label does not reflect reality for most social groups of dogs living together in homes. By subscribing to a belief in strict hierarchical relationships between dogs, owners are tempted to force their dogs into rigid and unsustainable roles which cause them to misinterpret their dogs' communication signals and use completely inappropriate "training" techniques. Karen Overall, a leading expert on canine and feline aggression, notes that most clients who had been advised to "support the dominant dog" caused the aggression between the two dogs to escalate.[7] This happens because owners erroneously assign a label of dominance with little or no evidence of support it. Even when one dog in the home is showing the majority of offensive signals and is consistently the dog who initiates aggression, it is questionable whether supporting an overly assertive dog who is bullying other dogs can be effective at reducing aggression between dogs. If the intent of supporting the more dominant dog is to encourage the victimized dog to show deference and appeasement, there are no good data showing that this in effect ever even happens. Also, it is important for owners to understand that in many cases of dominance-related aggression between dogs, the victimized dog fights back because of fear and often has begun to offer preemptive strikes, thus appearing to be the aggressor. This further complicates the owner's ability to understand the inter-dog dynamics and to assign labels of "dominant" and "submissive" to the fighting dogs. Moreover, when one dog is motivated by fear, supporting the aggressor will *not* change a response of defensive aggression because the victimized dog will still have no reason to begin to trust the aggressor. Supporting the more assertive dog is therefore not recommended and should be avoided because of the risk of causing an increase in aggression between the dogs.

PROBLEM AGGRESSION IN CATS

As in other species, aggression is a normal part of the cat's behavioral repertoire. Successful defensive behaviors that include aggression prevent an animal from being injured or becoming a meal for a predator. Cats may also use aggression to establish and defend territory. This is most commonly seen in multiple-cat homes when space is limited and cats may feel crowded. In intact animals, inter-male aggression occurs frequently during competition for mates, while maternal aggression allows a mother cat to protect her kittens. Just as in dogs, because aggressive responses are a part of every cat's behavioral makeup, individual cats cannot be classified as being either "aggressive" or "non-aggressive." Simply labeling a cat as "aggressive" provides no information about the context in which the aggression occurs and so is not helpful in developing effective treatment.

Inter-Cat Aggression in Multiple-Cat Homes

Inter-cat aggression is a common problem between cats who share the same household. Underlying causes include a lack of socialization, inappropriate play behaviors, fear, redirected aggression, and territorial aggression. In intact cats, inter-male aggression is common and increases in frequency during the breeding season. Aggressive play may be seen in cats who were hand-raised as orphans, were single kittens in a litter, or were weaned at too young an age. Cats who have not learned to inhibit their bite or keep their claws retracted during play can also cause fights to occur during play sessions with other cats. Adult cats who never developed normal feline communication skills may react defensively to any close encounter with another cat, even after being acquainted for a long period of time. Fear-induced aggression is seen in cats who lack socialization as well as in cats who are just naturally less social than others. Defensive body postures will be evident in these cats, and a cat who is fearful or defensive may spend much of his time hiding in the home to avoid interaction with other cats. Finally, redirected aggression occurs when a cat who is aggressively stimulated is prevented from directing his aggression toward the causative agent. A common example is a cat who is staring out of a window and sees a neighboring cat approach may turn and attack a house mate. If this redirection occurs with regular frequency, the interaction becomes a pattern of behavior that defines the relationship between the two cats. Redirected aggression is also seen in multiple-cat homes in which cats are overcrowded and one or more cats are unable to maintain a comfortable personal space.

Treatment: In cases of established inter-cat aggression in household cats, both behavior modification and home management can help to reduce aggression. Counter-conditioning and desensitization are often successful if the aggression has not been occurring for a long period of time and if both of the cats are relatively well-socialized to other cats. Meals can be used to counter-condition aversive emotional reactions. The cats are initially separated in different rooms using a baby gate and the cats are gradually fed in increasingly closer proximity to one another (Figure 11.4). During the period of reintroduction, fights and agonistic responses must be prevented by separating the cats completely except during feeding times. Each cat should be provided with his own sleeping area and litter box. In severe cases, pharmacotherapy can help to reduce aggression and anxiety during the first few weeks of treatment. However, when drug therapy

figure 11.4 Reintroduction to reduce or prevent aggression: Cats eating on each side of a baby gate

is needed for long periods, the associated side effects and health risks make this treatment unacceptable to many owners. In some cases, managing the home so that the two cats can each have separate territories in which they feel secure and which allow them to avoid contact with each other is the best solution for all involved. When the density of cats is contributing to aggression and one or more cats cannot tolerate living in a multiple-cat home, the owners may decide to rehome one or more of the cats to a home with no other cats.

Aggression toward a Newly Introduced Cat

Aggression between cats who have been recently introduced in a home is usually caused by territorial defense and fear. Territorial aggression is most commonly seen when caretakers are attempting to rapidly introduce a new cat into the home. The resident cat may react by simply threatening the new cat or by showing offensive threat postures followed immediately by a physical attack. In most cases, the new cat reacts either fearfully or with defensive aggression.

Prevention and Treatment: Territorial aggression can be prevented in most cases through careful and gradual introduction of the new cat. Regardless of age or gender, the new cat should be confined to a carrier or small area within a room for the first several days or weeks. During this time, the resident cat is allowed access to most of the house, with the exception of the room that contains the newly adopted cat. The resident cat is gradually allowed to approach the area that contains the new cat. Baby gates and closed doors are used to control the pace of these introductions. The new cat's territory can be gradually increased, but only when calm and accepting behaviors are observed in the resident cat.

Cat caretakers should always be advised to show patience when introducing a new cat, since some resident cats may need several weeks or even months to become acclimated to the new arrival. As the new cat gains more freedom in the home, the resident cat's reactions should be carefully monitored. Feeding the cats in the same room and at the same time, and gradually moving their bowls closer together, can be used to enhance this acclimation. If multiple litter boxes are not already available, more should be added. In addition, providing a variety of elevated resting and hiding places in several rooms of the home allows cats in multiple-cat homes to have control over the amount of contact that they have with other cats. Owners should understand that while some cats eventually begin to groom, sleep with, and play with each other, others are more solitary by nature and may simply prefer to be left alone. Understanding and accepting this variability in cat personality is imperative when attempting to keep two or more cats peacefully together in the same home.

Aggression toward Humans

Cats may attack, scratch, and bite human caretakers or visitors for several reasons. Play behaviors may be directed at a person who is running or walking by. For example, some cats ambush human housemates from behind doors and may severely attack ankles or legs. Others simply play too roughly, scratching or biting as they become stimulated during play sessions. A similar type of behavior is petting-induced aggression, which occurs when a cat who is being handled and petted suddenly turns to claw and bite. Defensive aggression can occur during handling for grooming or medical treatment. Finally, redirected aggression occurs when a human interferes with or is near a cat who is directing aggression toward another cat or stimulus.

Fear-Related Aggression: Cats who are nervous or fearful toward one or more family members or toward visitors to the home are quite common. Underlying causes include a lack of socialization when young or inadequate habituation to various types of people (common in cats who live exclusively indoors), the use of excessive punishment, or a past traumatic experience. In addition, genetics play an important role in feline fear behaviors. Individual cats vary tremendously in their degree of sociability and adaptability. While some are highly social and habituate easily to multiple situations and people, others are naturally more timid and become frightened easily, despite socialization and exposure. The majority of cats who are fearful choose avoidance, provided an escape route is available to them. However, cats who feel cornered or who have had severely traumatic experiences will resort to aggression to drive away the stimulus that is causing the fear. Cats with fear-related aggression show defensive body postures. They have a lowered body position, legs are tucked under the body, and they exhibit flattened ears, piloerection, and widened eyes with dilated pupils. The "hiss" is the characteristic defensive vocalization of cats, although some cats will also growl.

If approached, a defensively posturing cat will often swat with a front paw (with claws extended) before attempting to bite. In cases in which a cat has learned from prior experiences to use aggression preemptively, these warning signals may be absent and the cat may attack and bite immediately upon feeling threatened. Treatment for fear-related aggression in cats is similar to treatment for dogs and focuses on counter-conditioning and desensitization. The cat should be removed from all situations in which aggression occurs. Confining the cat to a wire crate both prevents aggression and allows control over the counter-conditioning process (Box 11.3). The use of drug therapy or synthetic feline pheromones such as Feliway may be helpful in reducing the cat's level of anxiety, especially during the early stages of treatment (see Chapter 8, p. 204).

Box 11.3   MARVIN'S REDIRECTED AGGRESSION

Background: Rebecca Taylor has two cats, Marvin (three years) and Walter (seven years) who until recently have lived very peaceably together. Last week, however, Marvin started viciously attacking Walter. He will chase him around the house, corner him and bite him, or pounce at him as he walks by. Rebecca is certain that this is not just rough play; Marvin is definitely attacking Walter. She does not understand this because the two cats have lived together for almost three years and have always been friendly and loving toward each other. Walter is becoming frightened of Marvin and has started to hide under the bed, refusing to come out for most of the day. Rebecca separates Marvin and Walter for 30 minutes after each episode, but as soon as Marvin is allowed out he begins to search for Walter, and attacks again. Rebecca also has tried squirting Marvin with a water bottle as soon as he starts to go after Walter and then locks him in another room until he calms down, but it does not seem to be helping. When asked about any recent changes in the household, Rebecca relates that she and her husband and their cats just moved to the area four months ago. They installed an Invisible Fence system around their yard about eight weeks ago. The two cats wear the e-collars and are able to have freedom in the yard. When the fence was first installed, Marvin was the most inquisitive, learning the boundaries very quickly and spending as much time as he could outdoors. Now, if allowed, he is outdoors most of the day. Rebecca states that they have seen a neighbor's cat sitting in the yard, but on the other side of the Invisible Fence, watching their cats. They have also observed Marvin staring at this cat and, on one occasion, they saw him chase the strange cat, but he stopped short of the fence boundary. The neighbor cat now seems to know to stay on the other side of the electronic fence and will sit staring at Marvin, sometimes for several hours at a time.

Cause of aggression: Marvin is most likely showing redirected territorial/offensive aggression toward Walter. This has occurred because Marvin is aggressively motivated toward the neighbor cat, but is repeatedly prevented from acting on this aggression. The pain of positive punishment may also be involved if Marvin experiences an electric shock when he attempts to chase the neighbor cat out of his yard.

Treatment: Treatment with redirected aggression involves removing the source of the aggression and then using counter-conditioning and desensitization to change the aggressive cat's association with the victimized cat.
- ***Removing the cause:*** In this case, this means either convincing the neighbors to keep their cat inside (unlikely), or restricting Marvin's and Walter's time in the yard to periods when the Browns are certain that other cats are not present.

- **Preventing repeated episodes:** Because redirected aggression can be very severe, Marvin and Walter must be confined to separate rooms or sections of the home during the treatment program.
- **Counter-conditioning and desensitization:** This is conducted by confining Marvin behind a baby gate or in a wire crate at meal times. Walter is then brought into the room, keeping him on the opposite side of the gate or far enough away from the crate to prevent a negative emotional reaction in either cat. Both cats are given their meals. As soon as they have finished eating, Walter is removed and taken back to his part of the home. This is practiced daily, gradually moving the cats closer to each other and always avoiding progressing too rapidly and risking an aggressive response in Marvin or a fearful response in Walter.
- **Reintroductions:** When both cats are eating peacefully together, Rebecca begins to allow them to visit though a baby gate, dividing two rooms. When this occurs with no aggression, supervised free time in the home is introduced. To ensure Walter's safety and continued security, he is provided with a "safe spot" in the home, a room that Marvin is not allowed to enter.

Play (Predatory) Aggression: Cats who bite or scratch while playing or who ambush people as they walk through a doorway are demonstrating uninhibited play behaviors and must be taught to play gently without biting or scratching. The most successful approach is to consistently redirect the cat's play toward a toy. This can be a stuffed toy or a hanging toy that is dangled from the doorway in areas that the cat most often sets up to ambush. Frequently providing new and novel types of toys will keep the cat interested and helps to distract the cat from focusing on feet and hands as toys. Providing consistent play periods and attention for the cat each day is also important. Using negative punishment is effective in some cases, especially if the cat is highly bonded to her human caretaker. This approach involves abruptly ending any interactive play that is inappropriate and giving the cat a "time out" by separating the cat from the owner for several minutes. Using this approach, the cat learns that play will only continue if her behavior remains playful and no biting or clawing occurs. However, for this approach to be successful, the owner's timing must be precise. In addition, repeated use of negative punishment should be avoided because of the risk for causing frustration and unintentionally increasing the cat's aggression. In cases that are refractory to previously described treatments, positive punishment in the form of a squirt bottle filled with water may be needed. The cat is sprayed in the face as she ambushes or begins to play too aggressively. In these cases, the risk of having the cat associate the human caretaker with the aversive stimulus (the water bottle) can lead to fear and avoidance, so any type of positive punishment should be used with caution.

Petting-Induced Aggression: Biting while being petted seems rather paradoxical to cat owners. The cat is sitting in the owner's lap and is apparently enjoying being petted when, suddenly, she turns and bites and then jumps off of the lap. In most, but not all cases, the bite is inhibited and does not break skin. In many situations, the cat actually solicits petting and attention and remains on the owner's lap for several minutes, only to turn suddenly and "bite the hand that pets it," so to speak. There are several theories that attempt to explain this behavior. Most likely, cats who use biting to end petting sessions have a limited tolerance for physical interaction and use the feline communication cue for terminating social grooming when they have had enough. For example, when two housecats are grooming one another, the interaction is often abruptly ended by one cat turning suddenly to bite the other. A second and related theory is that some cats have a limited tolerance for petting and find touch irritating, and bite to stop the interaction. Although owners are not always aware of the signs, most cats show subtle but distinct signals that they are becoming agitated and intolerant of petting. These include tail twitching, inhibited bites (nibbling), pivoting ears, and restlessness. Treatment for this type of biting involves learning to recognize the cat's signals and terminating petting sessions before the cat shows signs of impending aggression. The owner must establish a habit of always stopping petting sessions before the cat's tolerance threshold is approached. Counter-conditioning can also be helpful, and is established by feeding the cat small tidbits of a highly desirable food during each session of petting.

Redirected Aggression: Redirected aggression occurs when the victim of the cat's aggression is not the actual stimulus that triggered the cat's response, but just happens to be close to the cat at the time that the cat becomes aroused. One of the most common examples of redirected aggression occurs when an indoor cat observes a cat in the yard through a window or door. This can trigger a territorial response, but the cat is prevented from directing the response to the invading cat and so redirects to the owner or another cat in the home. Other common triggers are any type of stimulus that causes extreme fear and a response of defensive aggression in the cat, such as the presence of unfamiliar people or animals, or a new environment. An owner who is severely attacked when she attempts to remove her frightened cat from the room when visitors arrive is a victim of redirected, defensive aggression. In many cases, redirected aggression in cats can be very intense and may include severe and uninhibited bites. Because owners often are completely unaware of the trigger that caused the attack, this type of feline aggression appears to be unpredictable and can be very frightening to owners. Treatment involves identifying the trigger for the aggression and managing the cat's environment to prevent exposure to that stimulus. When possible, counter-conditioning can be used to reduce the cat's arousal state when exposed to the trigger. For example, when fear of visitors is the cause, counter-conditioning is used to reduce the cat's fear response and management techniques such as confining the cat before visitors arrive are used to prevent aggression. When an outdoor cat is the cause, blocking the cat's access to windows is often the most effective management technique. Owners should be advised that because redirected aggression in cats can be very severe, management to prevent future bites and behavior modification and/or anti-anxiety medications to reduce risk to people and other pets in the home are very important components of treatment.

REVIEW QUESTIONS

1. Identify the primary behavioral signs of human-directed, dominance-related aggression in dogs.
2. Describe the principal steps in a treatment program for dominance-related aggression occurring between two dogs living in the same home.
3. Define possessive aggression in dogs and provide an example.
4. Describe a protocol for properly introducing a new cat into a household that includes one or more resident cats.
5. Provide a set of tips for preventing human-directed play aggression in cats.

REFERENCES AND FURTHER READING

Appleby DL, Bradshaw JWS, Casey RA. **Relationship between aggressive and avoidance behavior by dogs and their experience in the first six months of life.** *Vet Rec*, 150:434–438, 2002.

Borchelt PL, Copopola MC. **Characteristics of dominance aggression in dogs.** Paper presented at Annual Meeting of the Animal Behavior Society. North Carolina, June, 1985.

Borchelt PL. **Aggressive behavior of dogs kept as companion animals: Classification and influence of sex, reproduction statues and breed.** *Appl Anim Behav Sci*, 10:45–61, 1983.

Borchelt PL, Voith VL. **Dominant aggression in dogs.** In: *Readings in Companion Animal Behavior*, VL Voith and PL Borchelt, editors, Veterinary Learning Systems, Trenton, NJ, pp. 230–239, 1996.

Campbell WE. **Behavior Problems in Dogs.** American Veterinary Publications, Inc., Santa Barbara, CA, 1975.

Donaldson J. **Culture Clash: A Revolutionary New Way of Understanding the Relationship between Humans and Domestic Dogs.** James and Kenneth Publishers, Oakland, CA, 221 pp., 1996.

Goodwin D, Bradshaw JWS, Wickens SM. **Paedomorphosis affects agonistic visual signals of domestic dogs.** *Anim Behav*, 53:297–304, 1997.

Guy NC, Luescher UA, Dohoo SE, Spangler E, Miller JB, Dohoo IR, Bate LA. **Demographic and aggressive characteristics of dogs in a general veterinary caseload.** *Appl Anim Behav Sci*, 74:15–28, 2001.

Guy NC, Luescher UA, Dohoo SE, Spangler E, Miller JB, Dohoo IR, Bate LA. **Risk factors for dog bites to owners in a general veterinary caseload.** *Appl Anim Behav Sci*, 74:29–42, 2001.

Heidenberger E. **Housing conditions and behavioural problems of indoor cats as assessed by their owners.** *Appl Anim Behav Sci*, 52:345–364, 1997.

Landsberg G. **The distribution of canine behavior cases at three behavior referral practices.** *Vet Med*, 86:1011–1018, 1991.

Levine E, Perry P, Scarlett J, Houpt KA. **Intercat aggression in households following the introduction of a new cat.** *Appl Anim Behav Sci*, 90:325–336, 2005.

Liinamo AE, van den Berg L, Leewater PAJ, and others. **Genetic variation in aggression-related traits in Golden Retriever dogs.** *Appl Anim Behav Sci*, 104:95–106, 2007.

Line S, Voith VL. **Dominance aggression of dogs toward people: Behavior profile and response to treatment.** *Appl Anim Behav Sci*, 16:77–83, 1986.

Lockwood R. **The ethology and epidemiology of canine aggression.** In: *The Domestic Dog: Its Evolution, Behavior, and Interactions with People*, Cambridge University Press, Cambridge, UK, pp. 131–138, 1995.

Love M, Overall KL. **How anticipating relationships between dogs and children can help prevent disasters.** *J Amer Vet Med Assoc,* 219:446–453, 2001.

Mertens PA. **The concept of dominance and the treatment of aggression in multi-dog homes: A comment on van Kerklove's commentary.** *J Appl Anim Welfare Sci*, 7:287–291, 2004.

Neilson JC. **How I treat food-related aggression in dogs.** *Vet Med*, April, pp. 247–252, 2007.

Netto WJ, Planta DJU. **Behavioural testing for aggression in the domestic dog.** *Appl Anim Behav Sci*, 52:243–263, 1997.

O'Farrell V, Peachey E. **Behavioral effects of ovariohysterectomy on bitches.** *J Small Anim Pract*, 31:595–598, 1990.

Orihel JS, Fraser D. **A note on the effectiveness of behavioural rehabilitation for reducing inter-dog aggression in shelter dogs.** *Appl Anim Behav Sci*, 112:400–405, 2007.

Orihel JS, Ledger RA, Fraser D. **A survey of the management of inter-dog aggression by animal shelters in Canada.** *Anthrozoos*, 18:273–287, 2005.

Overall KL. **Working bitches and the neutering myth: Sticking to the science.** *Vet J,* 173:9–11, 2007.

Overall KL. **Overview of canine aggression.** Invited paper. *World Congress WSAVA*, 2006.

Overall KL. **Dog bites to humans – demography, epidemiology, injury, and risk.** *J Amer Vet Med Assoc*, 218:1923–1034, 2001.

Overall KL. **Feline aggression. Part II. Common aggressions.** *Feline Pract*, 22:16–17, 1994.

Ozanne-Smith J, Ashby K, Stathakis VZ. **Dog bite and injury prevention – analysis, critical review, and research agenda.** *Injury Prevent*, 7:321–338, 2001.

Polsky RH. **Factors influencing aggressive behavior in dogs.** *California Vet*, 37:12–15, 1983.

Radosta-Huntley, L, Shofer F, Reisner I. **Comparison of 42 cases of canine fear-related aggression with structured clinician initiated follow-up and 25 cases with unstructured client initiated follow-up.** *Appl Anim Behav Sci*, 105:330–341, 2007.

Reisner IR, Houpt KA, Shofer FS. **National survey of owner-directed aggression in English Springer Spaniels.** *J Amer Vet Med Assoc*, 227:1594–1603, 2005.

Roll A, Unshelm J. **Aggressive conflicts amongst dogs and factors affecting them.** *Appl Anim Behav Sci*, 52:229–242, 1997.

Sherman CK, Reisner IR, Taliaferro LA, Houpt KA. **Characteristics, treatment, and outcome of 99 cases of aggression between dogs.** *Appl Anim Behav Sci*, 47:91–108, 1996.

Vas J, Topal J, Gacsi M, Miklosi A, Scanyi V. **A friend or an enemy? Dogs' reaction to an unfamiliar person showing behavioral cues of threat and friendliness at different times.** *Appl Anim Behav Sci*, 94:99–115, 2005.

Virga V, Houpt KA, Scarlett JM. **Effect of Amitriptyline as a pharmacological adjunct to behavioral modification in the management of aggressive behaviors in dogs.** *J Amer Anim Hosp Assoc*, 37:325–330, 2001.

Wilson F, Dwyer F, Bennett PC. Prevention of dog bites: Evaluation of a brief educational intervention program for preschool children. *J Commun Psychol*, 31:75–86, 2003.

Footnotes

[1] Hart B, Hart L. *Canine and Feline Behavioral Therapy*, Lea and Febiger, Philadelphia, PA, 275 pp., 1985.

[2] Overall KL. Feline aggression. Part II. Common aggressions. *Feline Pract*, 22:16–17, 1994.

[3] Lockwood R. The ethology and epidemiology of canine aggression. In: *The Domestic Dog: Its Evolution, Behavior, and Interactions with People,* Cambridge University Press, Cambridge, UK, pp. 131–138, 1995.

[4] Overall KL. Overview of canine aggression. Invited paper. *World Congress WSAVA*, 2006.

[5] O'Farrell V, Peachey E. Behavioral effects of ovariohysterectomy on bitches. *J Small Anim Pract*, 31:595–598, 1990.

[6] Overall KL. Understanding and treating canine dominance aggression: An overview. *Vet Med*, 94:976–978, 1999.

[7] Overall KL. Understanding dogs that fight. *World Congress ASAVA/FECAVA/SSAVA*, pp. 148–150, 2006.

appendix 1

Recommended Books

Abrantes R. *Dog Language: An Encyclopedia of Canine Behaviour*. Wakan Tanka Publishers, Naperville, IL, 263 pp., 1997.

Abrantes R. *The Evolution of Canine Social Behavior*. Wakan Tanka Publishers, Naperville, IL, 79 pp., 1997.

Abrantes R. *Dogs Home Alone*. Wakan Tanka Publishers, Naperville, IL, 50 pp., 1997.

Alexander MC. *Click for Joy*. Sunshine Books, Inc, Waltham, MA, 208 pp., 2003.

Aloff, B. *Canine Body Language: A Photographic Guide.* Dogwise, Wenatchee, WA, 372 pp., 2005.

Beaver B. *Feline Behavior: A Guide for Veterinarians*. WB Saunders Company, Philadelphia, PA, 276 pp., 1992.

Beaver B. *Canine Behavior: A Guide for Veterinarians*. WB Saunders Company, Philadelphia, PA, 276 pp., 1999.

Book M, Smith CS. *Right On Target – Taking Dog Training to a New Level.* Dogwise, Wenatchee, WA, 162 pp., 2006.

Bradshaw JWS. *The Behaviour of the Domestic Cat*. CAP International, Oxford, UK, 219 pp., 1992.

Burch MR, Bailey JS. *How Dogs Learn*. Howell Book House, New York, NY, 188 pp., 1999.

Case LP. *The Dog: Its Behavior, Nutrition and Health,* second edition, Blackwell Publishing, Ames Iowa, 479 pp., 2005.

Case LP. *The Cat: Its Behavior, Nutrition and Health*. Blackwell Publishing, Ames Iowa, 391 pp., 2003.

Clutton-Brock J. *A Natural History of Domesticated Mammals,* second edition. Cambridge University Press, Cambridge UK, 238 pp., 1999.

Coppinger R, Coppinger L. *Dogs: A Startling New Understanding of Canine Origin, Behavior and Evolution.* Scribner Publishing, New York, NY, 352 pp., 2001.

Csanyi V. *If Dogs Could Talk: Exploring the Canine Mind*. North Point Press, New York, NY, 334 pp., 2005.

Domjan M. *The Essentials of Conditioning and Learning*, third edition. Wadsworth/Thompson Learning, Belmont CA, 368 pp., 2005.

Donaldson J. *Dogs are From Neptune: Candid Answers to Urgent Questions about Aggression and Other Aspects of Dog Behavior*. Lasar Multimedia Publications, Montreal, Quebec, 162 pp., 1998.

Donaldson J. *Culture Clash: A Revolutionary New Way of Understanding the Relationship Between Humans and Domestic Dogs*. James and Kenneth Publishers, Oakland, California, 221 pp., 1996.

Hetts S. *Pet Behavior Protocols: What to Say, What to Do, When to Refer*. AAHA Press, Lakewood, CO, 335 pp., 1999.

Horwitz D, Mills D, Heath S. *BSAVA Manual of Canine and Feline Behavioral Medicine*. BSAVA, Gloucester, UK, 288 pp., 2002.

Landsberg G, Hunthausen W, Ackerman L. *Handbook of Behavior Problems of the Dog and Cat*, second edition. Saunders, New York, NY, 554 pp., 1997.

Lindsay S. *Handbook of Applied Dog Behavior and Training (Volumes 1–3)*. Iowa State University Press, 2002.

McConnell PB. *For the Love of a Dog: Understanding Emotion in You and Your Best Friend*. Ballantine Books, 332 pp., 2006.

McConnell PB. *The Other End of the Leash*. Ballantine Books, 272 pp., 2003.

Miklosi A. *Dog: Behavior, Evolution and Cognition*. Oxford University Press, Oxford, UK, 274 pp., 2007.

O'Heare J. *The Canine Aggression Workbook*, third edition. DogPsych Publishing, Ottawa, Canada, 201 pp., 2004.

Overall K. *Clinical Behavioral Medicine for Small Animals*. Mosby, St. Louis, MO, 544 pp., 1997.

Pryor K. *Don't Shoot the Dog*. Bantam Books, New York, NY, 187 pp., 1984

Pryor K. *Karen Pryor on Behavior*. Sunshine Books, North Bend, WA, 1995.

Pryor K. *Clicker Training for Cats*. Sunshine Books, Waltham, MA, 84 pp., 2001.

Reid P. *Excel-Erated Learning: Explaining How Dogs Learn and How Best to Teach Them*. James and Kenneth Publishing, Oakland, CA, 172 pp., 1996.

Robinson I (editor). *The Waltham Book of Human-Animal Interaction: Benefits and responsibilities of pet ownership*. Pergamon Press, Oxford UK, 148 pp., 1995.

Rogerson J. *Your Dog: Its Development, Behaviour, and Training*. Popular Dogs Publishing Company, London, UK, 174 pp., 1990.

Sanders CR. *Understanding Dogs: Living and Working with Canine Companions*. Temple University Press, Philadelphia, PA, 201 pp., 1999.

Serpell J (editor). *The Domestic Dog: Its Evolution, Behaviour, and Interactions with People*. Cambridge University Press, Cambridge, UK, 268 pp., 1995.

Stafford K. *The Welfare of Dogs*. Springer Publishing, Dordrecht, Netherlands, 280 pp., 2007.

Thurston ME. *The Lost History of the Canine Race: Our 15,000 Year Love Affair with Dogs*. Andrews and McMeel, Kansas City, MO, 301 pp., 1996.

Turner DC, Bateson P (editors). *The Domestic Cat: The Biology of Its Behaviour,* second edition. Cambridge University Press, Cambridge, UK, 244 pp., 2000.

Wendet LM. *Dogs: A Historical Journey: The Human/Dog Connection Through the Centuries*. Howell Book House, New York, NY, 258 pp., 1996.

appendix 2

Resources and Professional Associations

American Veterinary Society of Animal Behavior
Membership Information
12249 73rd Court North
West Palm Beach, FL 33412
http://www.avsabonline.org

Animal Behavior Resources Institute
P.O. Box 27348
Golden Valley, MN 55427
http://abrionline.org

Animal Behavior Society
Indiana University
2611 East 10th Street
Bloomington IN 47408-2603
http://www.animalbehavior.org

Association of Companion Animal Behavior Counselors
P.O. Box 104
Seville, FL 32190-0104
http://animalbehaviorcounselors.org

Association of Pet Behaviour Counselors (UK)
P.O. Box 46
Worcester, WR8 9YS,
England
http://www.apbc.org.uk

Association of Pet Dog Trainers (APDT)
150 Executive Center Drive
P.O. Box 35
Greenville, SC 29615
http://www.apdt.com

Canadian Association of Pet Dog Trainers
18700 Chief Lake Road
Prince George, BC
V2K 5K4
http://www.cappdt.ca

Certification Council for Professional Dog Trainers (CCPDT)
Professional Testing Corporation
1350 Broadway, 17th Floor
New York, NY 10018
http://www.ccpdt.org

Companion Animal Sciences Institute
1333 Rainbow Crescent
Ottawa, ON, Canada K1J 8E3
http://www.casinstitute.com

Etologisk Institute
DK-4270 Hong, Denmark
http://www.etologi.dk

International Association of Animal Behavior Consultants (IAABC)
P.O. Box 1458
Levittown, PA 19058
http://www.iaabc.org

International Institute for Applied Animal Behavior
1333 Rainbow Crescent
Ottawa, ON, Canada K1J 8E3
http://www.iiacab.com/

National Association of Dog Obedience Instructors (NADOI)
PMB 369; 729 Grapevine Hwy
Hurst, Texas 76054-2085
http://www.nadoi.org

San Francisco SPCA Academy for Dog Trainers
The San Francisco SPCA
2500-16th Street
San Francisco, CA 94103-4213
http://www.sfspca.org/academy

The Society of Veterinary Behavior Technicians
Membership Information
310 Olympic Lane
Mount Vernon, WA 98274
http://www.svbt.org

appendix 3

Sample Dog Behavior History Profile

STUDENT'S NAME _____

DOG'S NAME _____ BREED _____

AGE _____ SEX _____ SPAYED/NEUTERED? _____

Dog/Owner Interactions (Circle Appropriate Answers)

1. My dog was approximately [less than 7 weeks; 7 to 8 weeks; 8 to 12 weeks; 12 weeks to 6 months; 6 months to 1 year; older than 1 year] when I obtained him/her.

2. I acquired my dog from a/an [pure-bred breeder; ad in the newspaper; pet store; animal shelter or rescue organization; friend/neighbor; relative; as a stray].

3. I have owned my dog for [less than 1 month; 1 to 6 months; 6 months to 1 year; 1 to 2 years; more than 2 years].

4. To my knowledge, I am my dog's [first; second; third; fourth or more] owner.

5. I have owned [zero; one; two; more than two] other dogs in the past.

6. My dog spends the largest part of his/her day [indoors with 1 or more people; indoors alone, loose in the house; indoors alone, crated or confined to a small area; outdoors in a yard or kennel; tied outdoors].

7. My dog spends most of his resting/relaxing time: [on his own bed; on the couch or my bed; at my feet; by an outside doorway; in the yard or kennel; other _____].

8. At night, my dog sleeps [on my bed; in my bedroom; in the house but not in the bedroom; crated in the house; in a garage or kennel; tied or kenneled outside].

9. The type of area that my dog has available for elimination purposes is: [a fenced yard that he is in most of the day; a fenced yard that we let him out into and leave him there for periods of time; a fenced yard that he goes into for elimination only;

an unfenced yard where we chain or tie him; an unfenced area where I walk him on-lead; no set area; I walk my dog for elimination].

10. The person who my dog most often plays with is: [myself; my partner; my roommate; my children; visiting friends; other dogs].

11. Toys that my dog most likes to play with include [tug toys; squeaky toys; squeaky toys minus the squeaker (she always rips them out); rawhide bones; Nyla-bones; tennis balls; stuffed toys; other _____].

12. My dog's toys can usually be found [lying around the house wherever he has left them; on his bed or in his crate; near his food/water bowl; put away in a special spot that he/she has free access to; put away in a special spot that he/she does not have access to; nowhere, we do not keep the dog's toys in the house].

13. The way in which I usually exercise my dog is [walks on-lead, several times a day; one walk on-lead per day; daily walks off-lead in a park or other open area; playing in the yard or house; running/jogging with my dog; allowing my dog to play with neighbor dogs; taking my dog to the dog park; none of the above].

14. My dog usually receives strenuous exercise (i.e. hard running and playing for a minimum of 40 minutes) [every day; about every other day; several times per week; on weekends; one time per week; rarely].

15. My dog interacts with and plays with other dogs [every day; about every other day; several times per week; on weekends; one time per week; rarely; my dog generally does not like interacting with other dogs].

16. The type of physical contact that my dog most enjoys is: [tummy rubs; cuddling on the couch; gentle petting; playful petting; scratching; rough/rowdy petting and patting].

17. Our household also includes [zero; one; two; three; more than three] additional dogs and [zero; one; two; three; more than three] cats.

18. In our household, the person who most often feeds, brushes, bathes, and takes the dog to the vet is [myself; my partner; my roommate; my children; we all share these responsibilities equally].

19. In our household, the person who most often walks, plays with and trains the dog is [myself; my partner; my roommate; my children; we all play and walk the dog].

20. I consider myself to be [extremely; very; moderately; slightly] emotionally attached to my dog.

21. When I run errands in the car, my dog [always; often; occasionally; never] accompanies me.

22. When I go on vacations, my dog [always; often; occasionally; never] accompanies me/us.

23. Methods that I use to discipline my dog include (circle all that apply): [telling him/her "No!" and using a harsh voice; using physical reprimands, i.e. hitting or grabbing his nose; putting him outside or confining him/her to a crate or small room; ignoring bad behavior; collar corrections, i.e. jerking on the collar; other _____].

24. Methods that I use to praise my dog include (circle all that apply): [petting and cuddling; using verbal praise ("Good dog!"); playing games; giving food treats; other _____].

25. The type of collar that I use with my dog is (circle all that apply): [buckle collar; Premier collar; head halter (i.e. Gentle Leader); choke chain collar; pinch collar; harness; other _____].

26. Regarding obedience class/lessons, goals for myself include (circle all that apply): [to increase my understanding of my dog's behavior; to develop a closer relationship with my dog; to learn how to train my dog to be a better companion and member of the household; to solve problem behaviors that my dog is exhibiting; other _____].

Behavior Profile Information

1. I would describe my dog's energy level as [hyperactive; very high energy; normal; normal to low; basic couch potato].

2. I would describe my dog's emotional attachment to me and/or my family as [excessively attached; very attached; moderately attached; slightly attached; not attached at all].

3. My dog greets visitors he does *not* know at our home in a(n) [hyper-excitable; exuberantly friendly; moderately friendly; aloof; hesitant but then is friendly; fearful at first but recovers; always fearful; slightly aggressive; very aggressive] manner.

4. My dog greets visitors who she *does* know at our home in a(n) [hyper-excitable; exuberantly friendly; moderately friendly; aloof; hesitant but then is friendly; fearful at first but recovers; always fearful; slightly aggressive; very aggressive] manner.

5. When out walking away from home, my dog greets strangers in a(n) [**hyper-excitable; exuberantly friendly; moderately friendly; aloof; hesitant but then is friendly; fearful at first but recovers; always fearful; slightly aggressive; very aggressive**] manner.

6. When out walking away from home, my dog greets other dogs in a(n) [**hyper-excitable; exuberantly friendly; moderately friendly; aloof; hesitant but then is friendly; fearful at first but recovers; always fearful; slightly aggressive; very aggressive**] manner.

7. At the veterinarian's office, my dog behaves in a(n) [**hyper-excitable; exuberantly friendly; moderately friendly; friendly but quiet; hesitant but then is friendly; fearful at first but recovers; always fearful; slightly aggressive; very aggressive**] manner.

8. My dog is most likely to bark or whine (circle all that apply): [**when he is left alone by himself; when he is greeting me; when strangers come to the house; when he wants attention; when he hears certain noises; when he sees cats, squirrels, or other small animals; when he is playing with me**].

9. If he/she is approached by myself or a family member while eating, my dog [**stops eating and wags his tail; continues eating and wags his tail; freezes over his food bowl; begins to wolf down his food; walks away from the bowl; growls quietly; growls or barks menacingly**].

10. When being groomed (brushed, bathed, ears cleaned, etc.) my dog usually [**sits quietly and enjoys the grooming; sits quietly but does not enjoy the grooming; struggles a bit but usually settles down after a few minutes; struggles a bit throughout the grooming; struggles fiercely throughout the grooming; struggles and attempts to nip**].

11. When my dog is asked to give one of his/her favorite toys to me he/she [**readily gives it to me; runs away; holds on to it tightly; growls and refuses to give it up; tries to bite**].

12. My dog's favorite games include (circle all that apply): [**chasing a ball and bringing it back to me; chasing a ball and then running with it; play tug of war; wrestling and rough housing with family members; playing chase games; playing with other dogs in the family; playing with dog friends who are not part of the family; playing hide and seek; sitting close for cuddling and petting; other _____**].

13. I would describe my dog as: [**extremely playful; somewhat playful; not playful at all**].

14. The characteristics that I like about my dog include his/her (circle all that apply): **playfulness; high energy; affection for family members; affection for other people; behavior with other dogs; ability to learn new commands; protectiveness; boldness; calmness; sweet and gentle nature; behavior with children; physical appearance (i.e. coat type, size, etc); other _____**].

Appendix 3

15. Have you attended any previous training classes with your dog? **Yes/No** If so, what type of class was it? _____.

16. What commands does your dog usually respond to? **[sit; lie down; come; stay; give; take it (fetch); other _____]**.

17. Please circle any of the following behavioral problems that you are currently experiencing with your dog:

house-training	excessive chewing	barking when alone
running away	not coming when called	pulling on the leash
growling at strangers	growling at family members	nipping during play
biting people	possessive over toys/food	chasing cats/squirrels
jumping up to greet	fear of strangers	fear of family member
unmannerly in the car	fear of other dogs	destructive when left alone
rushing out of doors	aggressive towards unfamiliar dogs	stealing objects/food
barking for attention	digging	
fighting with other dogs in the home	playing too roughly with other dogs	

PLEASE ADD ANY ADDITIONAL COMMENTS REGARDING YOUR DOG'S BEHAVIOR AND YOUR GOALS FOR BOTH YOU AND YOUR DOG:

_____

_____

_____

_____

_____

(Form provided by permission; Linda P. Case, AutumnGold Consulting and Dog Training Center)

appendix 4

Sample Cat Behavior History Profile

STUDENT'S NAME _____ CAT'S NAME _____

BREED _____ AGE _____ SEX _____ SPAYED/NEUTERED? _____

Cat/Owner Interactions (Circle Appropriate Answers)

1. My cat was approximately [less than 7 weeks, 7 to 8 weeks; 8 to 12 weeks; 12 weeks to 6 months; 6 months to 1 year; older than 1 year] when I obtained him/her.

2. I acquired my cat from a/an [pure-bred breeder; ad in the newspaper; pet store; animal shelter or rescue organization; friend/neighbor; relative; as a stray].

3. I have owned my cat for [less than 1 month; 1 to 6 months; 6 months to 1 year; 1 to 2 years; more than 2 years].

4. To my knowledge, I am my cat's [first; second; third; fourth or more] owner.

5. I have owned [zero; one; two; more than two] other cats in the past.

6. Places that my cat enjoys resting include (circle all that apply): [on his own bed; in windowsills; on the couch; on my bed; in my lap; outdoors; other_____].

7. At night, my cat sleeps [on my bed; in my bedroom but not on the bed; in the house but not in the bedroom; outdoors].

8. I would describe my cats as: [extremely playful; very playful; occasionally playful; not playful at all].

9. The person or pet who my cat will play with includes (circle all that apply): [myself; my partner; my roommate; my children; visiting friends; the dog; other cats in the family].

-1-

10. Toys that my cat enjoys include [small stuffed toys; laser beams; balls that roll; toys that can be batted; catnip toys; lure toys (i.e. cat dancer); other_____].

11. The way in which I usually interact with my cat includes (circle all that apply) [playing with toys; going outdoors together for walks or playing; cuddling and petting in my lap; talking to my cat; brushing or combing; other _____].

12. The type of physical contact that my cat enjoys includes (circle all that apply): [gentle petting; sitting on my lap; playful or rough petting; chasing and pouncing play; brushing/combing].

13. Our household also includes [zero; one; two; three; more than three] additional cats and [zero; one; two; three; more than three] dogs.

14. In our household, the person who most often feeds, brushes, bathes, and takes the cat to the vet is [myself; my partner; my roommate; my children; we all share these responsibilities equally].

15. In our household, the person who most often walks, plays with and trains the cat is [myself; my partner; my roommate; my children; we all play and walk the cat].

16. In the household, my cat avoids or shows fear toward [myself; my partner; my roommate; my children; other cat(s); the dog(s); no one, our cat gets along with everyone in the home].

17. In the household, my cat avoids or shows aggression toward [myself; my partner; my roommate; my children; other cat(s); the dog(s); no one, our cat gets along with everyone in the home].

18. I consider myself to be [extremely; very; moderately; slightly] emotionally attached to my cat.

19. Methods that I use to discipline my cat include (circle all that apply): [telling him "No!" and using a harsh voice; using physical reprimands, i.e. hitting or swatting him; putting him outside or confining him/her to a crate or small room; ignoring bad behavior; squirting him with a water bottle; remote correction (scat matt or air canister); other _____].

20. Methods that I use to praise my cat include (circle all that apply): [petting and cuddling; using verbal praise ("Good girl!"); playing games; giving food treats; other _____].

21. The type of litter box and litter filler that my cat has are: _____.

22. Our household has [one; two; three; more than three] litter boxes available for our cat(s).

23. The litter boxes are located (identify location of each box in the home): _____.

24. I remove feces (and litter clumps if applicable) from my cat's litter box: [daily; every other day; twice a week; weekly; every other week; never].

25. I completely replace the litter filler in my cat's box: [daily; every other day; twice a week; weekly; every other week; once a month; never].

Behavior Profile Information

1. I would describe my cat's energy level as [hyperactive; very playful; normal; normal to low; basic couch potato].

2. I would describe my cat's emotional attachment to me and/or my family as [excessively attached; very attached; moderately attached; slightly attached; not attached at all].

3. When being petted, my cat usually [enjoys the petting for as long as we will pet her; enjoys it for a bit and then walks away; enjoys it for a bit and then becomes somewhat aggressive; avoids being petted].

4. My cat's favorite games include (circle all that apply): [chasing my hands or feet and biting at them; chasing a toy or ball; rough "wrestling" play with my hands; playing stalk and pounce games; playing with other cats or dogs in the family; sitting close for cuddling and petting; other _____].

5. My cat will usually or always respond to the following commands (circle all that apply): [comes to his name; gets off of counters when commanded; retrieves a toy; sits or lies down on command; meows on command; tricks _____].

6. My cat lives: [exclusively indoors (never allowed outside); mostly indoors (taken outdoors on a lead or harness); mostly indoors (allowed outside freedom with supervision); mostly indoors (allowed outside without supervision); indoor/outdoor; mostly outdoors].

7. My cat greets visitors to our home in a(n) [friendly; hesitant but then is friendly; fearful at first but recovers; always fearful (hides); aggressive] manner.

8. My cat uses his litter box [always (no house soiling); almost always; about 50 percent of the time; less than 50 percent of the time; hardly at all; never].

9. When my cat does not use the box it is: [urine only and a large amount; urine, but a very small amount; urine and feces; feces only; N/A].

10. If your cat does not use the box, list all locations that waste has been found: _____.

11. The characteristics that I like about my cat include his/her (circle all that apply): [playfulness; curiosity; affection for family members; affection for other people; intelligence; calmness; sweet and gentle nature; behavior with children; physical appearance (i.e. coat type, size, etc); other _____].

12. Please circle any of the following behavioral problems that you are currently experiencing with your cat:

house soiling (not using litter box)
scratching furniture/draperies
fear of one or more family members
playing too roughly with people (biting/scratching)
fear of other cats in home
hunting mice/birds

hyperactive
playing too roughly with other cats
fear of visitors
biting during/after petting
fighting with other cat(s) in home
excessive fear in many situations

excessive meowing
aggressive biting
excessive kneading

PLEASE ADD ANY ADDITIONAL COMMENTS REGARDING YOUR CAT'S BEHAVIOR AND YOUR GOALS FOR BOTH YOU AND YOUR CAT:

_____
_____
_____
_____
_____
_____
_____

(Form provided by permission; Linda P. Case, AutumnGold Consulting)

glossary of terms

agonistic — any behavior that is elicited during conflict; most commonly refers to aggressive behaviors, but can also involve fear, flight, or pacifying behaviors

agoraphobic — fear of new or unfamiliar places

allogrooming — grooming (preening) another individual

allorubbing — rubbing on another individual as a form of scent marking and expression of affiliation

altricial — requiring parental care and feeding for survival at birth; refers to a species whose young is born at a relatively immature stage of development

anorexia — loss of appetite for food

anxiety — anticipation of harm or danger

associative learning — learning through associations between two or more stimuli or events

aversive stimulus — a stimulus or event that an animal finds unpleasant or painful and wishes to avoid

behavior chain — multiple behaviors that are trained together in sequence and which eventually can be elicited using a single cue

bunting — head rubbing on objects or caretakers; form of marking behavior in the cat

castration — surgical removal of the testes in the male

classical conditioning — type of associative learning in which a neutral stimulus becomes conditioned to elicit a response through repeated pairing with an unconditioned stimulus

clicker — toy noisemaker commonly used as a conditioned reinforcer in dog and cat training

cobby body type — heavy-boned, sturdy, round, and compact body type with short legs and broad shoulders and hindquarters

cognitive dysfunction — age-related neurodegenerative disorder resulting in a decline in higher brain functions of learning and memory

commensalism — a relationship in which an individual (or species) benefits from associations and interactions with another individual, group or species on a temporary or permanent basis

conditioned reinforcer — neutral stimulus that repeatedly and reliably precedes a primary reinforcer, eventually taking on the properties of the primary reinforcer

conspecific — belonging to the same species or social group

contiguity — the relationship between two events in time and place; paired events

contingency — the predictability of two events occurring together

continuous reinforcement — provision of reinforcement every time a behavior is offered

coprophagia — stool (feces) eating

crepuscular — active during early morning and dusk hours cue — a stimulus that elicits a behavior; examples include verbal commands and hand signals

desensitization — behavior modification technique that involves gradually increasing an animal's tolerance to a particular stimulus by gradually increasing the intensity of the stimulus

differential reinforcement — provision of reinforcement only for those behaviors that meet a predetermined criterion of form, frequency, or duration

diurnal — being awake and active during the daylight hours and sleeping during the nighttime hours

dominance — relationship determined by behaviors that allow repeated access to and control over multiple types of desired resources

ecological niche — specific micro-habitat in nature to which populations or organisms gradually adapt; usually described in terms of providing food, nesting sites, opportunities to mate, and protection from predators

ethology — the study of the behavior of animals in their normal environment

extinction — the diminishing and eventual termination of a behavior when no reinforcement occurs or is provided for the behavior

extinction burst — a temporary increase in behavior in response to the removal of reinforcing stimuli; precedes extinction (termination) of the behavior

familial — observed within a family lineage

flooding — behavior modification technique in which the animal is repeatedly exposed to an aversive stimulus without opportunity to escape with the intent of desensitizing the animal to the stimulus; highly risky and not recommended for the majority of behavior problems

free-ranging — a non-captive (and often un-owned) animal who is living in a natural habitat, largely free from constraints imposed by humans

genotype — genetic makeup of an individual

habituation — non-associative learning involving reduction in an innate response (such as fear) through repeated exposure to a non-harmful stimulus

inguinal — pertaining to the groin region of the body

instrumental learning — trial and error learning (a type of operant learning)

intermittent reinforcement — occasional provision of reinforcement in response to a targeted behavior

luring — inducing part or all of a behavior by guiding (for example with food in the hand)

marking — deposition of urine (and sometimes feces) to leave an olfactory signal; unrelated to normal physiological elimination

negative punishment — the removal or withholding of a pleasurable or desired stimulus in response to a behavior, decreasing in a decrease in the frequency of that behavior

negative reinforcement — the removal or withholding of an aversive (unpleasant) stimulus in response to a behavior, resulting in an increase in the frequency of the behavior

neophobia — fear of the unknown or unfamiliar

neoteny — retention of infantile (neotenous) characteristics which persist into adulthood

nocturnal — being awake and active when it is dark and sleeping during the day

observational learning — a form of social learning in which one individual learns a new behavior simply by observing another engage in the behavior (and is not directly exposed to relevant stimuli)

Glossary of Terms

olfactory — involving the nose and sense of smell

ontogeny — development of an individual from birth to adulthood

onychectomy — surgical procedure for declawing cats in which the distal portion of the front and sometimes back digits are amputated

ovariohysterectomy — surgical removal of the female reproductive tract (ovaries, oviducts, and uterus); spaying

phenotype — set of observable characteristics that are determined by genotype

pheromones — hormone-like substances secreted by an individual and which affect the behavior of another individual of the same species; often related to reproductive behavior

pica — consumption of non-nutritive items

piloerection — reflex in which the muscles at the base of the dog's hairs contract and causes the hairs on the back to stand up

positive punishment — addition of an aversive (unpleasant) stimulus in response to a behavior that results in a decrease in the frequency of the behavior

positive reinforcement — addition of a pleasant or desired stimulus in response to a behavior that results in an increase in the behavior in the future

precocial — capable of a high degree of independent activity at birth

primary reinforcer — a stimulus that is inherently reinforcing when provided as a consequence of a behavior (i.e. requires no prior learning)

punishment — a consequence that causes a reduction in the frequency of the behavior that it follows

reinforcement — a consequence that causes an increase in the frequency of the behavior that either attains a desired stimulus (R+) or avoids an aversive stimulus (R-)

release word — a verbal cue that signals the end of a behavior; the word "okay" is commonly used in dog and cat training

sensitive periods — developmental stages during which an animal is especially sensitive to learning particular types of associations and social behaviors

sexual dimorphism — the occurrence of anatomical differences between males and females of the same species

shaping — modifying behavior by sequentially reinforcing behaviors that progressively approximate the target behavior; can be used to train dogs and cats to perform behaviors that would rarely occur spontaneously

social facilitation — effect of group interactions upon individual behavior

socialization period — developmental stage during which young animals form primary social relationships and develop species-specific behavior patterns

spraying — marking posture used by cats to deposit urine on vertical surfaces

stimulus — change in the environment that evokes a behavioral response (also called discriminative stimulus)

stressor — a stimulus that elicits a stress response in an animal

unconditional response — an unlearned response (occurs without prior learning or conditioning)

unconditional stimulus — stimulus that provokes an unconditioned response

xenophobia — fear of strangers; fear of the unfamiliar

index

Note: Page numbers followed by "f" refers to Figures.

A

Affiliative behaviors
  cats, 13, 15, 54, 59–60
  dogs, 10, 15
  playing, 280
African wildcats, 8, 12, 13, 70
Aggression, 73–75. *See also* Agonistic behaviors
  breed-specific, 19
  conflict and, 290
  defensive (cats), 68
  defensive (dogs), 67
  defined, 279–281
  desensitization and, 129
  dog bite prevention in children, 278–279
  between dogs, 284–285, 287–289, 297–301
  between dogs and cats, 14–15
  dominance-related (cats), 39, 70
  dominance-related (dogs), 73–76, 282–289
  eye-contact and (dogs), 57–58
  fear-related (cats), 304–306
  fear-related (dogs), 65–66, 295–300
  fighting, 266
  food bowl (dogs), 11, 70, 289–293
  free-ranging dogs, 10
  human-directed (cats), 304–306
  human-directed (dogs), 283–284, 294–295
  ineffective treatments for (dogs), 300–301
  inter-cat, 302–304
  petting-induced (cats), 306–307
  play-induced (cats), 306
  possessive (cats), 33
  possessive (dogs), 154–155, 283, 284, 289–293
  punishment and, 88–89, 227
  redirected (cats), 307
  territorial (cats), 302–304
  territorial (dogs), 70, 293–295
  wagging behaviors and (dogs), 64
Agonistic behaviors, 10, 63, 280. *See also* Aggression
Agoraphobia, 267
Allelomimetic behaviors, 97–98
Allorubbing, 12, 49, 59, 76
Alpha dogs, 9–10, 11. *See also* Dominant/submissive behaviors
Alprazolam, 257
Altricial (immature) offspring, 27
American Pet Product Manufacturers Association, 2
Amitriptyline, 256–257
Anti-bark devices, 227–229
Anxiety. *See* Separation anxiety
Appeasement behaviors. *See* Submissive behaviors
Arousal, 280
Associative learning, 82
Attention-seeking behaviors
  cats, 235, 237
  dogs, 221–222, 225–226, 231

Aversive stimuli. *See also* Avoidance behaviors; Punishment
  aggression (cat), 306
  barking training (dog), 227–229
  collars as, 174
  dog's name used as, 178
  down-stay as, 251
  elimination problems, 189, 202, 205
  excess vocalizations (cat), 237
  fear-related behaviors, 268, 295–297
  Flooding, 130–131
  jumping behavior training (cat), 153, 181, 234
  jumping up behavior training (dog), 214
  operant conditioning, 87–89, 92–95
  positive *vs.* negative control of behavior, 106–109
  in separation anxiety, 259
Avoidance behaviors. *See also* Aversive stimuli
  barking, 227
  chewing, 152, 217
  fear-related, 268–269, 304
  flooding and, 130
  house-training, 140, 143, 188–189, 202
  jumping up, 215
  play aggression, 306
  response to aversive stimuli, 87, 89, 106–108, 181

B

Back-chaining, 122–124
Bad behaviors. *See* Disruptive behaviors
Bark-activated devices, 227–229
Barking behaviors, 50–52, 222–229, 236–238
Barnyard cats, 12
Behavior chains, 122–124
Behavior modification. *See also* Counter-conditioning; Desensitization; Operant conditioning
  nothing in life is free (NILIF) procedure, 286–287
  positive reinforcement use, 109
  in separation anxiety problems, 254, 258–259, 261–262
  techniques of, 127–131
Bekoff, Mark, 47
Benzodiazeines, 257
Bite prevention, children, 278–279
Biting behaviors
  inhibition of, in puppies, 30–31, 34
  mouthing, 211–212, 287
  nipping in puppies, 147–149
  teaching children about, 278–279
Black, Janet, 170
Black, Steve, 170
Blackwell, E., 262
Boredom-related behaviors, 219, 223, 225, 229
Brambell Committee, 220
Breed clubs, 16
Breeds of dogs and cats, 15–22
Breed-specific behaviors, 18–19, 21–22, 75

British Shorthair cats, 20
Brown, Brad, 190
Brown, Sue, 190
Bunting, 49, 50f

C

Canadian Cat Association, 20
Canid species, 3
Canis familiaris, 4
Canis lupus, 7, 9
Carnivores, 3–4
Castration, 287
Cat Fanciers Association, 20
Chasing behaviors, 69–70
Chew bones and toys for dogs, 151–152, 217
Chewing behaviors (dogs), 151–152, 215–219, 221–222
Children, teaching, 278–279
Chirrups, 53
Citronella anti-bark collars, 227–229
Classes, socialization, 34–35
Classical (Pavlovian) conditioning, 82–87, 104–106
Classification of dogs and cats, 4
Clawing behaviors, 50, 153–154, 232–233
Clicker training, 95, 113–116, 166–167
Click-treats (CT), 114
Climbing up behaviors (cats), 152–153, 180–181, 233–235
"C" lip position, 63
Clomipramine, 256–257, 258–259
Clorazepate, 257
Cognitive dysfunction, 40
Collars, 174, 227–229
Come when called commands (cats), 181
Come when called commands (dogs), 176–179
Commensalism, 75
Communications
  auditory (cats), 28–29, 50, 53–55, 76, 236–238
  auditory (dogs), 50–52, 222–227, 236–238
  between dogs and cats, 14–15
  olfactory (*See* Olfactory communications)
  visual (*See* Visual communications)
Competitive interactions, 70, 280, 282, 284, 289, 293, 301
Conditioned (secondary) reinforcers, 95, 112–113
Conflict aggression, 290
Contiguity, 84
Contingency, 84
Continuous reinforcement schedules, 117
Coprophagia, 70, 99
Counter-conditioning. *See also* Desensitization
  aggressive dogs, 287
  defined, 129–130
  fear-related behavior problems, 269–271, 296, 298–299
  food bowl aggression, 292–293
  inter-cat aggression, 302, 305–307
  separation anxiety, 253–254, 261–262
  territorial aggression, 294–295
Counter crawling in cats, 152–153, 180–181, 233–235
Crane, Paul, 177
Crate training, 134–135, 256
Cues
  fading, 124–125
  pre-departure, 247, 249–250, 253–255, 263
  safety, 255–256, 260, 262, 263
  stimulus control, 124–127, 162
  verbal, 86

D

DAP (Dog Appeasing Hormone), 145–146, 258–259
Declawing, 233
Defensive threat behaviors, 67–68
Dependency reduction, 251–252. *See also* Separation anxiety
Desensitization. *See also* Counter-conditioning
  defined, 129
  fear-related behavior problems, 269, 271, 298
  food bowl aggression, 292–293
  inter-cat aggression, 302
  isolation, 254–255
  pre-departure cues, 253
  redirected aggression, 305–306
  separation anxiety, 261–262
  territorial aggression, 294–296
Differential reinforcement schedules, 117, 128–129
Digging behaviors, 229–230
Disruptive behaviors
  barking (dogs), 222–229
  chewing (dogs), 151–152, 215–219, 221–222
  digging (dogs), 229–230
  excessive vocalization (cats), 236–238
  furniture clawing (cats), 50, 153–154, 232–233
  hyperactive (dogs), 230–232
  jumping up (cats), 152–153, 180–181, 233–235
  jumping up (dogs), 170, 211–215
  nocturnal activity (cats), 235–236
  owner responsibilities, 220–221
  plant eating and pica (cats), 239
Distance-inceasing visual signals, 63–67
Distance of stays, 169
Distance-reducing visual signals, 56–62
Distractions, 169–170
Distress calling, 28, 29
Dog Appeasing Hormone (DAP), 145–146, 258–259
Doggy day care, 219
Domestication, 5–8
Dominant/submissive behaviors. *See also* Submissive behaviors; Submissive urination (dogs)
  cats, 12, 39, 70, 75–76
  dogs, 56, 72–76, 282–289
  dogs *vs.* wolves, 9–12
Double-sided tape, 153, 181
Down-stay training (dogs), 251–252, 255, 263
Down training (dogs), 120, 121f, 124–125, 163–164
Drug therapy, 206, 256–259, 263–264, 305
Duration of stays, 168–169

E

Eating houseplants (cats), 239
Ecological niches, 5
Elimination problems
  excitable urination (dogs), 194–195
  inappropriate (cats), 195–202
  incomplete house-training (dogs), 188–191
  kittens, 142–144
  learned avoidance in, 140, 143, 188–189, 202
  litter box aversion (cats), 195–197, 199, 201, 206
  location preferences, 188, 196–199
  marking behavior (cats), 202–204
  marking behavior (dogs), 192–193
  medical causes, 187, 196
  pre-adoption counseling for owners (dogs), 189–190
  puppies, 139–142
  submissive urination (dogs), 193–194

substrate preferences, 188, 191–192, 195–196, 198–199, 201
  urine spraying (cats), 46, 48, 202–205
Ellis, Mark, 298
Escape/avoidance behaviors. *See* Avoidance behaviors
Ethology, 27
Evolutionary history, 2–4, 6–7, 55–56
Excitable urination (dogs), 194–195
Exploratory play, 62
Extinction, 109, 127–129
Extinction burst, 128
Eye contact in dogs, 57–58

F
Fading cues, 124–125, 162
Fear behaviors, 65–67, 295–300
  aggression (cats), 304–306
  aggression (dogs), 65–66, 295–300
  common fears, 266–268
  counter-conditioning, 269–271, 296, 298–299
  desensitization to, 269, 271, 298
  learned avoidance in, 268–269, 304
  normal *vs.* problematic, 245–246
  treatment, 269–272
  visual communications (cats), 67
  visual communications (dogs), 64–66
Fear imprint stage, 32
Feces marking, 49
Feeding behaviors
  cats, 71–72
  dogs, 70
  food bowl aggression, 11, 70, 289–293
  food delivery toys, 216–217
  food preference learning, 98–99
Felid species, 3
Feline facial pheromones (FFP), 204–205
Felis catus, 4
Feliway, 305
Feral cats, 12–13
Feral dogs, 10
FFP (feline facial pheromones), 204–205
Fighting reactions, 266
Fillers for cat litter boxes, 144, 196–197
First day training, 134–135
Five freedoms, 220
Flagging, 63
Fleeing reactions, 266
Flehmen response, 48
Flooding, 130–131, 272
Fluency, 162
Fluoxetine, 257
Food bowl aggression, 11, 70, 289–293
Food competition, 70
Food delivery toys for dogs, 216–217
Food preferences, 98–99
Foreign body type, 20
Free-living cats, 12–13
Free-ranging dogs, 10
Free-shaping *vs.* prompting, 120–122
Freezing reactions, 266
Furniture scratching, 50, 153–154, 232–233

G
Gape response, 48
Generalization training, 162
Gentle play training, 147–151

Give command, 152, 219
Goals and objectives for training, 110
Go to bed command, 146, 167, 224, 251
Greeting behaviors
  aggressive, 294–295
  barking, 222–225
  bunting, 49, 50f
  between cats and dogs, 14–15
  distance-reducing signals (cats), 59–60
  distance-reducing signals (dogs), 56–59
  excessive, 248, 255, 299
  excitable urination, 194–195
  grunting, 52
  jumping up (dogs), 170, 211–215
  of puppies, 37
  sniffing, 45, 299–300
  submissive urination, 193–194
  tail wagging, 64
  trilling, 53
Groveling behaviors, 57
Growling, 52, 55, 63
Grunting, 52

H
Habituation, 30, 96–97
Heeling, 173, 175
Hissing, 55
History, breeds, 15–22
Hit it command, 167
House-training
  excitable urination (dogs), 194–195
  inappropriate elimination (cats), 195–202
  incomplete (dogs), 188–191
  kittens, 142–144
  learned avoidance in, 140, 143, 188–189, 202
  litter box aversion (cats), 195–197, 199, 201, 206
  location preferences, 188, 196–199
  marking behavior (cats), 202–204
  marking behavior (dogs), 192–193
  medically-caused soiling problems, 187, 196
  pre-adoption counseling for owners (dogs), 189–190
  puppies, 139–142
  submissive urination (dogs), 193–194
  substrate preferences, 188, 191–192, 195–196, 198–199, 201
  urine spraying (cats), 46, 48, 202–205
Howling, 52
Human-directed aggression, 283–284, 294–295, 304–306
Human socialization of pets, 29, 33–34, 136, 138
Hunting behaviors. *See* Predatory behaviors
Hyperactive dogs, 230–232
Hyper-attachment behaviors, 250, 258, 261.
  See also Separation anxiety

I
Imprinting behavior, 27
Incomplete house-training, 188–191
Inguinal area, 45
Instrumental learning. *See* Operant conditioning
Interactive toys for dogs, 216
Intermittent reinforcement schedules, 117–119
International Cat Association, 20
Introducing pets, 154–156
Isolation training, 145–147, 254–255.
  See also Separation anxiety

J

Johnson, Peter, 298
Jumping up behaviors
  cats, 152–153, 180–181, 233–235
  dogs, 170, 211–215
Juvenile period, 38–39

K

Kittens
  anxiety in, 32
  classes for, 34–35
  gentle play training, 149–151
  introducing, 155–156
  jumping up behavior training, 152–153
  litter box training, 142–144
  play behaviors, 33
  scratching post training, 153–154
  socialization, 32, 34–35, 138
Kitty condos, 233, 234f
Kneading behavior, 28, 29

L

Learned avoidance. See Avoidance behaviors
Leash pulling, 172–176
Leave it command, 152, 218
Leonard, Jennifer, 6
Litter box aversion (cats), 195–197, 196, 199, 201, 206
Litter box hygiene, 196, 198, 200–201
Litter box training (kittens), 142–144
Litters for cat litter boxes, 144, 196–197
Local enhancement, 99
Location preferences for elimination, 188, 196–199
Lorenz, Konrad, 27
Lupfer-Johnson, Gwen, 98–99
Lure training, 120–122, 161, 163, 165–166

M

Marking behaviors
  cats, 202–204
  dogs, 192–193
  feces, 49
  scent, 46–47
  scratching, 50
  urine, 39, 46–48, 192–193, 202–204
Mating calls, 55
Medical problems, 187, 196
Meowing, 28–29, 50, 53–55, 76, 236–238
Miacines, 3
Miles, Sandy, 236
Mills, Daniel, 204
Monoamine oxidase B inhibitors, 257
Motion-activated compressed air cylinders, 153, 181, 234–235
Mouthing behaviors (dogs), 211–212, 287
Murmer patterns, 53

N

Natural weaning, 37–38
Negative behaviors. See Disruptive behaviors
Negative punishment, 87, 89–92, 109, 148–149
Negative reinforcement, 87, 89–90, 92, 95–96, 107–108
Neonatal behavior patterns, 28–29
Neotenized features and behaviors, 19, 75, 76
Neoteny, 5

NILIF (nothing in life is free) procedures, 286–287
Nipping, 147–149
Nocturnal activity (cats), 235–236
Noise phobias, 267–268
Nose touch, 14, 15f, 59f, 166–167
Nothing in life is free (NILIF) procedures, 286–287

O

Object play, 62
Obligate carnivores, 4
Observational learning, 32, 99–101
Offensive threats, 63–64, 65f, 74f
Olfactory communications
  allorubbing, 12, 49, 59, 76
  feces marking, 49
  neonates, 29
  scratching, 50
  sniffing, 45–46, 48, 49, 299–300
  urine marking, 39, 46–48, 192–193, 202–204
  urine spraying, 46, 48, 202–205
Onychectomy, 233
Operant conditioning. See also Aversive stimuli; Behavior modification
  link with classical conditioning, 104–106
  prompting vs. free-shaping, 120–122
  reinforcement, 87, 89–96
  schedules of reinforcement, 116–119
  successive approximation, 109, 119–120, 164
  terminology, 87
Overall, Karen, 301
Owner's responsibilities, 220

P

Pack behavior, 9–12
Paedomorphism, 5
Pain shrieks, 55
Parker, Anna, 199
Parker, Joe, 199
Paroxetine, 257
Pavlovian theory, 82–87, 104–106
Paws up command, 212–213
Paw touch training, 167
Pedigrees, 16–18, 20
Personalities of cats, 35–36
Petting-induced aggression, 306–307
Pharmacotherapy, 206, 256–259, 263–264, 305
Phenothiazines, 257
Phenylpropanolamine, 195
Pheromones, 49, 204–205
Pica in cats, 239
Piloerection, 55, 63
Play behaviors. See also Toys
  affiliative, 280
  aggressive (cats), 306
  cats, 62
  dogs, 60–62
  between dogs and cats, 14–15
  kitttens, 33, 149–151
  puppies, 30–32, 147–149
  puppies vs. kittens, 33
Play bows, 61–62
Pointers, 18
Positive punishment, 87, 88–90, 89, 92, 107–108
Positive reinforcement, 87, 89–92, 95–96, 109
Positive vs. negative control of behavior, 106–109

Index

Possessive aggression, 33, 154–155, 283, 284, 289–293
Powers, Mike, 238
Pre-adoption counseling, 189–190
Precocial birds, 27
Predatory behaviors, 69–71
Pre-departure cues, 247, 249–250, 253–255, 263
Premack, David, 122
Premack Principle, 122–124
Primary reinforcers, 94, 112, 115
Prompting vs. free-shaping, 120–122
Proto-dog, 7
Pulling behaviors, 172–176
Punishment, 87–92, 107–109, 148–149, 259, 283–284, 300–301. See also Aversive stimuli
Puppies
　accepting isolation, 145–147
　chewing appropriately, 151–152
　classes for, 34
　gentle play training, 147–149
　house-training, 139–142
　introducing, 154–155
　play behaviors, 30–33, 147–149
　preventing nipping, 147–149
　socialization, 30–32, 34, 137
Puppy play vs. kitten play, 33
Purebred dogs, 16–18
Purring, 28, 53
"Pushing the go button," 173
Puzzle toys for dogs, 216–217

Q
Quiet command, 224–225

R
Redirection training techniques, 148–149, 151, 307
Reinforcement, 87, 89–96, 107–109, 112–114, 116–119
Releasing from commands, 161, 166, 168–169
Reliability training, 162
Repetitive barking, 223–225
Resource guarding. See Possessive aggression
Reversing the contingency, 165–166
Rochlitz, Irene, 220
Rolling in cats, 60, 62
Rooting reflex, 28, 29
Ross, Julie, 98–99
Royal Institute of Technology, 6

S
Safety cue toys, 255–256, 260, 262–263. See also Separation anxiety
Savolainen, Peter, 6
Scavenger theory, 5, 8
Scent glands, 49
Scent marking, 46–47
Schedules of reinforcement, 114, 116–119
Scratching behaviors, 50, 153–154, 232–233
Scratching posts, 153–154, 232
Secondary reinforcers, 95, 112–113
Selegiline hydrochloride, 257
Senior pets, 39–40
Sensitive periods
　defined, 26–27
　juvenile, 38–39
　nenatal, 28
　socialization, 30–38
　transitional, 28–30
Sensitization, 97
Separation anxiety
　behavioral modification techniques (dogs), 254, 258–259, 261–262
　behavioral signs, 249–250
　counter-conditioning (dogs), 253–254, 261–262
　dependency reduction, 251–252
　diagnosing, 246–247
　down-stay training for, 251–252, 255, 263
　drug treatment (dogs), 256–259, 263–264
　ineffective treatments (dogs), 259
　management approaches (dogs), 256
　non-drug treatment (dogs), 250–256
　normal vs. problematic, 245–246
　pre-departure cues and, 247, 249–250, 253–255, 263
　prevention of (dogs), 260
　risk factors and predisposing temperament traits, 247–249
　safety cue toys, 255–256, 260, 262–263
　treatment (cats), 264–265
Separation training, 145–147, 254–255
Serotonin uptake inhibitors, 257
Shaping, 91, 119–122
Side steps in cats, 62
Sit-stay training (dogs), 90–92, 116, 162–163
Sit training (dogs), 161–163, 213–214
Sloan, Sally, 288
Snarling, 63
Sniffing, 45–46, 48, 49, 299–300
Social facilitation, 69, 98
Socialization
　defined, 30
　with humans, 29, 33–34, 136, 138
　kittens, 32, 34–35, 138
　puppies, 30–32, 34, 137
　puppy play vs. kitten play, 33
Social learning, 97–100
Social play, 62
Social relationships
　cats, 12–13, 75–76
　dogs, 72, 74–75
Spaying, 282
Stay training (dogs), 164–166, 168–170
Stealing behaviors, 237–238
Stimulus (cue) control, 124–127, 162
Stimulus/response, 28, 29
Submissive behaviors, 52, 56, 65–66, 72–76, 74f. See also Dominant/submissive behaviors; Submissive urination (dogs)
Submissive urination (dogs), 193–194
Substrate preferences for elimination
　cats, 195–196, 198–199, 201
　dogs, 188, 191–192, 196
Successive approximation, 109, 119–120, 164
Suckling behavior, 28
Surface preferences for elimination
　cats, 195–196, 198–199, 201
　dogs, 188, 191–192, 196
Systematic desensitization. See Desensitization

T
Takeuchi, Y., 262
Target training, 166–167
Taxonomy, 4
Taylor, Rebecca, 305

Index

Territorial behaviors, 70, 223–225, 237, 293–295, 302–304. *See also* Marking behaviors
Timing of reinforcement, 116–119
Tongue flicking, 48
Touch sticks, 166–167
Toys. *See also* Play behaviors
  cats, 33, 62, 71, 236
  dogs, 33, 74, 123, 151–152, 216–217, 255–256, 260, 262–263
Training. *See also* House-training; Operant conditioning; Separation anxiety
  barking, 227–229
  chewing (dogs), 151–152, 215–219, 221–222
  classical conditioning in, 85–87
  clicker, 95, 113–116, 166–167
  collars, 174
  come when called (cats), 181
  come when called (dogs), 176–179
  crate, 134–135, 256
  digging behaviors (dogs), 229–230
  down (dogs), 120, 121f, 124–125, 163–164
  down-stay (dogs), 251–252, 255, 263
  excessive vocalization (cats), 236–238
  first day, 134–135
  furniture clawing (cats), 50, 153–154, 232–233
  gentle play, 147–151
  give command, 152, 219
  goals and objectives, 110
  go to bed, 146, 167, 224, 251
  hyperactive behavior (dogs), 230–232
  isolation, 145–147, 254–255
  jumping on counters and tables (cats), 152–153, 180–181, 233–235
  jumping up (dogs), 170, 211–215
  kittens, 34–35, 138, 142–144, 149–151, 152–154
  leave it, 152, 218
  lures, 120–122, 161, 163, 165–166
  mouthing, 211–212, 287
  nocturnal activity (cats), 235–236
  paws up command, 212–213
  paw touch, 167
  planning, 110–111
  plant eating and pica (cats), 239
  primary reinforcers, 94, 112, 115
  progress evaluation, 111
  puppies, 34, 137–142, 145–149, 151–152
  quiet command, 224–225
  redirection, 148–149, 151, 307
  releasing from commands (dogs), 161, 166, 168–169
  reliability, 162
  schedules, 111
  scratching post, 153–154
  sit (dogs), 161–163, 213–214
  sit/stay (dogs), 90–92, 116, 162–163
  stay (dogs), 164–166, 168–170
  stealing behaviors, 237–238
  target, 166–167
  turn to your name (dogs), 177–178
  wait command (dogs), 171–172
  walk on a loose lead (dogs), 172–176
  walk on harness (cats), 181–182
Transition period, 28–30
Tricyclic antidepressants, 256–257
Trilling, 53
Turn to your name command, 177–178

U

Umbilical cord technique (dogs), 190, 192, 193
Urine marking, 39, 46–48, 192–193, 202–204
Urine spraying, 46, 48, 202–205

V

Verbal cues, 86
Vila, C., 6
Visual communications
  defensive threat (cats), 68
  defensive threat (dogs), 67
  evolution of, 55–56
  eye contact (dogs), 57–58
  fear (cats), 67
  fear (dogs), 64–66
  greetings (cats), 59–60
  greetings (dogs), 56–59
  offensive threats (cats), 63–64, 65f
  offensive threats (dogs), 63, 74f
  piloerection, 55, 63
  play solicitation (cats), 62
  play solicitation (dogs), 60–62
  wagging (dogs), 64
Viveravines, 3
Vocalizations
  cats, 28–29, 50, 53–55, 76, 236–238
  dogs, 50–52, 222–227, 236–238
Vomeronasal organs, 48

W

Wagging behavior, 64
Wait training, 171–172
Walk on a loose lead training, 172–176
Walk on harness (cats), 181–182
Weaning, 36–38
Wells, Deborah, 228
Whining, 52
Wilson, Fiona, 279
Wilson, Jan, 211
Wolf packs, 9

Y

Yelping, 52
Yin, Sophia, 51